The Early History of Embodied Cognition 1740–1920

Internationale Forschungen zur Allgemeinen und Vergleichenden Literaturwissenschaft

VOLUME 189

The titles published in this series are listed at *brill.com/favl*

The Early History of Embodied Cognition 1740–1920

The Lebenskraft-*Debate and Radical Reality in German Science, Music, and Literature*

Edited by

John A. McCarthy
Stephanie M. Hilger
Heather I. Sullivan
Nicholas Saul

BRILL
RODOPI

LEIDEN | BOSTON

Cover illustrations: Image 1: "Der Mensch als Industriepalast" (Stuttgart, 1926. Chromolithograph. National Library of Medicine). Image 2: "The Sound of Seven Chakras" by Melanie Tee. Brill has made all reasonable efforts to trace all rights holders to any copyrighted material used on the cover. In cases where these efforts have not been successful the publisher welcomes communications from copyright holders, so that the appropriate acknowledgements can be made in future editions, and to settle other permission matters.

Library of Congress Control Number: 2015955684

ISSN 0929-6999
ISBN 978-90-04-30902-9 (hardback)
E-ISBN 978-90-04-30903-6 (e-book)

This book is printed on acid-free paper.

Table of Contents

Preface

This project evolved from a series of panels at the 2011 German Studies Association Annual Meeting which focused on writers and scientists from the long nineteenth century. The authors treated ranged on the literary side from Johann Wolfgang von Goethe to Friedrich Theodor Vischer and Wilhelm Bölsche and on the science side from Johann Friedrich Blumenbach (physical anthropology), Johann Christian Reil (chemistry / physiology / psychology), and Joseph Gall (phrenology) to Ludwig Büchner (physics), Ernst Haeckel (ecology), and Alfred Noll (chemistry / biology). Of course, not all could be covered, and the original concept inevitably evolved further. Other authors and scientists were added. A selection of the conference presentations was subsequently solicited for the volume and an editorial committee was formed, consisting of John A. McCarthy (Vanderbilt U) in the lead, Stephanie Hilger (U of Illinois, Urbana-Champaign), Heather Sullivan (Trinity U, TX), and Nicholas Saul (Durham, UK) to ensure transdisciplinary expertise. Each of the editors was active in soliciting several new contributions to provide coverage of important developments in the history of science and medicine at their intersections with humanistic inquiry. All contributions were evaluated by all the editors.

In the call for papers several key nineteenth-century works had figured prominently: Ludwig Büchner's *Kraft und Stoff* (Energy and Matter, 1855), with its focus on the inorganic, and Charles Darwin's *Origin of Species* (1859) with its parallel rethinking of the organic. Both works embedded human development in a context of recursive interaction with the broader environment. Both should therefore be seen as participating in the reconceptualization of the inner dynamism of the human and non-human launched for example by David Hartley's theory of vibrations in his *Observations on Man, his Frame, his Duty, and his Expectations* (1749) and Ruggiero Guiseppi Boscovich's atomistic theory propounded in his *Theoria Philosophiae Naturalis* (Theory of Natural Philosophy, 1763) – works which were reprinted and re-edited several times in the eighteenth century. As Darwin followed his early work with *The Descent of Man* in 1871, so Büchner published *Der Fortschritt in Natur und Geschichte im Lichte der Darwinischen Theorie* (Progress in Nature and History in the Light of Darwinian Theory) in 1884. While the latter applies a Darwinian paradigm to a reordering of historical-cultural developments that recalls the decentered anthropology of Johann Gottfried Herder's cultural history, both Friedrich August Lange and Karl Eugen Düring captured the radical mood of the era in their respective assessments, *Geschichte des Materialismus und Kritik seiner Bedeutung in der Gegenwart* (History of Materialism and Critique of its Meaning for the Present, 1865) and *Cursus der Philosophie als streng wissenschaftlicher Weltanschauung und Lebensgestaltung* (The Course of Philosophy as Rigorously Scientific World View and Formation of Life, 1875). Both had a direct impact on Friedrich Nietzsche's concept of vitalism. More than them, however, Nietzsche underscored the connection between scientific and aesthetic creativity, when he speculated that there is an innate relationship between chemical and artistic processes. Indeed, in the wake of L. Büchner, Julius Robert Mayer, and Hermann von Helmholtz, Nietzsche insisted that we should in this context fundamentally reconceptualize the notion of force and the conservation of energy in the universe. All these perspectives are in one way or another traceable to the grand *Lebenskraft* debate that stretched from 1770 to 1900 and which Alfred Noll reconstructed in *Die "Lebenskraft" in den Schriften der Vitalisten und ihrer Gegner* ("Life Force" in the Writings of the Vitalists and their Opponents, 1918).

The present book is a response both to Noll's account and to a modern counterpart, Howard Tuttle's *Human Life is Radical Reality* (2005). Tuttle encapsulates the neutral monistic trend of *Lebenskraft* in contemporary terms. His radical reality of human experience is similar to Anthony Chemero's thesis in *Radical Embodied Cognitive Science* (2009), but is closer to the current project's emphases. Radical reality is the result of an ongoing recursive process: human life derives from and reconstructs itself in response to circumstances encountered in the environment. This continuous feedback loop gives rise to a dynamic "system of importances" in human existence that takes its cue from the circumstances themselves. Humans are fated to become human by incorporating the non-human. Noteworthy is Tuttle's emphasis on the interconnections between the organic and inorganic, between the human and non-human. But equally significant is his highlighting of the qualitative, performative aspects of human existence, for they alone give rise to meaning. Meaning, as Tuttle knew from his study of Wilhelm Dilthey, Martin Heidegger, and Ortega y Gasset, properly belongs not to scientific practice or discourse, but to the realms of philosophy, religion, the arts and literature.

Andrew Pickering provided a further noteworthy prompt to reevaluate the historical contours of the soul-body-mind inquiries of the long eighteenth and nineteenth centuries. His treatise, *The Mangle of Practice: Time, Agency and Science* (1995), published ten years earlier than Tuttle's work, had already shifted the perspective from cognition to performance. In doing so, he intuitively picked up on Johann Wolfgang von Goethe's earlier brand of science that factors both the subject and the object into scientific practices. Pickering's subsequent work is focused ever more on engaging nature in a directly performative fashion; that is, one that does not center itself on knowledge acquisition through the mechanisms of scientific inquiry, but rather on the ways that humans are framed by their environment and give expression to their lives. He speaks of a "dance of agency" between people and things. This too surely belongs in the context of the *Lebenskraft* debate as an inevitable result of the expression of energy.

The present volume, then, highlights explorations of the nexus of scientific investigation, medical theory, aesthetic thought, and literary production in the cultural discourse mode of Elinor Shaffer's *The Third Culture: Literature and Science* (1998) and George Levine's perhaps more aptly entitled *One Culture: Essays in Science and Literature* (1987). Both eschew the false divide between science and literature associated with the C.P Snow/F.R. Leavis controversy because upon closer inspection both science and literature can be seen as discursive: each grows and evolves in a larger social and cultural environment. Such qualities moreover relate our project to the process science proposed by Alfred North Whitehead in *Process and Reality* (1929). Science and literature prove to be continually emergent "cultures," that is, as an irreducible whole which is always caught up in becoming more than the sum of its constituent parts.

Thus from a very wide variety of standpoints the contributors to this volume consistently focus on the emergent truth construction evident in the interplay of aesthetic form, theories of vital force and advances in empirical science. Themes common to all the individual chapters thus include the concept of energy and its relation to matter, the notion of narrative in rendering what some see as a monistic *Weltanschauung*, reflections on what constitutes "radical" reality, and the relevance of aesthetics in forming and evaluating scientific experimentation and its resulting paradigms.

John McCarthy's Introduction provides a general framework for locating the individual case studies that follow. Their focus ranges from the physical museal space of the *Hallesche Kabinetsammlung* to the interior spaces of body and mind in music, literature, and psychology. In the concluding section, Heather Sullivan's exploration of the Anthropocene serves as a prism through which one can view Goethean science around 1800 and contemporary ecocritical, holistic approaches to understanding the pervasive nature of reality throughout human life. And Monica Ledoux's inquiry into the role of the microbiome adds a further layer of complexity to the radical nature of mediation as primary and recurrent in human interactions with self, other, and environment. It provides a fitting conclusion to the tangled web of human physiology, medical treatment and philosophical speculation, from which the complex yet radical concept of the human soul emerged.

John A. McCarthy, Brentwood TN, January 2015

Establishing Parameters: *Lebenskraft* and Artifact

Introduction: Life Matters

John A. McCarthy

> "'Body am I, and soul' – thus speaks the child. And why should one not speak like children? But the awakened and knowing say: body am I entirely, and nothing else; and soul is only a word for something about the body."[1]

Abstract: In 1914 Alfred Noll published a seminal volume, *Die Lebenskraft Debatte* (*The Life-Force Debate*), which traced key stages in the development c.1770-c.1920 of one of the most significant and enduring controversies in human knowledge between the two cultures of humanities and science: the constitution of life itself. This introduction provides essential background information to the so-called *Lebenskraft* debate and in so doing aims to contextualize the case studies included here. The discourse of life force, or vitalism, sought a unitary origin of biological life, human consciousness, and psychophysical well-being. Life-force conceptuality manifested itself across the entire spectrum of discourses on medicine, philosophy, creative arts, and physical and mental health. Leibniz and Spinoza are identified as pivotal early modern contributors to the emergence of vitalist discourse, while the energies of the *Lebenskraft* debate are argued to resonate still in contemporary neurology, psychology, literature and film. Finally, meme theory in the wake of Richard Dawkins is seen as a logical heir of this historical debate.

Setting the Stage

1973 saw the republication of one of the most striking autobiographies of the mentally ill: Daniel Paul Schreber's *Denkwürdigkeiten eines Nervenkranken* (Remarkable Experiences of a Psychiatric Patient, 1903).[2] Schreber is notable as one of the most frequently cited psychiatric patients. After Sigmund Freud scrutinized Schreber's biography in his "Psychoanalytische Bemerkungen über einen autobiographisch beschriebenen Fall von Paranoia" (Psychoanalytical Comments on an Autobiographical Case of Paranoia, 1911) that emphasized the castration and father complexes, psychiatrists have turned to *Denkwürdigkeiten eines Nervenkranken* as an exemplary case study. For instance, New York psychoanalyst William G. Niederland

1 Friedrich Nietzsche, *Thus Spoke Zarathustra: A Book for All and None*, trans. with a Preface by Walter Kaufmann, The Modern Library (New York: Random House, 1995), "On the Despisers of the Body," p.34

2 Daniel Paul Schreber, *Denkwürdigkeiten eines Nervenkranken*, ed. by Samuel M. Weber (Frankfurt: Verlag Ullstein, Frankfurt, 1973). See also *Der Fall Schreber 1: um Neunzehnhundert, mit Aufsätzen von Franz Baumeyer*, ed. by Peter Heiligenthal and Reinhard Volk (Wiesbaden: Focus-Verlag, 1973).

(1904-1993), who had been studying the case since 1951, published his *The Schreber Case: Psychoanalytic Profile of a Paranoid Personality* in 1974. He considered Freud's interpretation to be the most important contribution to the study of psychosis.[3] London psychotherapist Morton Schatzman examined Schreber's biography in *Soul Murder: Persecution in the Family* (1973); the German translation appeared a year later.[4] Schatzman adopted an "anti-Freudian" approach by emphazing the role played by Schreder's repressive father, Daniel Gottlob Moritz Schreber (1808-61), offering an interpretation that differed markably from Freud's. In contrast to Schatzman, Niederland used a kind of psychogeography to highlight how body and soul are intertwined; that is, his interest was in the way issues, experiences, and processes that result from growing up in a human body play out in the mind and are symbolized in the wider social and natural worlds. The difference in approach by Schatzman and Nederland prompted a reviewer in *Der Spiegel* to draw the following poignant conclusion about changing theories of altered mental states:

> Gerade am 'Fall Schreber' nämlich ist abzulesen, wie radikal sich die Theorien über Psychose, Schizophrenie und Paranoia verändert haben und weiter verändern – von der klinischen Psychopathologie, die in der Geisteskrankheit ein geheimnisvolles organisches Leiden sieht, bis hin zu einer gegenwärtig sich neu orientierenden Schizophrenie-Forschung, die den Wahn vor allem als historisch-erkenntnistheoretisches, anthropologisches und soziales Problem begreift.[5]
>
> (Precisely in the Schreber case, namely, can we discern how radically theories of psychosis, schizophrenia, and paranoia have changed and continue to change, ranging from clinical psychopathology, which sees in mental illness a mysterious organic malady, to contemporary new orientations in research on schizophrenia that understand mania above all as a problem with historical-epistemological, anthropological, and social dimensions.)

In the forty years since then, major strides in pharmaceutical treatment and promising developments in gene therapy underscore the interconnectedness of body and spirit.

But what was so remarkable about Schreber's experiences in relation to the title of this introductory essay: Life Matters? The double entendre is intended. Life is about the quality of one's mental state, but it is also about the

---

3 William G. Niederland, *Der Fall Schreber : Das psychoanalytische Profil einer paranoiden Persönlichkeit* (Frankfurt a. M.: Suhrkamp, 1978), p.11.

4 Morton Schatzman, *Soul Murder. Persecution in the Family* (New York: Random House, 1973); Morton Schatzman, *Die Angst vor dem Vater* (Rowohlt, Reinbek bei Hamburg 1974).

5 anon., "Der Dutt, der keiner war," *Der Spiegel*, 36 (03.09.1973), 129-32.

physiological condition of the material body that encapsulates the mind. Ultimately, life matters for both mind and body; it is a question both of mental operations and chemical-organic ones. The question of how they are related is a core issue.

For nine years (1893-1902) Dr. Schreber, one-time President of the Saxon Senate in Dresden, was hospitalized in the Saxon state asylum Schloß Sonnenstein, diagnosed with *dementia paranoides.* Eventually, Schreber sued for having been committed to the asylum against his will and that, hence, his right to self-determination had been violated. Because of his own keen powers of objective analysis aimed at himself as well as his environment in defending his right to self-determination – during the final years of his internment Schreber had scrupulously written down his detailed self-observations – the court ruled in his favor. Upon his release, he published those observations. To be sure, Schreber would rant and rave about God, who sought to emasculate and then impregnate him; he did rail at the sun as the great whore and swear homunculi were trying to suck the marrow from his bones; and he would lose himself in transgendered sexual fantasies. But his doctors confirmed that he was otherwise well-behaved, a pleasant conversationalist, a superb chess player, and an accomplished pianist. What, in this close-woven tissue of consciousness, was real, what imagined? Indeed, can the entwinement of the two realms be explained physiologically?

In his monograph, Niederland spins out connections between the child's exposure to the father's nigh-on sadistic corporal training and the hallucinations the adult son rendered in his biography. The father Moritz Schreber was not only the initiator of the "Schreber Gärten," he was also a noted orthopedist, naturopath, gymnast, and inventor of mechanical devices to correct a child's posture ("Schreberscher Geradehalter," "helmartiger Kopfhalter"). Niederland concludes that the father's rigid educational methods traumatized his son by the age of ten. These corrective devices appear in morphed form in the paranoid illusions of the *Denkwürdigkeiten.* This strange case stands at the end of the time frame examined in the following chapters of this book. The critical attention accorded it as a fount of insights into the interactions of body and mind qualify Schreber's stories and medical treatments applied as late manifestations of the *Lebenskraft* debate that dominated much scientific discussion in the long nineteenth century and peaked around 1900 (cf. Stephanie Hilger's study in this volume).

Similarly, another more recent work that has enjoyed wide recognition establishes the opening frame for this undertaking: Daniel Kehlmann's *Die Vermessung der Welt* (Measuring the World, 2004). In his best-selling novel, Kehlmann renders a fictive conversation between the brothers Alexander (1769-1859) and Wilhelm von Humboldt (1767–1835) during a walk through a park following a visit to the renowned salon of Henriette and Marcus Herz, the intellectual hub of Berlin society, she a leading woman intellectual, he a leading physician of the day.[6] Their topic is the medical doctor and philosopher Julien Offray de La Mettrie's (1709-51) *L'homme machine* (1748). Alexander remarks:

> Es ist ein bemerkenswertes Buch. Der Autor behaupte ernstlich, der Mensch sei eine Maschine, ein automatisch agierendes Gestell von höchster Kunstfertigkeit. … Und ohne Seele, antwortete der Ältere. Sie gingen durch den Schloßpark; der Schnee lag dünn auf kahlen Bäumen. ... Nein, widersprach der Jüngere. Mit Seele. Mit Ahnungen und poetischem Gespür für Weite und Schönheit. Doch sei die Seele selbst ein Teil, wenn auch der komplizierteste, der Maschine. Und er frage sich, ob das nicht der Wahrheit entspreche.[7]
>
> (It is a remarkable book. Its author seriously asserts that the human being is a machine, a self-activated, finely crafted contraption. … And without a soul, the older brother commented. They made their way through the castle grounds; light snow lay on the barren trees. … No, the younger one countered. With a soul. With intuitions and a poetical appreciation of vastness and beauty. Nonetheless the soul was still a part of the machine, albeit the most complex. And he asked himself whether that was not the truth of the matter.)

A few pages later Kehlmann narrates young Alexander von Humboldt's epiphany about his life's calling: "Er wolle das Leben erforschen, die seltsame Hartnäckigkeit verstehen, mit der es den Globus umspanne. Er wolle ihm auf die Schliche kommen!" (He wanted to investigate life, to fathom that strange tenacity with which it embraced the globe. He wanted to read its mind; Kehlmann, *Vermessung*, p.26; my translation). Later in a cave he is amazed to see life forms thriving deep under the earth and exclaims that life seems to know no boundaries (p.30). Quite by accident he then encounters Luigi Galvani's (1737-98) book on animal electricity with its description of the frog-leg experiment, *De viribus electricitatis in motu musculari commentarius* (Commentary on the Electrical Energies in Muscular Dynamics, 1791) and decides to conduct such an experiment using his own body as a conductor in order to determine whether the *Lebenskraft* is a substance native to the living organism or the extrinsic result of electro-chemical reactions (p.31). The brutal experiment led to Humboldt's treatise on "Muskelfaser als leitende

---

6 On the Herzes in Berlin see Debora Hertz, *Jewish High Society in Old Regime Berlin* (New Haven: Yale UP, 1988).

7 *Die Vermessung der Welt*, pp.23-24. See also de La Mettrie's earlier *Histoire Naturelle de l'Âme* (A. Oxford,1747)

Substanz" (muscle fiber as conductive substance) which established his scientific reputation.[8]

This section of the novel is remarkable because it recalls in rapid succession monist de La Mettrie's assertion that the human body is a machine, tells us that Wilhelm von Humboldt wondered whether the soul might, in fact, be a highly complex component of the corporal machine after all, then recounts Alexander von Humboldt's Galvanic experiments on his own body to determine whether *Lebenskraft* is intrinsic or extrinsic to the living organism. Kehlmann's narration drives to the heart of this current project on fathoming the nature of life and the relationship between body and soul. In addition to focusing on the inquisitive geographer and botanist Humboldt, Kehlmann's novel also highlights the pioneering work of the mathematician Carl Friedrich Gauß. Kehlmann uses the two historical figures to draw emphatic attention to revolutionary efforts to measure scientifically the world mathematically and organically, to assess its heights and depths and to classify its contents. In similar fashion, the present volume, in however tenuous a manner, seeks to delineate major trends in understanding *Lebenskraft*; that is, to look more deeply into what constitutes life. The effort is an historical take on what preceded and led to our own age of the Anthropocene with its understanding of distributed agency in contrast to the classical binary notion of a single subject acting on passive matter (see the chapter by Heather Sullivan). Ultimately, the historical essays in this volume prepare the way for a radicalized view of the human body – each with its own microbiome – as already and always mediated in the world it inhabits (see the chapter by Monica Ledoux). Our own epoch of revolutionary neuroscience can be viewed as a logical extension of the creative force behind Gauß's number theory and differential geometry as well as Alexander von Humboldt's geobiological studies.

Seeking the Core

More specifically, the present volume offers a particular take on how we arrived at the present thinking about body and soul. Sounding very much like Goethe's Faust, who sought that which holds the universe together at its very core, Northrop Frey expressed the need in his *Anatomy of Criticism* (1957) to identify an "organizing principle, a central hypothesis" – analogous

8 Kehlmann, *Vermessung*, pp.31-33. The *Göttingischen Anzeigen von gelehrten Sachen* published a rave review on 26 September 1795 of Galvani's anonymously published *Dell' Uso e dell' Attivita dell' Arco donduttore nelle contrazioni dei Muscoli* (On the Use and Activity of the Conductive Arch in the Contraction of Muscles, 1794), the same time frame as Kehlmann's narration. The study confirmed his earlier recognition of an "extrinseca sostanza deferente" (p.1558).

to evolutionary theory in biology or the significance of gravity in explaining motion in the universe – that would allow a view of disparate parts actually consisting of a whole. Frey longed for a way of thinking that would allow the stages of development in literary endeavors across epochs and disciplines to become manifest. "We begin to wonder," he mused, "if we cannot see literature, not only as complicating itself in time, but as spread out in conceptual space from some kind of center that criticism could locate."[9] The approach in this volume resembles that search for some kind of center toward which researchers from different disciplines tend to gravitate. That center is life, and our contributors ask: what is life, and where do we locate value in life's plenitude of multiple forms? Is there a ghost in the machine, or is the soul merely the most finely wought part of the housing?[10]

The long nineteenth century – for our purposes 1770-1914 – experienced a radical redrawing of the lines between inanimate and animate matter. With the emergence of the energy table by century's end – today known as the energy matrix – the age-old body-soul debate took a new turn. To be sure, it was not a radical and sudden turn. It was, as Wolfgang Pauli was to put it at the end of our time frame, more a matter of persistent inquiry, of "continuously taking up an issue, thinking about the object, then putting it back down again, then gathering new empirical material, and continuing this [...] for years. In this way, the subconscious is stimulated by the conscious, and only in this way, if at all, can something come of it."[11] Eventually, "patterns of significance" (Frey, *Anatomy of Criticism*, p.16) and patterns of matter come into alignment. Atoms and archetypes seem to be related via deep structures. In seeking to emphasize the concept of patterns, this introduction offers a framework for aligning the essays in the following pages.

At least a partial reconstruction of the particular role that the *Lebenskraft* debate played in the emergence of "radical reality" will prove helpful in assessing how the individual contributions collected here fit together like pieces of a puzzle, the kind of endeavor that Daniel Kehlmann means with

9 Northrup Frey, "Polemical Introduction," in N.F., *Anatomy Of Criticism Four Essays* (Princeton UP, 1957), pp.3-29 (pp.15-16). (source: http://northropfrye-theanatomyofcriticism.blogspot.com/2009/02/polemical-introduction.html)

10 Gilbert Ryle coined the phrase, "ghost in the machine" in rejecting Cartesian dualism. See his "Descartes's Myth," in *The Concept of Mind* (Hutchinson, London, 1949). See also Arthur Koestler, who explored Ryle's notion further in his *The Ghost in the Machine* (Hutchinson: London, 1967).

11 Cited by Ernst Peter Fischer, "Going Bravely into the Unexplored," *Max Planck Research: Science Magazine of the Max Planck Society*, 1 (2006), 15-18 (p.18). *Atom and Archetype: the Pauli/Jung letters, 1932-1958* ed. by C.A. Meier and translated from the German by David Roscoe with an introductory essay by Beverley Zabriskie (Princeton UP, 2001).

his references to Alexander von Humboldt in *Measuring the World*. Equally important in this regard is the chapter by Jennifer Hellwarth. She traces seminal developments in the evolution of the multivalent concept of πνεῦμα (*pneúma*: vital, unifying force) from ancient Persia, Arabia and Greece through medieval Christian times to the new, empirical sciences of the seventeenth century. It also impacted the theory of humors propounded by Aristotles and elaborated upon by Galan. The term is rich with physiological, cosmological, and cultural meanings that shaped conceptions of reality beyond medical practice itself. Variously referring to vital heat, active principle, aer, cause, soul – *pneúma* formed the implicit (if not explicit) basis of the *Lebenskraft* debate that emerged in the early eighteenth century and ran through the nineteenth. This introduction draws attention to the contours of that debate over the span of ca. 150 years. The chapter by Ingo Uhlig augments these two historical surveys with keen insights into the remains of lived life, aesthetically structured to convey an appreciation of the wholeness from which they sprang and to reinvigorate them with motion or *vis activa primitiva*, a concept Uhlig traces to one of Gottfried Wilhelm Leibniz's four ontological expressions of substantial force (primitive active force, primitive passive force, derivative active force, and derivative passive force). Leibniz's focus on dynamism marked a break with Cartesian ontology and helped lay the ground work for process philosophy that Alfred North Whitehead labelled organic realism in his *Process and Reality* (1929).[12] Some contributors, but not all, explicitly argue against the backdrop of the historical and thematic structuring manifest in the opening chapters. One goes well beyond our main time frame to show how the supposedly self-sufficient human body is actually host to billions of symbiotic microbes (Monica Ledoux). Some but not all have recourse to literary examples of the forms *Lebenskraft* can assume.

Literary historians are well aware of the vital forces that figure prominently in literary works of the era foregrounded here. Popular ones immediately come to mind: Kleist's *Über das Marionettentheater*, E. T. A. Hoffmann's *Der Sandmann*, Georg Büchner's *Woyzeck*, Friedrich Nietzsche's *Also sprach Zarathustra*, Gerhart Hauptmann's *Bahnwärter Thiel*, or Max Nordau's *Seifenblasen* – not to speak of non-German examples such as *Frankenstein*, *Dracula*, and *Dr. Jekyll and Mr. Hyde*. Cinematic classics include Carl Boesse and Paul Wegener's *Der Golem*, Robert Wiene's *Das Cabinet des Dr. Caligari*,

12 Alfred North Whitehead, *Process and Reality: An Essay on Cosmology*, corrected edition, ed. by David Ray Griffin and Donald W. Sherburne (New York: The Free Press, 1978). Originally published as *Process and Reality. An Essay in Cosmology. Gifford Lectures Delivered in the University of Edinburgh During the Session 1927–1928* (New York and Cambridge UK: Macmillan and Cambridge UP, 1929).

Ridley Scott's *Alien* and *Prometheus* diptych, and Andy and Lana Wachowski's *The Matrix*. But how much 'note do humanists generally take of the philosophical and scientific inquiries into the origins of life and the rise of human consciousness that underlie these artistic treatments? Without life, without consciousness, no art, no awareness of beauty, tragedy, or profundity. In the preface to his new cosmology of relations Alfred North Whitehead saliently remarked: "[…] it must be one of the motives of a complete cosmology to construct a system of ideas which brings the aesthetic, moral, and religious interests into relation to those concepts of the world which have their origin in natural science."[13] And where in the new radical reality do we plug in the creative impulse? For Whitehead creativity is inherent in the process of becoming; is it the novelty in reality itself; a "lure to creative emergence" in which the perceiving subject is a participant (Whitehead, p.21, 263; cf. also 28, 128, 187, 222, 346). Or it might, as physicist Ernst Peter Fischer suggests, be a matter of archetypes that we seekers project onto reality to tease out of nature mirror images of our imaginings?[14]

While providing the background in this introduction for a renewed look at literary and artistic (especially musical) productions of life energy, I do not analyze any of them directly, although several of the contributors offer valuable insights (e.g., Alice Kuzniar, Cate Reilly, Nicholas Saul, Alexis Smith, Ann Schmiesing,). Noteworthy for its encyclopedic expanse is James Kennaway's survey of reverberations of the vitalism debate between the history of medicine and musical theory throughout the time frame of this volume. Ultimately, models of will power and self-control emerged as crucial markers of health (or feebleness), culminating in Nietzsche's medical-based critique of Wagner's music as both symptom and cause of pathologically weak will power.

Radical Reality

"Radical reality" raises suspicions about the essence of existence by essentializing naturalism. Naturalism posits a single world knowable by science, a world that is both caused and itself an agent for change. The term

---

13 North Whitehead, *Process and Reality* (1978), p.xii. Accessed 6 Dec. 2014. https://archive.org/stream/AlfredNorthWhiteheadProcessAndReality/Alfred%20North%20Whitehead%20-%20Process%20and%20Reality#page/n9/mode/2up.

14 E. P. Fischer, "Going Bravely into the Unexplored," p.18. See also John A. McCarthy, *Remapping Reality: On Chaos and Creativity in Science and Literature* (Amsterdam & New York: Rodopi, 2006). To be sure, Northrop Frey accorded archetypes a central role in his *Anatomy of Criticism* (Princeton UP, 1957), pp.95-114 and 131-242: http://northropfrye-theanatomyofcriticism.blogspot.com/2009/02/source.html. Accessed 21 Aug. 2014

"radical reality" betrays a humanistic orientation, for talking about radical reality is not something that occurs to nanophysicists, microbiologists, or neuroscientists. Rather, the concept stands in opposition to traditional rationalism that had led to the construction of metaphysical systems (e.g., Kant, Hegel). In an expanded, holistic sense, "radical reality" places the body at the center of deliberations without seeking to privilege the inquiring human mind as an anomaly in the physical world. Howard Tuttle gives apt expression to the concept in his *Human Life is Radical Reality* (2005):

> Through the absorbing of circumstances the human being constructs the world. By the use of a life form which ancient Greek philosophy designated as pragmata, human beings perform the actions which 'must be done' for the uses and needs of human life. Such humanization of circumstances allows us to develop our lives in the 'system of importances,' which surrounds us. Our life is a continual reabsorption of circumstances, a continual working out of our lives through them. In this respect human beings are doomed to achieve themselves through something else, to become human through the absorbing of the non-human.[15]

Noteworthy in the current context is the emphasis Tuttle places not only on the interconnections between the organic and inorganic, between the human and non-human, but also on the qualitative aspects of human existence that give rise to meaning. These qualitiative aspects include, of course, mental agility and the creative imagination which result from chemical-electrical processes involved in human thought. This mental energy is part and parcel

---

15 Howard N. Tuttle, *Human Life is Radical Reality: An Idea Developed from the Conceptions of Dilthey, Heidegger, and Ortega y Gasset* (New York: Peter Lang, 2005), pp.97-100, 153-59 (p.158). The Center for Naturalism provides a pithy definition of naturalism that resonates with Tuttle's "radical reality": "Naturalism is the understanding that there is a single, natural world as shown by science, and that we are completely included in it. Naturalism holds that everything we are and do is connected to the rest of the world and derived from conditions that precede us and surround us. Each of us is an unfolding natural process, and every aspect of that process is caused, and is a cause itself." http://www.centerfornaturalism.org/descriptions.htm. Accessed 17 Aug. 2014. Thelma Lobel, *Sensation. The New Science of Physical Intelligence* (New York: Atria Books, 2014), speaks of the "tangled web our senses weave," presenting her views as a new investigation of the intimate connection between body, mind, and our sense of self. Other typical works from a cognitive-science perspective include Lawrence Shapiro's *Embodied Cognition* (New Problems in Philosophy) (New Brunswick NJ: Routledge, 2010) and Anthony Chemero's, *Radical Embodied Cognitive Science* (Cambridge: MIT Press, 2009). They are interested in mental representational rather than in the nexus of the inanimate, animate, and mental. Thus, they are unaware of Whitehead and do not seem to realize that the roots of cognitive science reach deeper than William James' pragmatism.

of the total energy of the universe. Despite its transformations from one form to another, the total energy of the system remains constant. In this light, Gottfried Wilhelm Leibniz's designation of God, or *Urmonade*, as the ultimate *puissance* reflected throughout creation seems rather foreward looking.[16]

A major impact on monist thinking in our time frame derives from Baruch de Spinoza. Jonathan Israel has argued persuasively in *Radical Enlightenment* (2001) that Spinoza emerged "as the supreme philosophical bogeyman of Early Enlightenment Europe." With his philosophical system he imparted "shape, order, and unity to the entire tradition of radical thought," by which Israel essentially meant a strong atheist bent. The Dutchman became the source for a subsequent systematic redefinition of man, cosmology, politics, social hierarchy, sexuality and ethics. With his theory – especially as it is expressed in the first two parts of his *Ethics* (1675) – that the universe consists of a single substance and that thought and extension are attributes of the world, Spinoza caused a revolution of thought. (Extension is a concept derived from the scientific understanding of the geometrical properties of matter.) Thus the seeming duality of his concepts *natura naturans* (i.e., creative potential, energy, thought) and *natura naturata* (i.e., realized potential, materialized energy) converge into one. Israel summarizes: "all manifestations of mind or minded-substance are part of Spinoza's single thinking-extended substance and therefore governed by the same set of rules – the laws of nature – as any other part of *Natura Naturata*."[17] One can see how Spinoza could provide the basis for the later materialism of Julien Offray de La Mettrie (1709-51), Denis Diderot (1713-84), Claude Adrien Helvétius (1715-71) and Paul Henri Dietrich Baron d'Holbach (1723-89).

Equally important for the later *Lebenskraft* debate was the work of the Swiss experimental physiologist, Albrecht von Haller (1708-77), whom Brian McInnis discusses in detail. Known to literary scholars as the author of the widely acclaimed didactic poem, *Die Alpen* (The Alps, 1732), he is best known for his *Icones anatomicae* (1756) and eight-volume study of the human anatomy, *Elementa physiologiae corporis humani* (1757–66) (Elements of Physiology of the Human Body). His *Icones anatomicae* was greatly admired for it rendered for the first time in exquisite detail the arterial network of the human body graphically illustrated by the artist C.J. Rollinus. As a summation

16 See for a detailed treatment of this theme J. A. McCarthy, "The Little Divine Machine: The Soul/Body Problem Revisited," in: *Neuroscience and Religion: Brain, Mind, Self, and Soul*, ed.by Volney P.Gay (Lanham & Boulder: Rowman & Littlefield, 2009), pp.241-269.

17 Jonathan Israel, *Radical Enlightenment: Philosophy and the Making of Modernity 1650-1750* (Oxford: Oxford UP, 2001), pp.159-74, 230-52 (pp.175, 230-31, 233).

of the state of knowledge about human anatomy and physiology, the *Elementa physiologiae* gave new meaning to Alexander Pope's secular commandment in his *Essay on Man* that "the proper study of mankind is man." It became a standard work in the history of medicine, with new editions well into the twentieth century.

With references to Haller and to Alexander Pope, Kant penned his precritical treatise on the origin and operation of the universe, *Allgemeine Naturgeschichte und Theorie des Himmels* (Universal Natural History and Theory of the Heavens, 1755), labelled "philosophische Träume" (philosophical dreams), probably because he did not offer new empirical evidence but only a critical examination of the logic of recent and ancient cosmologies against the backdrop of Newtonian laws of motion.[18] Together with his earlier *Gedanken von der wahren Schätzung der lebendigen Kräfte* (Thoughts on the True Estimation of Living Forces) (1747) that examined theories of *vim activam* (ebullient energy) and *vim morticem* (decreasing energy) – especially those of Leibniz – Kant early on signaled the revolution of thought that came to characterize the importance of his later critical writings. That mental revolution ranged for Kant far beyond the scope of systematic philosophy to include the cosmic and microscopic scale along with the meso or human one. Actually, he began his career with natural science questions. As a result Kant seems to have been a kind of universal "cosmic constant" in the background of the debate on *Lebenskraft*. In his debate with Leibniz's concept of an inherent "lebendige Kraft" that causes movement, logic led Kant to conclude that all motion is essentially driven by external causes. And that insight provided the impetus for the *Allgemeine Naturgeschichte*. Kant explained that his philosophical ruminations were far from being "willkürliche Erdichtungen" (arbitrary fictions):

> Ich habe mich in der Tat mit größester Behutsamkeit aller willkürlichen Erdichtungen entschlagen. Ich habe, nachdem ich die Welt in das einfachste Chaos versetzt, keine andere Kräfte als die Anziehungs- und Zurückstoßungskraft zur Entwickelung der großen Ordnung der Natur angewandt, zwei Kräfte, welche beide gewiß, gleich einfach und zugleich ursprünglich und allgemein sind. Beide sind aus der Newtonischen Weltweisheit entlehnet. (Kant, *Werke*, vol. I, p.242)

> In fact, I have rejected with the greatest care all arbitrary fictions. After I placed the world in the simplest chaos, I have applied to it no forces other than the powers of attraction and repulsion, so as to develop the great order of nature.

18 Immanuel Kant, *Allgemeine Naturgeschichte und Theorie des Himmels* (1755), in *Werke in zehn Bänden*,, ed. Wilhelm Weischedel, 10 vols (Darmstadt: Wissenschaftliche Buchgesellschaft, 1983), vol. I, p.242. Kant cites, among others, Epicurus, Lucretius, Johannes Kepler, Christiaan Huygens, James Bradley, Thomas Wright, Edmond Halley, Pierre Louis Moreau de Maupertuis.

> These two forces are both equally certain, equally simple, and at the same time equally primal and universal. Both are taken from Newtonian philosophy.[19]

An early step in Kant's revolutionary thinking is his theory of the role nebulae play in the formation of the universe. The forces of attraction and repulsion caused widely dispersed, thin clouds of dust and gasses – atoms according to the ancient Greeks – to collapse in upon themselves via a spinning action. Denser bodies are located closer to the center of attraction, whereas bodies grow ever lighter and more diffuse the further removed from the center they are. The gravitational spinning action can also explain the rotation of the earth and other planets, he mused.[20] Ultimately, particularly after Pierre Simon Laplace (1749-1827) provided the mathematical evidence in support of Kant's ideas on the role of nebulae as the nurseries of the cosmos, Kant's theory became the accepted model for star and planetary formation. Moreover, in *Universal Natural History and Theory of the Heavens* Kant also theorized that distant nebulae, which many thought to be circular, were elliptical figures when viewed from the side. His insightful suggestion was to view the nebulae as actually constituting separate galaxies similar to our own Milky Way. Emphasizing the role of gravitational forces, he thus conjectured: "all these endorse perfectly the idea that these elliptical figures should be taken as exactly the same world systems and, so to speak, as Milky Ways."[21] Thus, he offered a multi-galaxy theory; the universe consists possibly of billions of separate galaxies. This he envisioned long before technological advances in telescope and lens construction in the early twentieth century provided Edward Hubble with the empirical evidence to confirm the theory.[22]

---

19 Immanuel Kant, *Universal Natural History and Theory of the Heavens Or An Essay on the Constitution and the Mechanical Origin of the Entire Structure of the Universe Based on Newtonian Principles*, translated 2008 by Ian Johnston., "Preface," accessed 5 August 2014: http://records.viu.ca/~johnstoi/kant/kant2e.htm#t5.

20 Kant, *Allgemeine Naturgeschichte und Theorie des Himmels*, I, pp.266, 326-44. English version: Kant, *Universal Natural History and Theory of Heavens,* trans. by I. Johnston, Part Two, sections 1 and 7: http://records.viu.ca/~johnstoi/kant/kant2e.htm. Accessed 5 August 2014

21 Weischedel, *Kants Werke* (1983), vol. I, p.266: "Alles stimmet vollkommen überein, diese elliptische [sic] Figuren vor eben dergleichen Weltordnungen, und, so zu reden, Milchstraßen zu halten." More elaborately explained in Zweiter Teil, 8. Hauptstück, pp.365-70. Kant, *Universal Natural History and Theory of the Heavens*, trans. Johnston, Part One, "On the Systematic Arrangement of the Fixed Stars."

22 See Owen Hempton, "Immanuel Kant: Discoverer of Nebulae and the Multi-galaxy Universe" (18 Feb. 2014): http://www.fromquarkstoquasars.com/kant/. Accessed 5 August 2014.

By arguing in his 1747 treatise, *Gedanken von der wahren Schätzung der lebendigen Kräfte,* against Leibniz's concept of internal "life forces" of movement, Kant would seem to be in opposition to the debate that would unfold in the years to follow. Yet when he subsequently turned his attention to the origins and structure of the universe resultant from two simple forces of gravity, Kant gestured toward humankind's embeddedness in the totality of nature, a realm, he insisted, that is marked by limitless reproductive ability that entailed no loss of energy. Kant concludes the pivotal eighth section of his *A Universal Natural History* with the words:

> Die Natur [...] fasset in dem Umfange ihrer Mannigfaltigkeit alle mögliche [sic] Abwechselungen, sogar bis auf die Mängel und Abweichungen, in sich. Eben dieselbe unbeschränkte Fruchtbarkeit derselben hat die bewohnten Himmelskugeln sowohl, als die Kometen, die nützlichen Berge und die schädlichen Klippen, die bewohnbaren Landschaften und öden Wüsteneien, die Tugenden und Laster, hervorgebracht. (Kant, *Werke*, vol. I, p.374)

> Regardless of the fact that it is essentially established for perfection and order, nature includes in itself in the range of its multiplicity all possible changes, even deficiency and deviation. Just this unlimited fecundity of nature has produced the inhabited celestial globes, as well as the comets, the useful mountains and the harmful cliffs, the habitable landscapes and barren deserts, the virtues and vices. (Johnston, *Universal Natural History*, end of Part II, section 8)

Do we not sense here an anticipation of the first law of thermodynamics, formulated a century later? And could we not read the preface to his Universal history as an "ecocritical" statement *avant la lettre* when, for instance, he explains:

> Die nach ihren allgemeinsten Gesetzen sich bestimmende Materie bringt durch ihr natürliches Betragen, oder, wenn man es so nennen will, durch eine blinde Mechanik anständige Folgen hervor, die der Entwurf einer höchsten Weisheit zu sein scheinen. Luft, Wasser, Wärme erzeugen, wenn man sie sich selbst überlassen betrachtet, Winde und Wolken, Regen, Ströme, welche die Länder befeuchten, und alle die nützlichen Folgen, ohne welche die Natur traurig, öde und unfruchtbar bleiben müßte. (Kant, *Werke*, vol. I, p.231)

> Matter, which organizes itself according to its most general laws, produces through its natural behaviour or, if we prefer, through a blind mechanical process, good consequences, which appear to be the design of a supremely High Wisdom. When we observe air, water, and heat left to themselves, they produce wind and clouds, rain, streams which moisten the lands, and all the useful consequences without which nature would have had to remain sad, empty, and barren. However, they produce these results not through mere chance or accident, which could just as readily have resulted in something detrimental. But we see that

these consequences are limited by its natural laws so as to work only in this way.[23]

Kant uses the concept of Divine Intelligence ("Entwurf einer höchsten Weisheit") not to explain that God creates order out of chaos, but to encourage his conservative religious readers (and censors!) to accept his argument that internal laws govern the universe and earth's biosphere. These laws are sufficient in and of themselves to transform "die Materie, die der Urstoff aller Dinge ist" into awe-inspiring order that is far from being accidental in his mind.[24] That is why Kant speaks of the "supremely Highest Wisdom" and not Deity or Divine Being. Order out of chaos due solely to the laws of nature is, in the eyes of the contemplating human observer, seemingly wise, indeed.

The *Lebenskraft* Debate

Far more widely acknowledged in the *Lebenskraft* debate are the views of the French naturalist Jean Baptiste Pierre Antoine de Monet, Chevalier de Lamarck (1744-1829), who proposed the notion of inherited characteristics in his *Philosophie zoologique* (*Zoological Philosophy*, 1809). The essence of Lamarckism is, as is well known, the insight that all changes in inorganic and organic matter are the result of laws, not of miraculous intervention.[25] He argued that the physiological needs of organisms resulted from their interactions with the environment, an idea that recurs with force in the philosophy of Arthur Schopenhauer and later that of Nietzsche (see Kennaway and Cate Reilly in this volume).

Although the roots of the debate reach back to the organicism of German philosopher G. W. Leibniz around 1700 and the hypothesis of a single "living filament" as the "cause of all organic life" of the English evolution-

---

23 http://records.viu.ca/~johnstoi/kant/kant2e.htm#preface.

24 "The material which is the primordial stuff for all things"; Weischedel, *Kants Werke*, vol. I, p.234. While Kant here pursued theoretical physics, what MIT physicist Alan Lightman has called the "the deepest and purest branch of science [...] the outpost closest to philosophy, and religion," Kant knew nothing of the "multiverse" concept derivative of such contemporary ideas as dark matter, "eternal inflation," and "string theory" all of which suggest that the universe we inhabit is, in fact, an "accidental" universe because its laws might be unique unto itself with no relevance for multiple others universes with their own fundamental principles of operation. Nonetheless, Kant's search for simple and fundamental principles of logic and mathematics to explain our universe is mainstream. Cf. Alan Lightman, "The Accidental Universe: Science's Crisis of Faith," *Harpers Magazine*, December 2011, 34-40 (pp.35-36, 38).

25 Cf., e.g., Ann Thomson, *Bodies of Thought: Science, Religion, and the Soul in the Early Enlightenment* (Oxford UP, 2008), rev. by Michael Printy in: *European History Quarterly*, 42 (2012), pp.374-375. http://ehq.sagepub.com/cgi/reprint/42/2/374

ary biologist Erasmus Darwin, I focus more narrowly on the long nineteenth century.[26] The internal German debate regarding the interconnections between the parts and the whole and the concept of "life force" that span the era from Lamarck to Darwin and Nietzsche has, for the most part, fallen off the radar screen.[27]

In physiology the German debate centered on the concept of *Lebenskraft* as a kind of corollary to the nature and function of gravity in astronomy. Terms such as *Kraft* and *Energie* were central to both dimensions of the debate; the terms began to turn up in various discourses by 1800, even playing a central role in Schiller's aesthetic theory.[28] To the classical understanding of force and energy belong such related ideas as creative wind/breath

---

26 Cf. Gottfried Wilhelm Leibniz, *Monadology* (1714) where he pithily formulates such concepts as God understood as the "Supreme Substance" ((§40); "each organic body is a kind of divine machine" (§64); "there is nothing waste, nothing sterile, nothing dead in the universe" (§69); "all bodies are in a perpetual flux like rivers, and parts are passing in and out of them continually" (§71); and "not only is the soul (the mirror of an indestructible universe) itself indestructible, but so also is the animal itself" (§77). Cited according to G. W. Leibniz, *Philosophical Writings*, ed. by G. H. R. Parkinson, trans. by Mary Morris and G. H. R. Parkinson, Everyman Library (London: J.M. Dent, 1997), pp.185, 189-91. Erasmus Darwin, *Zoonomia or The Laws of Organic Life*, vol I (London: J. Johnson, 1794), pp.505-507 (Sect 39.4.8 of Generation). Surely not by coincidence Erasmus Darwin began his *Zoonomia* with the assertion that motion is endemic to all life and concluded his work with the query: "would it be too bold to imagine, that in the great length of time, since the earth began to exist, perhaps millions of ages before the commencement of the history of mankind, would it be too bold to imagine, that all warm-blooded animals have arisen from one living filament, which THE GREAT FIRST CAUSE endued with animality, with the power of acquiring new parts, attended with new propensities, directed by irritations, sensation, volitions, and associations; and thus possessing the faculty of continuing to improve by its own inherent activity, and of delivering down those improvements by generation to its posterity, world without end!" He next applied the same argument to other organisms such as gnats, worms, leeches, snails, shell-fish, coralline insects, and sponges (p.506). And to these he added vegetation before concluding the section with: "shall we conjecture that one and the same kind of living filament is and has been the cause of all organic life?" (p.507)

27 There are notable exceptions. Timothy Lenoir, *The Strategy of Life* (Chicago: U of Chicago P, 1989), discusses the "teleomechanical school" of thought that argued the principles of physics and chemistry do not apply to biology. Whereas physical and chemical processes are purposeless, the interactions of cause and effect in an organism are teleological (pp.7-35). While the origins of such purposiveness are unknown, early thinkers subscribed to Leibnizian and Kantian concepts of Supreme Substance and Highest Wisdom. E. Darwin clearly focussed on the purely biological.

28 Cf. John A. McCarthy, "Energy and Schiller's Aesthetics from the Philosophical to the Aesthetic Letters," in: *Who is this Schiller Now? Essays on his Reception and Significance*, ed.by Jeffrey L. High, Nicholas Martin, and Norbert Oellers (Rochester NY: Camden House, 2011), pp.165-86.

(πνεῦμα/pneúma; see Jennifer Hellwarth's essay in this volume), vitality (*vim*) and energy pulse or bundle (*Kraftzentrum*) that cause motion whether in terms of the vital source of life (*Lebenskraft*) in animals and plants or in gravitational, electromagnetic, electrochemical, and thermodynamic forces on the cosmic-, meso-, micro- and nano scales (that is, galactic, solar, human, molecular). In sum, energy is any source of useable power; that is, has the capacity to do work. By the mid-1790s Johann Gottlieb Fichte could claim that all reality is action.[29] By the turn of the century, F. W. J. Schelling could posit gravity as the central force in his *Einleitung zu seinem Entwurf eines Systems der Naturphilosophie* (1799).[30]

Alfred Noll, sometime professor in Jena, offers examples of the physiological debate in his slim essay collection, *Die Lebenskraft* (Life Force, 1914), with excerpts from eleven authors seeking to define the term and phenomenon. Texts range from Friedrich Casimir Medicus's *Von der Lebenskraft* (On Vitality, 1774) to Ludimar Hermann's *Lehrbuch der Physiologie* (Introduction to Physiology, 1910).[31] Of the eleven selections arguing for vitalism and the eleven objecting to the traditional concept of a separate life force, I can discuss only a representative sample. Physician Christoph W. Hufeland (1762-1836), medical doctor Joachim Dietrich Brandis (1762-1846), Halle physiologist Johann Christian Reil (1759–1813), Austrian ophthalmologist and physiologist Georg Prochaska (1749-1820), and philosopher, physiologist, physician Ludwig Büchner (1824-99) must stand for a large group of commentators whom Noll presents.[32] In all these writers *Lebenskraft* was a postu-

---

29 Johann Gottlieb Fichte, *Grundlage der gesammten Wissenschaftslehre* (Foundations of the Entire Science of Knowledge) (Jena: Gabler, 1794–95), p.37: "Alle Realität ist *thätig*, und alles *thätige* ist Realität" (emphasis in is in the original). Text cited according to http://www.zeno.org/Lesesaal/N/9781484031100?page=0

30 F. W. J. Schelling, *Einleitung zu seinem Entwurf eines Systems der Naturphilosophie*, ed. by Wilhelm G. Jacobs (Stuttgart: Reclam, 1988), §2, pp.29, 64. He later restates his purpose as: "Die allgemeine Aufgabe der spekulativen Physik läßt sich jetzt so ausdrücken: *die Konstruktion organischer und anorganischer Produkte auf einen gemeinschaftlichen Ausdruck zu bringen*" (The general task of speculative physics can be summarized as follows: to discover the common basis for the construction of organic and inorganic products (p.58; emphasis in original).

31 Alfred Noll, ed., *Die "Lebenskraft" in den Schriften der Vitalisten und ihrer Gegner*, Voigtländers Quellenbücher 69, 4th ed. (Leipzig: R. Voigtländer Verlag, 1914). Also published as *Was die Gelehrten einst von der sogenannten Lebenskraft hielten* (Leipzig: R. Voigtländer Verlag, o.J.) which is available as an e-text or PDF. For further reading on issues central to the *Lebenskraft* debate see, e.g., "International Bibliographic Guide to the History of Pathology (IBGHP)," ed. by Axel W. Bauer at http://www.umm.uni-heidelberg.de/ag/gte/IBGHP.pdf

32 Pathologist August Hirsch recounts the *Lebenskraft* debate in the first half of the nineteenth century and includes details on magneticism. See A. H., *Geschichte der medi-*

late subject to observational inquiry, serving primarily as an organizing principle. All examined the nature and origin of *vis vitalis* understood as either the capacity of organic matter to receive external stimuli and to react to them independently of chemical and mechanical processes or understood as the proactive capacity to alter or even to counteract chemical processes.[33]

Noll provides a brief introduction to trace the debate as it came to be more widely discussed in the second half of the eighteenth century and then dominated the scene through the first half of the next century. At that point, Knoll switches to the anti-Vitalists, beginning with François Magendie and continuing through Ludwig Büchner to Rudolf Virchow. The move amounts to a shifting of the search for the soul/anima in the body itself. The early vitalists took their cues initially from chemist doctor Georg Ernst Stahl (1659-1734), who posited phlogiston as the key element in chemical reactions transmitted throughout the four elements leading to life.[34] They also responded to physician Jean Baptist van Helmont (1580-1644), who introduced the term "gas" into chemical terminology, distinguishing various gases distinct from air and located the seat of life and the soul in the stomach, which he called "a sack, or membrane" analogous to "the Membrane of the womb."[35] For him water and air were the two primitive elements. None-

---

zinischen Wissenschaften in Deutschland (Bd. 22 der *Geschichte der Wissenschaften*) (München & Leipzig: R. Oldenbourg, 1893), esp. pp.462-79.

33 On the transfer from clinical studies to more popular writing, see *Medizinische Schreibweisen. Ausdifferenzierung und Transfer zwischen Medizin und Literatur (1600-1900)*, STSL 117, ed. by Nicholas Perthes and Sandra Richter (Tübingen: Niemeyer, 2008).

34 Georg Stahl, *Opusculum Chymico-Physico-Medicum* (Nuremberg, 1715), pp.334-39. Phlogiston appears in its simplest form in the air. Through admixture with the other elements, it enters into the plant world, from there into the animal one: "His omnibus circumstantiis permotus, statuo, quod phlogiston hoc principium, in pura sua simplicitate, maxime naturaliter ipsi aeri infit, & ex illo in Mixtiones varias derivetur" (p.338). Hence, through interaction with other elements, phlogiston becomes a more complex phenomenon, what he calls "pinguedo una intertexta" (an interwoven richness, p.339).

35 John Baptista van Helmont, *Oriatrike or, Physick refined. The Common Errors therein Refuted, and the whole art Reformed and Rectified, Being a new Rise and Progress of Phylosophy and Medicine, for the Destruction of Diseases and Prolongation of Life*, trans. by J. C. Sometime of M. H. Oxon (London: Lodowick Loyd, 1662), p.289. Like Haller and Erasmus Darwin, each proved to be influential in shaping the later discourse. When we consider that Justus von Liebig propounded a view of fermentation that resembled in part the phlogiston theory and that Nietzsche emphasized explosiveness and accorded the stomach the major role in constructing life in its fullest manifestations, a continuity of sorts becomes manifest despite the repeated assertions of investigators that theories come and go with frequent regularity. Friedrich Nietzsche spoke of the "Bauch des Seins" (belly of being), "Bauch der Dinge" (belly of reality) and equated spirit with the stomach ("der Geist *ist* ein Magen!") in Friedrich Nietzsche, *Sämtliche Werke*,

theless, according to Noll, the early vitalists used the term *Lebenskraft* loosely to talk about both *vis vitalis* (vitalistic force) and *spiritus vitalis* (spiritual force) without drawing any clear distinction between the two or locating the soul in a particular organ (Noll, p.78). (Early Renaissance thinkers located the soul either in the head or the heart.)

The first German to use the term *Lebenskraft* according to Noll was Friedrich Casimir Medicus (1736-1808), who used it (along with "Lebensgeist" and "Seele") in a lecture at the Kurpfälzisch-Theodorischen Akademie der Wissenschaften in Mannheim in 1774 (Noll, p.8). This life force (*Substanz*) is present throughout the animal and plant kingdoms as well as being the dominant force in man's physical or "mechanical" nature. He locates the life spirit in the brain and spinal chord of man and animals and in the marrow of plants (Noll, p.12). The world of the spirit is the situs of reflective capacity, the ability to think being wholly independent of the life "substance" that regulates the organic world.[36] The soul is independent of the *Lebensgeist* and hence has no control over bodily functions such as breathing, digestion, and the nervous system. In considering why the soul has no direct control over the body's mechanisms, Medicus conjectures that the *Lebensgeist* can pass through the central nervous system without hindrance, whereas the soul's movement, which also uses the nervous system, is obstructed by the "Nervenknoten" (nerve centers) between the head and the heart, for example (Noll, p.13). Nonetheless, in exceptional cases of extreme joy or migrane headache, the soul is capable of getting through those obstructions. Moreover, the difference between the *Lebensgeist* (life of the spirit) and *Lebenskraft* (life of the body) is that critical reflection varies according to exercise, whereas the life substance is the same in every human being from conception. Medicus also asserts that "der Lebensgeist keine neue Fertigkeiten mehr erlangen könne" (the life of the spirit cannot acquire any new competencies; Noll, p.10). By that he seems to mean that the quantity and quality of the "substance" is determined at conception and cannot renew itself, as is observable with the rejuvenating processes of the *Lebenskraft*. The develop-

---

Kritische Studienausgabe, ed. by Giorgio Colli and Mazzino Montinari, 15 vols. (München and Berlin: Deutscher Taschenbuchverlag and de Gruyter, 1993), vol. IV, pp.36, 170, 258 (hereafter cited as KSA); English version cited according to Nietzsche, *Thus Spoke Zarathustra*, trans. by Kaufmann (1995), pp.32, 132, 206. Joachim Dietrich Brandis, *Versuch über die Lebenskraft* (Hannover: Hahn'sche Buchhandlung, 1795), argued: "[...] so wie schon über diesen Gegenstand [die Lebenskraft] so manche Hypothese eine kurze Zeit gegolten hat und wieder ungültig geworden ist" (p.xi).

36 *Substanz* in Medicus' usage is analogue to its philosophical meaning of not being further reducible; it is the energy pulse, so to speak, of all that follows from it, not unlike the *Strahlkraft* of Leibniz's monad. See G. W. Leibniz, *Monadologie*, ed. with introduction and notes by Hermann Glockner (Stuttgart: Reclam, 1963), p.56.

mental drive (*Kunsttrieb*) of each species varies, moreover, because each species has a particular function. Only the *Lebensgeist* is capable of further development so that a monkey will always remain different from a human being (Noll, p.12). The monkey is presumed not to have *Lebensgeist.* In sum, *Lebenskraft* designates for Medicus autonomic processes ("die vegetativen Functionen") and is separate from the capacity for thought (*Lebensgeist*) and the function of the soul (*Seele*).

Noll mentions Christoph W. Hufeland (1762-1836) only in passing. He cites Hufeland's *Ideen über Pathogenie und Einfluß der Lebenskraft auf Entstehung und Form der Krankheiten* (Jena, 1795) as an example of the turn toward pathology as a means for defining *Lebenskraft* by its loss (Noll, p.14). Yet many consider his *Makrobiotik, oder die Kunst, das menschliche Leben zu verlängern* (Macrobiotics, or the Art of Prolonging Life, 1794) to be his most significant legacy (see Kuzniar's essay below). Although Hufeland laid down major characteristics of *Lebenskraft* (e.g., irritability, what we today call the immune system, dependency on external influences, and its omnipresence in all parts of the body), Noll does not seem to know it. Moreover, in reaction to the flood of publications on the topic that introduced all kinds of confusing terminology, Hufeland decided to publish a separate essay on efforts to define *Lebenskraft.* Noll does not reference this essay either, "Mein Begriff von der Lebenskraft," that provides salient information. I accord the latter closer attention.

According to Hufeland, the most appropriate connotation for *Lebenskraft* is its denotation of chemical reactions: "Auch hierzu ist das Wort Lebenskraft am schicklichsten, indem es nicht bloss das Reizverhältnis des Lebens, sondern auch das chemische Verhältnis des Lebens (d.h., die chemische-vitale Beschaffenheit der Materie und ihre in der Vitalität gegründeten Ursachen) umfasst" (In this case too the term *Lebenskraft* is best suited because it not only entails the ratio of stimuli to life but also the chemical conditions of life, i.e., the chemical-vital quality of matter and its causes rooted in vitality).[37] Hufeland considers *Geist* and *Seele* to be poor substitute terms for *Lebenskraft. Geist* is so broad in connotation that it is unsuitable; *Seele* is less appropriate because it is rooted in the ability to think and reflect (*Denkkraft*). "Denken und Leben [seien] zweyerlei Dinge" (reflection and life are two different things; "Begriff," pp.792-93). Hence, Hufeland rejects those frequently used terms. He also rejects the designations *Caladum innatum*, *po ernogon*, and *impetum faciens* as inaccurate ("Begriff," p.794). *Reizbarkeit* and *Empfindlichkeit* are merely characteristics of

37 Christoph Wilhelm Hufeland, "Mein Begriff von der Lebenskraft," *Journal der practischen Heilkunde*, 6 (1798), 785-96 (p.791). Cited according to http://books.google.com/books?id=RhAr8i9Fj1cC&pg=PP3#v=onepage&q&f=true.

Lebenskraft, he avers, and cannot be used to denote *Lebenskraft* itself ("Begriff," pp.793-94). To be sure, *Incitabilität* is more encompassing than *Reizbarkeit* because it denotes the whole range of sensibilities, but is still not an appropriate synonym because it is restricted to the dynamic relationship ("dynamisches Verhältnis") and does not include the chemical dimension ("chemisches Verhältnis") within the animal-chemical ratio ("animalisch-chemisches Verhältnis"). Hufeland writes: "das Vermögen der Mischung und Darstellung der Materie den animalisch-chemischen Karakter zu geben" ("Begriff," p.795). Because researchers have more recently posited *Lebenskraft* as being the result of chemical reactions, Hufeland is averse to accepting the term chemische Lebenskraft" ("Begriff," p.796) as the appropriate designation of what *Lebenskraft* is designed to express. Why? Because it seems to him that the grounding of the life force in chemical processes has not been proven but remains an hypothesis that seeks evidence. Thus, Hufeland ultimately opts for "unbestimmte Lebenskraft" as the appropriate designation in the ongoing debate. Any and all specific qualifiers can only be added after the science has provided the irrefutable evidence ("Begriff," pp.795-96). I added "irrefutable evidence" which Hufeland clearly implies.

Early in his remarks Hufeland asserted that no one really knows what the essential basis of life is. Nonetheless, we sense that one does exist. Therefore, he argues, we need some way to talk about it in order to advance scientific research. When he speaks, therefore, of *Lebenskraft* he means the inner cause of the life of a body or its constituent parts. In other words, he uses *Lebenskraft* to designate *ratio vitalitatis.* But he is ready to switch terms as soon as someone can demonstrate what the actual cause of life is.[38]

J.D. Brandis, a student of the physicist and satirist Georg Christoph Lichtenberg (1722-79) and naturalist/physiologist Johann Friedrich Blumenbach (1752-1840), is important in this survey because he urged investigators to put aside the traditional preconceived notion of matter consisting only of bodies and of two main "substances" consisting of body and soul. He also does not accept the notion that life-energy is fluid in nature, passing easily through the arteries and veins, miniscule muscle fibers or nerve sin-

---

38 Hufeland, "Begriff," p.788: "Niemand wird läugnen, dass wir von dem wesentlichen Grund des Lebens noch nichts wissen und doch existirt einer, und wir müssen ein Wort dafür haben, um ihn in unsere practischen Untersuchungen und Bearbeitungen mit aufnehmen zu können. Diesen unbekannten *innern* Grund des Lebens eines Körpers oder seiner Theile, diese *ratio vitalitatis* will ich andeuten, wenn ich sage *Lebenskraft.* Es ist also ein blosses Wortzeichen, und ich gelobe hiermit, das Wort sogleich aufzugeben, sobald mir jemand die *Sache* sagt, d.h. worinn dieser innere Grund des Lebens besteht."

ews. The investigator should follow the evidence without preconceptions.[39] His purpose is not to disprove previous theories of *Lebenskraft* but to provide the practicing physician with information that might prove vital to making decisions that are often a matter of life or death. Key to Brandis' own way of thinking about the essence of life is his emphasis on the inherent relationship between life and movement. In following this course, he adopts Blumenbach's theories of cell structure and cell formation ("Reproductionskraft").

Brandis' first attempt to define *Lebenskraft* is straightforward and simple enough: in order for the human body as a machine to be able to interact with its environment, it must possess a force (or forces) that enables the material parts of the machine to become a communicative instrument ("welche den Körper fähig macht, uns zum Communicationswerkzeug mit der Körperwelt zu dienen;" Brandis, *Versuch*, p.2). The state that allows for this intercommunication to happen is labeled life. And the impulses ("Kräfte") which actively carry out the interactions are called vital impulses ("Lebenskräfte") (Brandis, *Versuch*, p. 2). Thereby, he drew out further connotations of *Kraft*/force, moving the model away from the traditional billiard-ball paradigm. He has no use for the presumed need of "Mittelkörper," as he calls them, capable of passing through the arterial system or causing oscillations in our central nervous system. They are pure fantasy.[40] Later he draws an analogy between the workings of *Lebenskräfte* and those of electricity: in fact, he is rather convinced that electricity is *Lebenskraft* itself (Brandis, *Versuch*, p.81).

To be sure, the life force manifests itself quite differently in different parts of the organism, in nerves, muscle fibers or cell membranes (Brandis, *Versuch*, pp.34-50). Others align this differentiation with the excitability of Albrecht von Haller's "*Irritabilitätslehre*" that posited "Irritabilität" und "Sensibilität" as phenomena inherent in living organisms (i.e., *vires insitae, innatae*) (cf. Brian McInnis in this volume). For Brandis, who adopts a phlogistic model of interaction, the key to the life force is its ability to combine and mix with other parts of organic matter (Brandis, *Versuch*, p.76-77). Due to this mixing function life is possible. Once *Lebenskraft* departs, the

39 See preface in Brandis, *Versuch* (1795), pp.ix-xxiv (xiv-xvii). The digital copy of the Digitale Bibliothek Braunschweig is cited, whenever the passage is not in Noll's excerpt.

40 Brandis, *Versuch*, pp.12-13: "Nervensaft, Schwingungen der Nerven, Blut oder irgend eine andere Flüssigkeit, welche die willkührlich gedachten Zellen in der Muskelfiber ausdehnen und so die Zusammenziehung der Muskelfaser bewürken sollen, sind alles Ideen, die uns bei der Erklärung der in der organischen Körperwelt vorgehenden Bewegungen unmöglich Genüge leisten können."

organic matter begins to decay because the other chemical elements, which had been kept in balance, reassert their own laws of interaction. The result is "fermentation" ("Gährung"). Hence, he confidently asserts: "die Lebenskraft ist der Materie eigen und wirkt den physischen Kräften der Verwandtschaft, Anziehung usw. entgegen" (life-energy is unique to matter and counters the physical forces of affinity and attraction; Noll, p.16). This life force is inherent in organic matter itself, it does not inhere in the blood alone. Analogue to a spark that ignites a reaction on a larger scale, Brandis supposes, a similar phlogistic process might occur on the microscopic level of fibers and filaments. Hence, *Lebenskraft* is dependent upon the phlogistic exchange of carbon and oxygen or phosphor in the filaments where arteries and veins interact with the surrounding tissue (Noll, p.17; on the foregoing see Brandis, *Versuch*, p.76-84). Brandis clearly moved the debate in a new direction.

Johann Christian Reil interpreted biology in physiochemical terms. With his popular journal, *Archiv für Physiologie* (1796–1815), he opened the way for a transdiscursive application of terminology. He devoted the entire first number of his journal to the *Lebenskraft* debate, summarizing recent arguments for his readers and formulating fundamental principles of physiological observation in pithy manner, to wit: concepts of animal physiology he regards as phenomena of a special kind of force that exist within the whole chain of natural forces. Hence, he rejects the notion of a separate life substance along with the concept of the soul as the source of life because there is no empirical evidence to assume its existence. Critical for the origination of life is the manner in which inorganic matter becomes interconnected under the influence of such attributes as heat and light, and elements such as electricity, oxygen, and carbon dioxide.[41] Noting that there are only a few general characteristics of empirical natural philosophy – "coherence, gravity, and expansion" (*Cohärenz, Schwere und Expansivkraft*; Reil, *Versuch*, vol. I, p.20) – he points out that the further one goes down the pyramidal structure of natural phenomena, the more matter reveals individually characteristic traits, so that the (organic) individual specimen has its unique form and qualities (Reil, *Versuch*, vol. I, p.21). He readily admits that scientific terminology and descriptions of such phenomena as *Kraft* are essentially subjective interpolations on the researcher's part. One such manifestation is muscle contraction in animals ("Zusammenziehung"). It results, he argues, from an altered coherence in their material substance. To be sure, he does not agree with Ernst Platner, who conflated this contraction ("Kontraktalität")

---

41 Johann Christoph Reil, *Von der Lebenskraft*, ed. Karl Sudhoff (Leipzig: Johann Ambrosius Barth, 1910). Originally published in *Archiv für Physiologie*, 1 (1796), 8–162, which is cited here (pp.8, 26, 32-40).

with the coherence or elasticity observable in inorganic nature.[42] But he does approvingly cite Kant on the inherent unity of the parts and the whole in organic matter when Kant argues that organic matter is more than just a by-product of nature and that it is also expressive of nature's purpose.

Reil argued against the chemist Georg Ernst Stahl (1660–1734), who had coined the term phlogiston. Three essences were thought to comprise all matter: sulfur or *terra pinguis*, the essence of flammability; mercury or *terra mercurialis*, the essence of fluidity; and salt or *terra lapida*, the essence of fixity and inertness. Stahl renamed *terra pinguis* phlogiston, considering it an element capable of existing in isolation in a relatively pure form. Three general principles dominated his thought. First, continuing Cartesian dualism, he had theorized an irreducible difference between mind (anima) and matter. The mind is driven by an immaterial vital principle, he thought, that distinguishes living forms from inanimate matter. (He did not distinguish between animal, plant, and miasma as living organisms with or without sentient abilities. Kant, of course, had posited the human mind – ratio – as the highest expression of *Lebenskraft*.) Second, the anima is teleological (in a Leibnizian sense of the monad). Stahl held that inorganic matter is totally dependent on external force, hence is mechanical, while organic matter is both self-directive and reactive. Third, the anima exerts itself through mechanical principles by causing a body to be in motion (here he thought most particularly of the circulatory system). The anima as source of motion explained, to Stahl's mind, how something immaterial can act upon something material (thus addressing a major issue in Cartesian dualism). By advocating a mechanical view of body and a teleological view of mind, Stahl combined an analytic approach to matter with a holistic approach to mind. Nevertheless, the duality of mind and body remained intact.

The continued dualism evident in Stahl was a bone of contention for Reil. In contrast to Stahl, Reil adopted a monist position, contending that powers and forces of motion are grounded in matter itself. Hence, *Lebenskraft* was not a separate, external force for him. Energy he posited as the core value common to both organic and inorganic matter:

> Alle Erscheinungen in der Körperwelt sind Resultate einer bestimmten Form und Mischung der Materie (§ 1). [...] *Das Verhältnis der Erscheinungen zu den Eigenschaften der Materie, durch welche sie erzeugt werden, nenne ich Kraft.* Kraft kann also so allgemein, besonders und individuell gedacht werden, als sich die Verhältnisse der Wirkungen zu den Ursachen und die Erscheinungen zu den Eigenschaften der Materie denken lassen. Kraft ist also etwas von der Materie unzertrennliches, eine Eigenschaft derselben, durch welche sie Erscheinungen hervorbringt.

42 Reil, "Lebenskraft," *Archiv*, vol. I, p.57. This passage was not taken over into Noll's compendium.

> Kraft ist ein subjectiver Begriff, die Form, nach welcher wir uns die Verbindung zwischen Ursach und Wirkung denken. Wäre es möglich, dass wir uns jeden Körper so wie er ist, die Natur seiner sämtlichen Grundstoffe und ihre Verbindung, ihre Mischung und Form, auf einmal deutlich denken könnten, so hätten wir den Begriff der Kraft nicht nötig, der zu manchen fehlerhaften Folgerungen Anlass giebt.[43]

> (All phenomena of the physical world derive from a specific form and a mixture of material reality (§1). [...]. I call the relationship of manifestations to their material characteristics, by means of which phenomena are produced, energy. Energy can be conceived of as broadly, specifically, or uniquely as the [nature of] the nexus of cause and effect and of [individual] manifestations to the qualities of matter. Hence energy is inseparable from matter; it is an inherent quality of matter, that produces phenomena.
> Energy is a subjective concept; it is the vehicle by means of which we conceive of the relationship between cause and effect. Were it possible for us to conceive clearly all at once the nature of each body together with its complete composite basic materials and their arrangement, then we would not need the concept of energy, which has led to so many false conclusions.)

A subjective concept inseparable from matter, *Kraft* uses nerves and muscle fiber as vehicles for its conveyance. Matter is, in fact, a form of energy or *Kraft* that is revealed in contraction and extension, growth and movement. Reil considered *Kraft* to be the means by which we conceive of the relationship between cause and effect and to be a subjective concept. What is life within this context, he asks? Signs of life, he concludes, are tied to the ability to procreate, to grow, to feed and be nourished, to replicate, and to take on a specific form. This process of chemical metabolism he labels "thierische Krystallation" (animal crystallization) analogue to crystallization in minerals.[44] More important for my purposes, however, is Reil's identification of five categories of energy: "physical energy" (*physische Kraft*) "life energy" (*Lebenskraft*) "autonomic energy" (*vegetative Kraft*), "fleshly energy" (*animalische Kraft*), and "capacity for reason" (*Vernunftvermögen*). In essence, he saw the animal body as a "large republic" consisting of many different parts but collaborating for the survival of the whole organism. The capacity for thought, however, is peculiar to humans.[45] His ideas resonated throughout the nineteenth century.

---

43 Reil, "Lebenskraft," *Archiv*, vol. I, pp.45-46, 71; compare Noll, *Lebenskraft*, pp.21–22. The reference to animal crystallization (p.71) is not in Noll. Emphasis in the original.

44 Reil, "Lebenskraft," *Archiv*, vol. I, p.65: "Zeugung, Wachstum, Ernährung, Reproduction, Bildung."

45 Reil, "Lebenskraft," *Archiv*, vol. I, pp.47-50 (= Noll, *Lebenskraft*, pp.22–23). On the body as a "republic" see Hirsch, *Geschichte der medizinischen Wissenschaften* (1893), p.464.

For his part, vitalist Viennese Professor of Anatomy and Ophthalmology Georg Prochaska (1749-1820) accorded electricity a larger role, albeit one still envisioned as a kind of fluid substance ("elektrisches Fluidum"), even when talking about sparks and spasms and citing its benefit in curing paralysis and convulsions. He also pointed to its role in sensitive reactions to changes in barometric pressure, tingling sensations of the tongue in speech acts, and internal flashes of light in the eye.[46] He saw in the central nervous system the key to answering all the other questions about life and movement (Prochaska, *Lehrsätze*, p.90 [p.108]).

Especially notable in our context, is the need Prochaska felt to account for the many unconscious functions of the brain. To be sure, he readily grants that we know preciously little about the architecture of the central nervous system because science still lacked the ability to see into cellular structure. Nonetheless, he was convinced that the nervous system was the means by which animals and humans are capable of sensory perception and motor movement ("das Werkzeug ihrer Empfindung und Bewegung [sey] das Nervensystem"; Prochaska, *Lehrsätze*, p.91 [109]). He concluded that the *sensorium commune* (the spinal cord, the medulla oblongata and the basal ganglia) reflects the transmission of sensory impressions into motor actions by definite laws peculiar to itself, and independently of consciousness (cf. Prochaska, *Lehrsätze*, pp.89-102 [107-20]).

Electrical impulses seemed to play a key role here too. As a consequence, Prochaska thought of energy as an aggregate for all parts of a body evident in its functions. Energy, then, was not derivative of a single, inherent and unique *Selbstkraft* (autonomous source). Prochaska, who pioneered the field of neurophysiology, is remembered for his comprehensive theory of reflex action involving the concepts of *vis nervosa* and *sensorium commune.*[47] *Vis nervosa* he described as a latent force possessed by the nerves themselves; *sensori-*

---

46 Georg Prochaska, *Lehrsätze aus der Physiologie des Menschen: zum Gebrauche seiner Vorlesungen* (Vienna: Wappler, 1797), pp.60-65 (p.62). Cited according to the digital version held by the Max Planck Institute for the History of Science Library (= pp.78-83: http://echo.mpiwgberlin.mpg.de/ECHOdocuView?url=/permanent/library/DDV3YCT2/pageimg&start=71&viewMode=index&pn=83&mode=imagepath.

47 See Georg Prochaska, *De Usu et Functione Systematis Nervorum* (Vienna, 1800). Together with Johann August Unzer's (1727-1779) earlier *Erste Gründe einer Physiologie der eigentlichen thierischen Natur thierischer Korper* (*The Principles of Physiology*, 1771), it is considered foundational in neuroscience. (Previously, Prochaska had published his *Tractatus de structrua nervorum* [Vienna, 1779]). In 1851 the York physician Thomas Laycock, an authority on the reflex actions of the brain, translated both into English as *The Principles of Physiology by John Augustus Unzer and a Dissertation on the Functions of the Nervous System by George Prochaska (1749-1820)*, trans. and ed. by Thomas Laycock (The Syndenham Society, London 1851). In these two authors Laycock found precedence for his own work.

um commune he rendered as the nexus of nerves involved in sensory perception and those conducive to motor action. Prochaska used the term *vis nervosa* in direct analogy to Isaac Newton's *vis gravitans*, because he believed it to be an elemental form of energy, not observable except through its effects in unconscious reflexes and mental reflection. Prochaska's early endeavors in understanding the connections between the mechanistic and the mental were a precursor of contemporary explorations of the neural basis of behavior, that is, of similarity, replication, causes, and agency.[48]

The implied (or asserted) connection between the inorganic and organic spheres via the concept of energy as movement was made explicit by the chemist Friedrich Wöhler (1800–82). In 1828 Wöhler synthetically converted ammonium cyanate NH4 (NCO) into urea (NH2)2CO, an isomer, that is, a compound with the same number of atoms but a different molecular structure. It was the first of a number of experiments using amino acids (polypeptides) that synthetically reproduced biological compounds in proteins.[49] Hence, Wöhler's synthesis of ammonium cyanate into similarly structured urea is considered the starting point of modern organic chemistry that demonstrated that life emerges from inorganic matter via a process of elective affinities. Given the wide ranging implications of Wöhler's work in the early nineteenth century, it is not surprising that Goethe used the attraction of similar compounds as a structuring device in his novel, *Die Wahlverwandtschaften* (Elective Affinities, 1809-10). Nonetheless, it took another generation or two before the vitalist argument was displaced as the dominate model.[50] Wöhler's legacy is evident in modern chemistry. For instance, chemist Dennis H. Rouvray has convincingly argued that all scientific concepts and classifications have a basis in similarity. Common in chemical applications are similarities such as analogy, complementarity, equivalence relation, scaling, and self-similarity.[51] Understandably, then,

---

48 Cf. Jeffrey D. Schall, "Actions, Reasons, Neurons, and Causes," in *Neuroscience and Religion* (2009), pp.175-98.

49 Friedrich Wöhler, "Ueber künstliche Bildung des Harnstoffs," *Annalen der Physik und Chemie*, 88.2 (1828): 253–56. Cited by Noll, *Lebenskraft*, p.82. The conversion is rendered as NH4 (NCO) Õ NH3 + HNCO Ö (NH2)2CO.

50 http://en.wikipedia.org/wiki/W%C3%B6hler_synthesis. This website offers a graphic rendering of the chemical reactions involved. Accessed 18 August 2014.

51 Dennis H. Rouvray, "Similarity in Chemistry: Past, Present and Future," *Topics in Current Chemistry*, 173 (1995), pp.1-30. Without analog comparisons to Greek cosmology, e.g., chemistry would not have been able to develop as a science. See especially his useful timeline of major atomistic properties (p.3), his diagram of the uses of Platonic polyhedra (p.4), and his dendogram of the hierarchy of similarities in the chemical properties of the twenty naturally occurring amino acids (p.22).

Wöhler's argument is critical to the present undertaking for both historical and contemporary reasons.

At the same time as these developments, the Italian mathematician, astronomer, and natural philosopher Ruggiero Giuseppi Boscovich (1711–87) transferred all this movement in chemicobiology and astronomy to the inner workings of molecules, positing dynamics as the fundamental principle of atomism. That opened the door to the insight that the nature of matter is not fixed. Hence one can speak of a Copernican Turn from the cosmic to the molecular scale of things.[52] All of these developments prompted a rethinking more generally of humankind's relationship to empirical reality. Kant's "Revolution in Thought" ("*Revolution der Denkart*") comes immediately to mind. It was a revolution that began with his pre-critical writings as noted above.

In the wake of such rethinking, Ludwig Büchner, who was an avid opponent of the theory of an autonomous "Lebenskraft," reduced every physical manifestation to the "operations of nature" without drawing a qualifying distinction between organic and inorganic matter.[53] His theory can also be seen as an echo of a key component in Kant's cosmology. Büchner was thus viewed as a chief exponent of nineteenth-century scientific materialism. It started with the dissemination of his ideas in *Kraft und Stoff* (*Force and Matter*, 1855). The purpose of his study was to demonstrate the ultimate reality of physical force, the interchangeability of matter and energy, and the indestructibility of energy. Nature for him was purely physical with no purpose, no will, and no grounding in an extraneous authority. In adopting this position, he took the next logical step after Kant had cautiously asserted that all of creation is autopoietic. (His caution was due, no doubt, to the religious conservativism of his era.) Matter is immortal, indestructible and is subject to constant and continuous metamorphosis. Like Kant before him and resonating with Darwin's contemporaneous *Origin of Species*, Büchner's *Kraft und Stoff* caused quite a stir.[54] Ultimately, it cost the author his professorship at

52 Adolf Müller, „Ruggiero Giuseppe Boscovich," in *The Catholic Encyclopedia*, vol. 2 (New York: R. Appleton, 1907). http://www.newadvent.org/cathen/02691a.htm. Accessed 8 August 2014.

53 Ludwig Büchner, "Die Lebenskraft," in Noll, *Lebenskraft*, pp.68–77. Cf. Ludwig Büchner, *Kraft und Stoff. Empirisch-Naturphilosophische Studien in allgemein verständlicher Sprache*, 13th ed. (Leipzig: Theodor Thomas, 1874), Chapter XVIII: "Lebenskraft," pp.242-55: https://openlibrary.org/books/OL23308316M/Kraft_und_Stoff. Accessed 9 July 2014.

54 Both Charles Darwin and Ludwig Büchner can be seen as participating in a reconceptualization of the inner dynamism of the human and non-human launched by David Hartley's theory of vibrations (*Observations on Man, his Frame, his Duty, and his Expectations* (1749) and Ruggiero Guiseppi Boscovich's atomistic theory (*Theoria*

the University of Tübingen because of its radical grounding of everything in natural processes. Büchner emphatically stated:

> *Der Stoff* ist unsterblich, unvernichtbar; kein Stäubchen im Weltall kann verloren gehen, keines hinzukommen. Es ist das große Verdienst der *Chemie* in den letzten Jahrzehnten, uns auf's Klarste und Unzweideutigste darüber belehrt zu haben, daß die ununterbrochene Verwandlung der Dinge, welche wir tagtäglich vor sich gehen sehen, das Entstehen und Vergehen organischer und unorganischer Formen und Bildungen nicht auf einem Entstehen und Vergehen vorher nicht dagewesenen Stoffes beruhen, wie man wohl in früheren Zeiten ziemlich allgemein glaubte, sondern daß diese Verwandlung in nichts Anderem besteht, als in der beständigen und unausgetzten Metamorphosierung *derselben Grundstoffe, deren Menge und Qualität an sich stets dieselbe und für alle Zeiten unabänderliche bleibt.* Mit Hülfe der Waage ist man dem Stoffe auf seinen vielfachen und verwickelten Wegen gefolgt und hat ihn überall in derselben Menge aus irgend einer Verbindung wieder austreten sehen, in der man ihn eintreten sah. Die Berechnungen, die seitdem auf dieses Gesetz gegründet worden sind, haben sich überall als vollkommen richtig erwiesen.
>
> (Matter is immortal, indestructible. There is not an atom in the universe, which can be lost. We cannot, even in thought, remove or add an atom without admitting that the world would thereby be disturbed and the laws of gravitation and the equilibrium of matter interfered with. It is the great merit of modern chemistry to have proved in the most convincing manner that the uninterrupted change of matter which we daily witness, the origin and decay of organic and inorganic forms and tissues, do not arise, as was hitherto believed, from new materials, but that this change consists in nothing else but the constant and continuous metamorphosis of the same elementary principles, the quantity and quality of which ever is, and ever remains, the same. Matter has, by means of the scales, been followed in all its various and complicated transitions, and everywhere has it been found to emerge from any combination in the same quantity as it has entered. The calculations founded upon this law have everywhere proved to be perfectly correct.)[55]

Philosophiae Naturalis, 1763). Both their works experienced several editions in the eighteenth century.

55 Büchner, *Kraft und Stoff* (1874), pp.10-11; Ludwig Büchner, *Force and Matter: Empirico-Philosophical Studies Intelligibly Rendered*, trans. J.F. Collingwood (London, 1870). Text scanned and modernized by Jerome S. Arkenberg. Both texts are available online: https://archive.org/stream/forceandmatterem00beucuoft/forceandmatterem00beucuoft_djvu.txt and https://openlibrary.org/books/OL23308316M/Kraft_und_Stoff. In his preface to the eighth edition of *Kraft und Stoff* in 1863 (included in the 13th edition of 1874), which served to express his delight at the imminent translation of the work into English for the first time, Büchner provided a long list of innovations in science disciplines that he could not have anticipated when he first penned his study in 1855. He specifically drew parallels between Darwin's theory of evolution and his own observations in the non-organic world. See his "Statt des Vorworts zur achten Auflage" (pp.lxxix-lxxxvii).

Several years earlier German physician and physiologist Emil du Bois-Reymond (1818-96), widely known for his pioneering work in electrophysiology and who has been called "the most important forgotten intellectual of the nineteenth century," had published a similar polemic against the concept of *Lebenskraft* that adds a novel and unexpected perspective.[56] In his "Glaubensbekenntnis über das Wesen der sogenannten Lebensvorgänge und die dabei tätigen Kräfte" (1848), du Bois-Reymond argued against the concept of *Lebenskraft* as a unique substance, asserting that even using terms such as *Kraft* and *Stoff* simply continued the old dualism of God and world, body and soul.[57] He explained that *Kraft* – especially in the combination of *Lebenskraft* implicit in his argument – is a personification of something we do not understand ("Lebenskraft," vol. II, p.13; Noll, p. 60). (This is the same point that Reil had made fifty years earlier.) For Bois-Reymond the evidence was overwhelming that electricity offered the key to understanding life and that nerves and muscles were the basis of thought and action. After citing considerable empirical evidence from the physical sciences that refute the idea of an autonomous *Lebenskraft*, he concluded by averring that the belief in a special *Lebenskraft* was nonetheless so persistent because it was for many essentially a matter of *faith*. Moreover, the belief in it served the purposes of certain institutions (e.g., the Church). Hence, analogous to doctrines of faith, the insistence on *Lebenskraft* as an autonomous force seemed ineradicable from public discourse ("Lebenskraft," vol.II, p.22; Noll, p.68). No wonder that Ivan Turgenev modeled his nihilistic protagonist physician Bazarov in *Fathers and Children* on du Bois-Reymond (Finkelstein, p.xii).

While Büchner, for his part, was not always clear on the relation between matter and force, he nonetheless posited natural and spiritual forces as inherent in matter. Drawing an analogy to the steam engine that produces motion via the energy of fire, water, and air pressure, he argued that motion in organic and inorganic nature results from the unseen dynamics of energy exchange that produce via the process of *Stoffwechsel* (metabolism) a total

---

56 Gabriel Finkelstein, *Emil du Bois Reymond: Neuroscience, Self, and Society in Nineteenth-Century Germany* (Cambridge MA: MIT UP, 2013), p.xv. Finkelstein offers a fascinating introduction to his broad significance.

57 Emil du Bois-Reymond, "Glaubensbekenntnis über das Wesen der sogenannten Lebensvorgänge und die dabei tätigen Kräfte" (1848), in *Untersuchungen über tierische Elektrizität* (Berlin, 1848), vol. I, p.xxxvi. Noll cites this source (pp.56-68). However, Du Bois-Reymond reissued his polemic against the special *Lebenskraft* theory in 1912, adding many explanatory footnotes to update his argument. Vgl. Du Bois-Reymond, "Über die Lebenskraft. Aus der Vorrede zu den 'Untersuchungen über tierische Elektrizität' vom März 1848," In: *Reden von Emil du Bois-Reymond in zwei Bänden*, edited by Estelle du Bois-Reymond (Leipzig: Veit & Comp., 1912), vol. II, pp.1-26 (p.14).

sum of effects.[58] He called these internal workings in animate beings mind, soul, or thought. One of his central postulates was that force/energy and mind emanate from original matter. Because the quality of atoms is indestructible, as it was then thought, and that all organisms are constructed of the same atoms that are found in inorganic matter, except for their arrangement, there cannot exist, L. Büchner concluded, any separate *Lebenskraft*: "kann es auch keine besonderen organischen Kräfte, keine Lebenskraft geben" (Noll, p.70). Organic life (itself a tautology) clearly originated for L. Büchner from the effects of molecular forces ("Wirkung der Molekularkräfte"; Noll, p.70). His view of nature, therefore, was radically materialistic. There was no room for a *spiritus vitalis*. Therein he echoed his contemporary, the Professor of Pathology and Clinical Medicine Rudolf Virchow (1821-1902) who had argued in similar fashion in his polemical response to the vitalists, "Alter und neuer Vitalismus" (1856).[59] Virchow was a major figure who, like du Bois-Reymond, was strongly influenced by physiologist Johannes Müller's (1801-58) cell theory. Müller used *Lebenskraft* to designate the complex chemical-physical interactions central to cell formation in the development of life forms. The physiologist Max Verworn based his encyclopedic *Allgemeine Physiologie* (1896), which is dedicated to Johannes Müller, on the cell as the elementary organism in all life.[60]

At other times, however, Büchner was not so sure what to do with consciousness. Mind and matter sometimes seemed to him to be two different aspects of the human body; the mind in that case was not necessarily rooted in matter (Büchner, *Kraft und Stoff*, pp. 97-138). This indecision recalls for us Nietzsche's comment in *Also sprach Zarathustra* cited in the masthead that the child believes s/he is body and soul, whereas the more aware adult comes to accept her/himself only as body: "'Leib bin ich und Seele' – so redet das Kind. Und warum sollte man nicht wie die Kinder reden? Aber der Erwachte, der Wissende sagt: 'Leib bin ich ganz und gar, und Nichts ausserdem; und Seele ist nur ein Wort für Etwas am Leibe" ("'Body am I, and soul' – thus speaks the child. And why should one not speak like children? But the awakened and knowing say: 'body am I entirely, and nothing else; and soul is

---

58 *Kraft und Stoff*, pp.11, 122; *Force and Matter*, pp.11, 132.

59 Rudolf Virchow, "Alter und neuer Vitalismus," *Archiv für pathologische Anatomie und Physiologie und klinische Medizin*, 9 (1856), 3-55 (p.8). Cited according to Noll, pp.77-81 (78). Virchow makes clear that his polemic is directed against the concept of *spiritus vitalis*, which he sees as the vitalist's approach; *vis vitalis* is the sole legitimate term for him (Virchow cited by Noll, p.78).

60 Cf. Max Verworn, *Allgemeine Physiologie. Ein Grundriss der Lehre vom Leben*, fünfte, vollständig neu bearbeitete Auflage, (Jena: Gustav Fischer, 1909), pp.64-141. (Noll cites very little: Noll, pp.83-84). See also Rudolf Virchow, *Die Cellularpathologie in ihrer Begründung auf physiologische und pathologische Gewebelehre* (Berlin: Hirschwald, 1858).

only a word for something about the body.'"[61] Nietzsche's conceptions of the eternal return and will to power also play a role here when seen as expressions of the laws of thermodynamics. It should prompt us to consider seriously Ludwig Büchner's own hesitation about how to describe the relationship between physical and mental operations. In any event, Büchner's ideas about the sameness of matter and energy live on in contemporary fields of cosmology and particle physics, even if Büchner has not entered the contemporary debate regarding organic matter by name as Darwin has.[62]

As is well known, Darwin's description of the natural system is "a genealogical arrangement" that foregrounded the interconnectedness of all organic matter and amounted to one grand natural system. Darwin identified specific laws of growth with reproduction that include inheritance and variability in the struggle for life:

> The fact, as we have seen, that all past and present organic beings constitute one grand natural system […]. The real affinities of all organic beings are due to inheritance or community of descent. The natural system is a genealogical arrangement, in which we have to discover the lines of descent by the most permanent characters, however slight their vital importance may be.[63]

Unlike Darwin, Lamarck allowed for cultural and intellectual inheritance along with genetic transferral. (The former is now commonly known as memetics.) For his part, Nietzsche spoke of the unity of the physical, physiological, and mental as conjointly constituting the "economy of the whole."[64] Received ideas for him could be just as (trans)formative as hereditary traits. This point Hugh Ridley underscores in his *Darwin Becomes Art: Aesthetic Vision in the Wake of Darwin* (2014). The transformative force of ideas is, of course, central to this current undertaking, manifest in its linking of

---

61 Friedrich Nietzsche, *Sämtliche Werke*, Kritische Studienausgabe, ed.by Giorgio Colli and Mazzino Montinari, 15 vols. (München and Berlin: Deutscher Taschenbuchverlag and de Gruyter, 1993), vol. IV, p.39 (hereafter cited as KSA); English version cited according to Nietzsche, *Thus Spoke Zarathustra*, trans. by Kaufmann (1995), "On the Despisers of the Body," p.34.

62 See, e.g., Lisa Randall, *Warped Passages: Unraveling the Mysteries of the Universe's Hidden Dimensions* (New York: HarperCollins, 2005); George Smoot and Keay Davidson, *Wrinkles in Time* (New York: William Morrow & Co., 1993).

63 Charles Darwin, *The Origin of Species* (1859), Chapter 14: Recapitulation and Conclusion: http://www.literature.org/authors/darwin-charles/the-origin-of-species/ chapter-14.html.

64 On this issue see my "'Live like a Man and not like a Monkey': Nietzsche's Philosophic Vitalism and Darwin's Theory of Evolution," in: *The Evolution of Literature: Legacies of Darwin in European Cultures*, ed. by Nicholas Saul and Simon J. James (Amsterdam & New York: Rodopi, 2011), pp.255-81 (pp.270-74).

scientific paradigms and aesthetic responses. Ridley's focus on intelligence rather than adornment as the chief characteristic of (enduring) art mirrors the analyses in this volume.[65] In any event, Büchner, Darwin, and Nietzsche each expressed versions of the so-called will to power as the fundamental force or energy in the universe, inherent in and between all its constituent parts. That will is expressed in the give-and-take of cooperation and productive interaction as well as in explosive singularities.

Central to my present purposes is the similarity between Darwin's idea of nature as a genealogical arrangement and Büchner's conception of nature as matter and energy that do not approach entropy and that require no authentication by an authority external to the process itself. In 1884 – at the very time that such issues occupied Nietzsche – Büchner published his treatise on the concept of progress in the wake of Darwin: *Der Fortschritt in Natur und Geschichte im Lichte der Darwinschen Theorie* (*Progress in Nature and History in the Light of the Darwinian Theory*, 1884). These concerns with natural selection, the nexus of matter and energy, and the notion of progress are dominant in Nietzsche's philosophical musings.

In 1866 the young Nietzsche hit upon Friedrich Albert Lange's *Geschichte des Materialismus und Kritik seiner Bedeutung in der Gegenwart* (*History of Materialism and Critique of its Present Importance*, 1865). The book proved decisive in awakening his interest in atomism, chemistry, and physiology. He viewed Lange's work as "undoubtedly the most significant philosophical work to have appeared in recent decades," calling it in a letter of 1868 "a real treasure-house" of the "materialist movement of our times."[66] Lange, in turn, had assiduously read Büchner's *Kraft und Stoff*, reacting positively to its interpretation of nature as being purely physical. Nietzsche also read and reacted to the French naturalist Lamarck. Büchner's treatise *Progress in Nature and History in Light of Darwinian Theory* is applicable to Nietzsche, that "profound physiologist."[67] And Nietzsche's own problematizing of human conscious-

---

65 Hugh Ridley, *Darwin Becomes Art: Aesthetic Vision in the Wake of Darwin, 1870-1920* (Amsterdam & New York: Rodopi, 2014), pp.212-13. On Darwin and Nietzsche see pp.68-85 and on aesthetics esp.chap.V: "Ojectivizing Feeling" (pp.165-208).

66 Quoted in Leiter, *Philosophy Guidebook to Nietzsche*, p.65, and George J. Stack, *Lange und Nietzsche* (Berlin & New York: de Gruyter, 1983), p.10. Original source: *Nietzsche Briefwechsel. Kritische Gesamtausgabe*, eds. G. Colli und M. Montinari (Berlin & New York: de Gryuter, 1975), vol.I, p.2. *Briefe an Hermann Mushacke (Nachschrift)*, November 1866, p.184 (*Archidatenbank des Goethe- und Schiller-Archivs*, GSA 71/BW 290, 7-8). For further orientation see Thomas H. Brobjer, "Nietzsche Reading and Knowledge of Science: An Overview," in *Nietzsche and Science*, ed. by Gregory Moore and Thomas Brobjer (Ashgate, 2004), pp.21-50.

67 Richard S. G. Brown, "Nietzche that Profound Physiologist," in *Nietzsche and Science* (2004), pp.51-70.

ness – hence of literature and art – as physiological should be seen in the context of the contagious *Lebenskraft* debate in the long nineteenth century.

It is all about energy. Or in Nietzsche's formulation in his effort to get to the bottom of things: will to power. Will to power – he declared, demanding total honesty when confronting the basic text of *homo natura* – is the expression of a healthy, growing organism: "life itself is *essentially* appropriation, injury, overpowering of the strange and weaker, suppression, severity, imposition of one's own forms, incorporation and, at the least and mildest, exploitation."[68] Exploitation is not something bad and to be eradicated as per the utopian conceptions of science and idealistic philosophy. Hence, applying the concept to society at large, he averred that a state without exploitation sounded to him "like promising a life in which there will be no organic functions" (BGE, p.260). Quite the contrary to common understanding, "exploitation" does not pertain to a primitive society; "it pertains to the essence of a living thing in a fundamental organic function." Consequently, the will *to* power is nothing more than "the will *of* life" (emphasis added; BGE, p.260). This Nietzsche considered the novelity of his theory. Indeed, it was explosive, opening up new perspectives on the relationship between mind and matter and renewing the old debate on the nature of *Lebenskraft* by essentializing the fundamental principle of life: absorption.

The connections between matter and spirit/energy led to new aesthetic renderings of human activity in a "mechanical" body situated in a "mechanical" world. Nietzsche underscored elsewhere the artistic quality of *Lebenskraft* when he conjectured that an innate relationship might exist between chemical and artistic processes: "Die chemischen Verwandlungen in der unorganischen Natur sind vielleicht auch künstlerische Prozesse" (Chemical transformations in the inorganic world are perhaps also artistic processes).[69] He urged us to rethink the concept of force and the conservation of energy in the universe in the wake of Büchner, Julius Robert Mayer, and Hermann Helmholtz. Clearly, with his concept of vitalism Nietzsche participated in this *Lebenskraft* debate, representing perhaps its apotheosis around 1900. It has morphed into the organically grounded will *to* power as will *of* life.

---

68 Friedrich Nietzsche, *Beyond Good and Evil*, trans. by R. J. Hollingdale (New York: Penguin Putnam, 1990), No. 259, p.194. Hereafter cited as BGE. The German version is found in KSA, vol.V, p.207.

69 Nietzsche, Nachlass I (Musarion); cf. also KSA, vol.7, p.395. I take a different tack from that of Claudia Crawford, *The Beginnings of Nietzsche's Theory of Language* (Berlin: de Gruyter, 1988), pp.115-27, where she contends that Nietzsche thought we could only understand the manifestations of life, but not its origins. To this point see my *Remapping Reality* (2006), chap.3: "Grounding Creativity – Nietzsche and the New Universe," pp.111-34 (pp.121-24).

To be sure, he could confound his readers by expanding will to power to cover all expressions of energy release and by asserting that natural science does not *clarify* natural processes ("Welt-Erklärung") but rather *interprets* them ("Welt-Auslegung und -Zurücklegung"). Consequently, he groups natural scientists of his day together with the class of idealists derivative of Plato.[70] This aversion we can perhaps explain as Nietzsche's rejection of dogma of any kind that hinders the eyes from seeing, the ears from hearing, and the mind from understanding. The belief in science as capable of answering all questions amounted to him to be no different from religious dogmatism. And we have seen how carefully Kant, Reil, du Bois-Reymond and others had to express themselves to avoid conflict with powerful institutions and individuals. The great iconoclast Nietzsche threw one more stick of dynamite at the mental block hindering further progress in the revolution of thought.

Nietzsche's insistence that even physicists are interpreters underscores the centrality of the hermeneutic act in human life. Coupled with the interconnection of the organic and inorganic as it has unfolded in the *Lebenskraft* debate begs for an explanation of the carriers of cultural phenomena (ideas, behaviors), the so-called memes. Reil had already recounted "sympathetic" reactions in animals, and lectured on the "capacity to make associations" (*Assoziationsvermögen*) that led to the coordination of various parts and systems in organic matter.[71] The physiologists and anatomists considered in this introduction readily admitted that they often had to take recourse to analogy and metaphor to fill in the gaps of reliable empirical evidence, although scientists after ca. 1850 waxed ever more confident in their prospects for completely understanding the inner workings of nature. We should heed Nietzsche's warning, take a further step, and expand our view of empirical reality to include mental or psychic phenomena that interact with biological functions. Such interactions already marked the work of Johann Wilhelm Ritter (1776-1810), Schelling, and the Romantics (see the essays by Kuzniar, Schmiesing, and Smith).[72]

Ultimately, the step would lead us to consider neurological cases such as that of the Daniel Paul Schreber or of intersex ones such as that of Mar-

---

70 Nietzsche, *Jenseits von Gut und Böse*, #14, KSA, vol.V, p.28. See also KSA vol.13, p.15[51]. Cf. Babette Babich, "Nietzsche's Critique of Scientific Reason and Scientific Culture," in *Nietzsche and Science* (2004), pp.133-54 (p.141).

71 See Reil's discussion of sympathy in Noll, *Lebenskraft*, pp.61–64 (§17. Sympathie – Consensus); pp.81–87 (§21. Drittes Gesetz; Assoziationen). Originally published in *Archiv*, 1 (1796), 109–14; 140–51.

72 Cf. Jocelyn Holland, *Key Texts of Johann Wilhelm Ritter (1776-1810) on the Science and Art of Nature* (Leiden: Brill, 2010) and Helmut Müller-Sievers, *Self-Generation: Biology, Literature, Philosophy around 1800* (Stanford: Stanford UP 1997).

tha/Karl Baer (see the chapter by Stephanie Hilger) or of Alfred Döblin's blurring of boundaries between chemical processes and psychological interiority (see the chapter by Cate Reilly) within a materialist context. Simultaneously with Nietzsche's call to view the body as the source for everything associated with mind and soul, for instance, William James premised his *The Principles of Psychology* (1890) on positivistic materialism. Eschewing any kind of dualism, he contended in full accord with Nietzsche: "The very Self or *ego* of the individual comes [...] to be viewed no longer as the *pre-existing* source of the representations, but rather *as their last and most complicated fruit*."[73] Nietzsche had read Büchner, who had included several chapters on brain-thought-soul in his *Kraft und Stoff* (pp.123-78). And Büchner had read Hudson Tuttle's (1836-1910) *Arcana of Nature* where he characterized the brain as "bridging over the chasm between matter and spirit."[74] The move to self and psyche would entail a logical further step: to consider the meme as a product of radical reality.

Memes and *Lebenskraft*

Coined by Richard Dawkins in *The Selfish Gene* (1976), the word derives from the Greek word "mimeme" (imitated) but it is also related to "mneme" (memory). It denotes a unit of cultural information that propagates, like a virus, from one mind to another. The analogy to the way a gene propagates from organism to organism as a unit of genetic information in biological evolution is clear and intended. Memes exert strong influence over our physical environment as well as our social networks.

In the wake of Dawkins, Daniel Dennett applied the idea to philosophy of mind in his *Consciousness Explained* (1991) and linked the concept to evolutionary theory in his *Darwin's Dangerous Idea* (1995). Susan Blackmore, *The Meme Machine* (1999), sought to place memes at the center of a radical and counter-intuitive naturalistic theory of mind and of personal identity; Lyall Watson, *Dark Nature: A Natural History of Evil* (1995), tried to make sense of natural and supernatural phenomena in biological terms. All agreed that

---

73 William James, *The Principles of Psychology,* 2 vols (New York: Henry Holt & Co., 1890), vol.I, p.3. My emphasis.

74 Hudson Tuttle, *Geschichte und Gesetze des Schöpfungsvorganges,* trans. by H. M. Achner (Erlangen: Ferdinand Enke, 1860), p.292 ("welches gleich einer Brücke über die Kluft zwischen Materie und Geist hinspannt"). In the masthead to his chapter on "Gehirn und Seele," Büchner cited Hudson's remark in the German translation: "Von der Materie erheben wir uns zum Geist über das Gehirn" (*Kraft und Stoff,* p.123; "From matter we ascend to spirit through the brain; Hudson, p.291). The English title is *Arcana of Nature, or the History and Laws of Creation,* 2 vols (first publication: 1860-63) (Boston: W. White & Co, 1870), citations pp.273, 274.

memes, like genes, do not purposely do or want anything; they either get replicated or not. Influenced by Rupert Sheldrake, *A New Science of Life* (1981) and *The Presence of the Past* (1988), Watson stated the essential law of memes pithily: "Vary, multiply, and let the fittest survive" (*Dark Nature*, p.255). The proximity of meme propagation to the biological strain is evident in his summarizing statement: "Culture is […] an organism taking a variety of forms, which compete against each other and change and evolve through the medium of their units of inheritance – the memes" (*Dark Nature*, p.256). Thus the old debate about *Lebenskraft* assumes in our contemporary era whole new significance. If nineteenth-century science ultimately concluded that *spiritus vitalis* inhibits scientific inquiry, today we could conclude that an emphasis on *Lebenskraft* as solely at home in a material realm of *vis vitalis* is "forschungshemmend" when confronted with growing insights into how the human organism has succeed in creating a parallel universe that seemingly abides by Darwinian principles of natural selection. In *The Mangle of Practice* (1995), Andrew Pickering took a different, yet similar approach to the ways that humans are enframed in their environment and give expression to their lives through performance. His metaphorical characterization of a "dance of agency" between people and things resonates with Nietzsche's use of the dance in *Also sprach Zarathustra* as a metaphor for the drive of the spirit to overcome the omniscent inhibiting force of gravity.[75] The two drives are of course inherently related.

No wonder, then, that the human brain appears to be the new frontier. Despite significant progress, we understand little about its architecture. And why would it be otherwise when "[m]ore than 80 percent of the roughly 30,000 genes in human DNA are active in the brain."[76] Indeed, those numbers together with the huge number of neurons are mind-boggling. Once one adds the possible interconnections among them, the numbers become astronomical. Science writer Carl Zimmer notes laconically:

> On its own, in its dish, the neuron can't do much. But join together 302 neurons, and they become a nervous system that can keep the worm *Caenorhabditis elegans* alive – sensing the animal's surroundings, making decisions and issuing commands to the worm's body. Join together 100 billion neurons – with 100 trillion

---

75 Andrew Pickering, *The Mangle of Practice: Time, Agency, and Science* (University of Chicago Press, 1995). Later he addressed the dance of agency as a pivotal concept in "Material Culture and the Dance of Agency," in: *Oxford Handbook of Material Culture Studies*, ed. by Dan Hicks and Mary C. Beaudry (Oxford UP, 2010), p.191-208, and in "The Dance of Agency," in: *Ideas on the Nature of Science*, ed. by David Cayley (Goose Lane Editions, 2009), pp.81-91. Nietzsche, *Zarathustra*, KSA, vol. IV, pp.139-40.

76 Carl Zimmer and Benedict Carey, "A Spark for a Stagnant Search," *NYT, Science Times*, Tuesday, July 22, 2014, pp.D1, D5.

> connections – and you have yourself a human brain, capable of much, much more.[77]

The neuron is the core component of the brain and of the nervous system more generally. Together with neotransmitters and glia cells it processes and transmits information via electro-chemical interactions with other neurons. Without them the body would not be able to interact with its environment. The layers of neural activity and the exact role(s) of glia cells that constitute at least as much mass as the neurons in the brain are so complex that researchers can focus only on a minute part of the architecture at a time. Yet, recent advances in molecular genetics and aetiology offer hope of locating the gene that sparks schizophrenia and hence of reversing the disorder.[78] In other words, of altering mental states.

Yet all this research into the human is fundamentally based on the view that the body is a machine. To be sure, a highly sophisticated one – one hosting, moreover, trillions of microbes necessary to its functioning such that we are 90% bacterial and just 10% human[79] – but one, nonetheless, with a soul. If by "soul", like Kehlmann's Alexander von Humboldt, we understand something associated with mind, then the soul too is a highly complex and finely crafted part of the body. Such is clearly Nietzsche's intent in asserting: "soul is only a word for something about the body." Then too, albeit in a seemingly quaint analogy to Leibniz's concept of the Divine understanding as the "region of eternal truths," the human mind has evolved into a force of agency, capable of materializing potential realities. Eternal truths refers to all the ideas and possible worlds God could have created, but exist only as potential in the Mind of God.[80]

---

77 Carl Zimmer, "100 Trillion Connections: New Efforts Probe and Map the Brain's Detailed Architecture," *Scientific American*, (January 2011).

78 Schizophrenia Working Group of the Psychiatric Genomics Consortium, "Biological insights from 108 schizophrenia-associated genetic loci," *Nature*, 511 (24 July 2014), pp.421-27.

79 Melinda Wenner, "Humans Carry More Bacterial Cells than Human Ones," *Scientific American* (30 Nov. 2007): http://www.scientificamerican.com/article/strange-but-true-humans-carry-more-bacterial-cells-than-human-ones/. Accessed 24 November 2014.

80 *Monadology*, §44: "C'est parce que l'entendement de Dieu est la région des vérités éternelles, ou des idées dont elles dépendent, et que sans lui il n'y aurait rien de réel dans les possibilités, et non seulement rien d'existant, mais encore rien de possible." Cited is the electronic version prepared by M. Daniel Banda of Gottfried Wilhelm Leibniz, *La Monadologie*, édition annotée par émile Boutroux (Paris: C. Delagrave, 1881):http://classiques.uqac.ca/classiques/Leibniz/La_Monadologie/La_Monadologie.html. See also G. W. Leibniz, *Theodicy*, §189: "In the region of the eternal verities are found all the possibles." G. W. Leibniz, *Theodicy: Essays on the Goodness of God the*

And Leibniz valued the human capacity for reflective and applied thought higher than the soul because the mind is itself "a little deity" capable of replicating the creative energy of the Divine understanding.[81] Even Nietzsche's awakened adult is apparently constrained to search out untruth, as he labels that which has not been claimed to be dogmatic Truth or not yet been thought of. Just as physicists have gained empirical evidence of the fabled and elusive theoretical Higgs boson (the so-called "God particle"), which provides mass to elementary particles and could explain the expansion of the universe following the Big Bang, mind researchers (i.e., neurologists, psychologists, pychiatrists, philosophers) are closing in on a neurological explanation as to why humans have a penchant to believe in God.[82] The term "God" (or earlier: "gods") might in fact be a nomenclature for that which causes motion: energy (*Kraft*). In the context of such theorizing, we do well to recall F.W.J. Schelling's dictum of 1799 that "*speculative* physics [...] has always been the mother of all great discoveries in nature."[83]

According to the current operative paradigm, then, matter, mind, and consciousness are inextricably intertwined, sharing as they do the electromagnetic spectrum. Their interactions are inherent to what amounts to a dynamic radical mediation; mediation in the sense that matter, mind, and consciousness are never really posited as an inalterable substance or state but

---

Freedom of Man and the Origin of Evil, ed. with an introduction by Austin Farrer, translated by E.M. Huggard from C.J. Gerhardt's Edition of the Collected Philosophical Works, 1875-90 (Routledge & Kegan Paul Ltd, 1951). Gutenberg e-edition used. (= Gerhardt; vol.VI, p.229).

81 Tellingly, he theorized: "que les âmes en général sont des miroirs vivants ou images de l'univers des créatures ; mais que les esprits sont encore des images de la Divinité même, ou de l'Auteur même de la nature : capables de connaître le système de l'univers et d'en imiter quelque chose par des échantillons architectoniques ; chaque esprit étant comme une petite divinité dans son département" ("that souls in general are living mirrors or images of the universe of created things, but that minds are also images of the Deity or Author of nature Himself, capable of knowing the system of the universe, and to some extent of imitating it through architectonic ensamples [echantillons], each mind being like a small divinity in its own sphere" (*Monadology*, §83; *Theodicy*, §147). English translation by Robert Latta available at: http://oregonstate.edu/instruct/phl302/texts/leibniz/monadology.html

82 Volney P.Gay, "Neuroscience and Religion: Brain, Mind, Self, Soul," in: *Neuroscience and Religion* (2009), pp.1-18; and Stephan Carlson, "The Neuroscience of Religious Experience: An Introductory Survey," in: *Neuroscience and Religion*, pp.153-73.

83 Schelling, *Einleitung zu seinem Entwurf* (1988): "Es ist daher begreiflich, daß *spekulative Physik* (die Seele des wahren Experiments) von jeher die Mutter aller großen Entdeckungen in der Natur gewesen ist" (§4, p.29).

experience continuous alteration filtered via multiple levels of interaction.[84] They constitute the kaleidoscopic, complexly distributed agency in the Anthropocene that Heather Sullivan addresses in her essay or the microbiome that Monica Ledoux renders in her contribution. While far from complete, current understanding of the architecture of mind and body and their interactions with the environement is far advanced over the insights of eighteenth- and early nineteenth-century researchers. A refrain resonating throughout the historical inquiries into how the world and *l'homme machine* work is the researcher's ready admission: I really cannot say definitely what the inner processes of life and movement are; but I am confident that future research can build upon what is offered here. This we heard Kant, Brandis, Reil, Prochaska, and others confess. By analogy the essays collected here – little energy bundles, if you will – seek not merely to detail historical ideas and events but to stimulate future thought and action. In any event, in whatever form it appears, life is a matter for contemplation, analysis, and wonder. It is matter and it matters.

---

84 Cf. Richard Grusin, "The Radical Mediation of Life in the Anthropocene" (22 October 2013), C21 Blog: http://www.c21uwm.com/2013/10/23/the-radical-mediation-of-life-in-the-anthropocene/. Accessed 24 November 2014.

Pneuma – Sexuality – Sex Difference: From Arabic to European Philosophy and Medical Practice

Jennifer Wynne Hellwarth

Abstract: The theory of *pneuma* (life force, vital heat)**,** marked largely by Galen's notion that there were two kinds, *pneuma zotikon* (life principle, vital spirit) and *pneuma psychikon* (or spiritus animalis, the mind, or psychic spirit), came to be translated into Arabic starting in the ninth century. Many Arabic texts included the addition of an earlier, Greek formulation of a third kind of *pneuma,* which is called "natural" (*spiritus naturalis*). The tripartite division between spirits and their functions, which also corresponded to the tripartite division of the soul as articulated in Plato and Aristotle's works, was largely accepted in the Middle Ages as being crucial to physical and mental function and quality of life – the primary moving cause. This essay seeks to contextualize and historicize the theory of *pneuma* and how the theory of "vital spirits" was shaped and transmitted from Greek to Arabic, and then into European philosophy and practices. This essay defines the concept of *pneuma* and its relationship to the soul, sexual desire, sexuality, sex difference, and sperm production. Further, it traces some of these principles to medieval literary examples.

Though the term *Lebenskraft,* describing the physical, chemical, and mechanical forces of organic matter, appears to have been first introduced into medicine (and then into colloquial speech) in about 1774, I would like to suggest that the history of this concept reaches back before the eighteenth-century coining of the term.[1] And, while the definition and the development of the term *Lebenskraft* and its use in German philosophical thought and literary production falls outside my bailiwick – I will leave the development of the concept of *Lebenskraft* to be carefully defined and explored by the scholars contributing to this collection – I would like to suggest that the term brings with it a history embedded with physiological, cosmological, and cultural meanings dating back to the ancient, then medieval and early modern concepts of *pneuma,* or life force or principle. Further, the debate about the relationship between the mind, soul, and body (recalling that the notion of *pneuma* –or "vital force" – was considered the primary vehicle of the soul) still continues, as we see in the increased studies in neuroscience, cognitive science, and neurophysiology. We also see continued discourses around sexuality and sexual arousal (physiological and psychological) that resonate with those emerging from pneumatology: one only need think of beliefs related to competitive sports or other typically masculine endeavors, and of

1 Thomas H. Broman, *The Transformation of German Academic Medicine, 1750-1820* (Cambridge: Cambridge UP, 1996), pp.87-89.

the idea of sperm as a vital or precious fluid that needs to be retained in order for a man to preserve his strength for his various endeavors and competitions.[2] This essay, then, is just one entry point into an historical context of the inquiries into the soul-mind-body connection and vitalism in the long eighteenth and nineteenth centuries; it also seeks to touch upon the ways in which scientific thought and debates intersected and informed literary production in the Middle Ages and beyond.

To begin, Aristotle, in his fourth century BCE *Peri pneumatos* (*De spiritu* in Latin), extrapolates on the general concept of innate *pneuma* – its nature, its characteristics, and its relationship to a living creature (animal or human) as a "primary moving cause." He also explores the connection of the *pneuma* to the soul, "referring to the unity of the soul and *pneuma* as its instrument"; that is, *pneuma* is the primary vehicle of the soul.[3] Aristotle's work, along with Hippocratic theories of humors, is considered to be one of the chief sources of Galen's (ca. 129-216 CE) work on *pneuma.* Though Galen differed with Aristotle on the heart as the controlling organ of the body – he argued for a tripartite relationship between heart, liver, and brain – he adhered to Aristotle's notion that mind and body are not separate faculties. Though there was debate about this separation of faculties, most specifically amongst the Stoics, Galen's concepts of *pneuma* – *pneuma zotikon* (or *vital pneuma*) and *pneuma psychikon* (or *spiritus animalis*) – were transmitted in the ninth century through medieval Arabic texts. At this time, the theory developed largely into a tripartite physiological pneumatology that included the recognition of a third *pneuma, spiritus naturalis*; this tripartite schema was then transmitted to Western medicine, where it prevailed for nearly eight centuries.[4] This essay seeks to give a brief overview of some of the key moments in this history to highlight the central concepts and complexities of this the-

---

2 A couple of popular culture examples from the last fifty years demonstrate this belief, including scenes from movies like *Dr. Strangelove* and *Bull Durham* in which a man's sperm is seen as a "precious" fluid that, when "spent," can reduce his power. In these films, sperm is also represented as being sought after by women. Both of these notions – sperm as a precious fluid that is desired by (heat seeking) women – are deeply embedded in the ancient theory of humors as well as Indian Ayurvedic medicine.

3 Abraham P.Bos and Rein Ferwerda, *Aristotle, On the Life-Bearing Spirit* (De Spiritu)*: A Discussion with Plato and his Predecessors on* Pneuma *as the Instrumental Body of the Soul* (Leiden, Boston: Brill, 2008), p.2. Aristotle's work "On *pneuma*" is "absent in the Greek lists of Aristotle's writings but it is mentioned in the Arabic catalogue of Ptolemy el-Garib" (Bos and Ferwerda, p.3). In addition, Pre-Socratic (sixth century BCE) philosopher Anaximenes is held to be one of the key originators of this theory.

4 Danielle Jacquart and Claude Thomasset, *Sexuality and Medicine in the Middle Ages* (Cambridge: Polity Press, 1988), pp.48-49.

ory, particularly those that connect to the mind-body-soul debate, sexual reproduction, desire, sex differentiation, and gender roles. This essay also seeks to briefly touch on the ways pneumatic theory influenced medieval – as well as later – imaginative literature.

Defining *Pneuma*

The idea and definition of *pneuma*, most often translated into English as "vital heat" or "life-bearing spirit," or even "vital power," was described by Aristotle in a number of his works.[5] In *De generatione animalium*, Aristotle explains the central role of innate *pneuma* to the "perception, mental activity, and physiological vigour" of all living creatures.[6] The text also connects *pneuma* to conception as well as to how the "faculties of the soul are transmitted to offspring."[7] The notion of innate *pneuma* is embedded in a system that connects the non-human, the "astral" (or *aether*) element of the stars, with the human, as it is drawn into the body through respiration; it is the source of vital heat. According to Aristotle, "*pneuma* is already present in semen and is an analogue of the astral element, which is responsible for the fertility and the life-generating power of semen."[8] Aristotle thus draws a close connection between these two elements, the role respiration plays in feeding the "internal fire" of living creatures, and the special role that *sperma* plays as a unique foam-like substance which provides the "form" in the act of conception.[9]

The role and purpose of *pneuma* reflects as well the humoral principles established by Hippocrates, Aristotle, and later in greater detail by Galen. Humoral theory informed the explanatory principles for medical practice for the Greeks and Romans, and then later the Muslims, and then Christian Europeans through at least the sixteenth century and in a more limited way into the nineteenth century. Humoral theory is a worldview based on the relationship between the macrocosm and the microcosm and the interaction between the four primary principles in the body. In brief, the theory held that the body contained four basic substances – humors – which, when in balance, made for a healthy individual. These bodily humors are black bile, yellow bile, phlegm, and blood, and they were thought to be directly related to the four earthly elements, earth, fire, water, and air (respectively). Each

5 Bos and Ferwerda, p.2. note 4; Friedrich Solmsen, "The Vital Heat: The Inborn Pneuma and the Aether," *The Journal of Hellenic Studies*, 77. 1 (1957), 119-123 (p.119).

6 Bos and Ferwerda, p.2.

7 Solmsen, p.119.

8 Bos and Ferwerda, p.1.

9 Bos and Ferweda p.9-10. This theory of sperm was established in earlier works by Plato, Diogenes, and Empedocles (Slomsen, pp.119-120).

individual had his or her own temperament (Melancholic, Choleric, Phlegmatic, or Sanguine) depending on his or her basic humoral makeup. This temperament might shift as the balance of one's humors shifts, but essentially one temperament was thought to be fundamentally more dominant than another.

Just as the humors dictated a person's temperament, all ailments and diseases were believed to be caused by an imbalance of humors in the body. These imbalances would be addressed through diet, exercise, and medicinal remedies designed to bring the body's humors into balance. This might include purging (blood, excrement, phlegm, etc.), fumigation, applying heat or cold, and the use of herbs and other substances thought to be associated with counteracting or restoring the deficit humor. Treatments would also be determined based on macrocosmic elements that were associated with the humors and the elements, including seasons and astronomical and astrological influence.[10]

Each humor is also associated with a particular organ and, in a tripartite pneumatic frame, certain organs were paired with one of the three *pneumas.* The theory held that food is digested and transmitted (from the stomach) to the liver, where it is transformed into blood with the natural *pneuma* (*spiritus naturalis,* natural spirit). It is then imbued with the vital *pneuma* (*pneuma zotikon,* vital spirit) as it passes to the heart; the vital spirit is imbued with air from the lungs, drawn in through air inhaled in breathing (also known as *aether* or astral air as described above). The psychic *pneuma* (*pneuma psychikon,* animal spirit) is formed as the vital *pneuma* moves to the base of the brain, and then the arterial blood moves through the brain's vessels in a kind of distillation process. The *pneuma* then moves outward in arteries and veins to all the organs.[11] In this schema, the natural *pneuma* is associated with yellow bile, fire, and the liver (or sometimes the gallbladder); the vital *pneuma* is associated with blood, air, and the heart; the psychic *pneuma* is associated with phlegm, water, and the brain. Furthermore, *pneuma* has associated with it rational, spirited, and appetitive "principles in the brain, the heart, and the liver respectively" as the function of *pneuma* changes and is elaborated depending on where it is located in the body.[12] It should be noted, however, that

---

10 Jacquart & Thomasset, pp.48-86. See also Peter E. Pormann and Emilie Savage-Smith, *Medieval Islamic Medicine* (Edinburgh: Edinburgh UP, 2007), pp.41-79.

11 Jacquart & Thomasset, pp.48-49; Jacalyn Duffin, *History of Medicine: A Scandalously Short Introduction* (Toronto: U of Toronto Press, 2000), pp.43-45; Geoffrey Lloyd, "*Pneuma* Between Body and Soul," *Journal of the Royal Anthropological Institute,* 13.1 (2007), 135-146 (p.143); C. E. Quin, "The Soul and the *Pneuma* in the Function of the Nervous System after Galen," *Journal of the Royal Society of Medicine,* 87 (July 1984), 393-395 (p.393).

12 Lloyd, pp.142-143.

there is some variation amongst theorists in these associations and pairings, as the different organs are sometimes connected with different humoral qualities – and note, too, that the humor associated with black bile has no specific (tripartite) *pneuma* associated with it.

Not surprisingly, the theory of humors was central to medieval medical practices and understandings of sex differentiation, conception, and sexual desire, and the theory supported each culture's view of women as the inferior sex by ascribing to the female (or feminine) the weaker or inferior temperaments. The physiology of the body was thought to consist of a combination of four attributes – hot, cold, dry, wet – and these elements were in turn associated with male or female characteristics. It was hypothesized that humans, as well as animals, consisted of the elements of fire and water (fire: hot/dry; water: cold/wet). Males were considered hotter and more perfect. Women were considered to be essentially colder than men, and those temperaments that were associated with colder elements were considered more feminine in nature. Likewise, the theory of conception generally, and the relationship of *pneuma* to humoral theory more specifically, hierarchized the role of the male both socially, in terms of gender roles, and biologically, in terms of their role in conception, particularly in the formation of sperm. The association of some elements with more masculine or hierarchized characteristics naturally extended to the nonhuman macrocosmic elements of earth, water, air, and fire, as well as the celestial bodies and seasons.

The relationship between *pneuma* – in the broadest sense of the term – and air in the functioning of the body, the relationship between *pneuma* and the soul, and the relationship between *pneuma* and conception are examples of ways in which the philosophical, physiological, medical, and even spiritual systems were shaped by a male discourse that viewed women and feminine characteristics as weak, imperfect, irrational, passive, and inferior. As explained above, the air (or *aether,* the cosmos) sets in motion *pneuma* throughout the body (including digestion). Air – variously described as *aer, aither, pneuma, phusa*, and *anemos* – brings with it a gendered quality, one that emerges in the role of *pneuma* in sexual desire, sex differentiation, conception, as well as in relation to the soul.[13] Most clearly this gendered connection is evident in a binary system famously attributed to Aristotle that delineates the two fundamental principles of active and passive (along with other

13 Lloyd, p.136; Lloyd does an excellent job articulating the history, subtle variations, and uses of these related terms referred to in this paper: *aer, aither, pneuma, phusa*, and *anemos.*

binaries). These basic principles apply to *pneuma* and its relationship to the soul and bodily function.[14]

In this schema, the passive is a "quality-less substance or matter", and the active is identified as "the cause, god, reason, fate, soul, or *pneuma*", where *pneuma* is the "active principle" that "pervades everything" and holds everything together.[15] *Pneuma* was also considered to be connected to the macrocosmic: "just as our soul, being air (*aer*), holds us together, so does *pneuma* or air (*aer*) enclose the whole world."[16] For example, the (clearly masculine) power of the wind, or air is, as Geoffrey Lloyd points out, "worked out especially in connection with sex [...] the winds were reputed to be able to impregnate mares."[17] This (metaphorical) image that connects the power of air/wind with conception is found in texts from Plato, Homer, and then later Aristotle, and Galen. The various meanings of *pneuma* also include "breeze and breath/respiration but stretch to spirit, inspiration, and eventually to [...] Holy Spirit" all of which are associated with masculine characteristics.[18] The ancient writers such as Hippocrates, Aristotle, and later theorists, saw these elements both in terms of sex differentiation as well as in relation to the soul. This combination of elements also accounted for the composition of the soul: "Soul, *psuché* [...] is a blend of fire and water," though the blending is different in the case of males and females.[19] Further, Aristotle viewed the soul and the body as interconnected, where the soul is the activity of the body.[20] Numerous philosophers would find divine presence in *pneuma,* which they saw as sharing both the cosmic, divine, and psychic principles.[21] Thus the role of *pneuma* in bodily function, reproduction, and the soul was intimately tied to concepts of active/passive, form/matter, male/female.

Similarly, the process of distillation noted above in relation to *pneuma* in the body had an analogy in the production of *sperma*; this theory privileged the role of male sperm.[22] There were, of course, debates over the centuries,

---

14 Lloyd also includes a discussion of the Stoics and *pneuma,* and how those theories connect to later Galenic and the Arabic texts (p.142).

15 Lloyd p.142.

16 As quoted in Lloyd, p.145, n. 9

17 Llyod, p.136.

18 Lloyd, p.137.

19 Lloyd, p.140.

20 Lloyd, pp.140-141.

21 Solmsen, p.122; the Stoics, Aristotle, Diocles, etc. shared this notion. In addition, we see this connection in Virgil's *Aeneid* VI, where Virgil connects *aether*, or *pneuma* as "the home or essence of the soul" and "employs freely one or the other of our concepts as a symbol of man's divine origin" (Solmsen, p.123).

22 Jacquart and Thomasset, p.49.

first amongst Greeks and Romans, then later in the Muslim, Jewish, and Christian worlds, about the nature of sperm and the role of female versus male seed in conception (which I explore more carefully below). But the general consensus saw sperm as "man's purest blood" and the "residue of the last coction" which, as it makes its way from the brain towards the sexual organs through veins and arteries, finally "undergoes one final metamorphosis" in which the purified "moist sperm accompanied by *pneuma* is ejaculated."[23] This view connected with later discussions in the Middle Ages, first by Arabic, Jewish, early Christian, and then later Christian European thinkers, philosophers, and scholars.

Notions of Sex and Reproduction and the Role of "Vital Heat"

The discourse on conception, sex differentiation, sexual desire, and the soul took various forms, and they were not at all universal. In the creation and generation of the embryo, so Aristotelian theory goes, the female "matter" that contributed to conception was largely made up of the menstrual blood, while the "form" in the semen was the male's contribution.[24] In terms of understanding sex differences and reproduction, the Medieval Islamic medical and philosophical discourses were not unified, but showed complexities

---

23 As quoted in Jacquart and Thomasset, p.60. This quotation is a translation of a late fourth-century Christian philosopher and Galenist, Nemesius, bishop of Emesa, from (what is now) Syria. Interestingly, Nemesius is noted by Philip van der Eijk as an early practitioner and theorist of addressing the problem of the relationship between mind and body through associating different parts of the brain with different functions, and by exploring "the structure of the human body, the position and function of its constituent parts, and the relation between the body and the soul (or the mind) in complex activities such as sense perception, voluntary movement, imagination, memory, thought, desire, and the emotions. The presupposition of the work is that the human soul-body composite is as good as it possibly can be: the body and its parts serve as an 'instrument' (the Greek word is organon) for the mind and its functions." Philip van der Eijk, "Nemesius of Emsea and Brain Mapping," *The Lancet*, 372.9637 (9 August 2008), pp.440–441 (p.440).

24 See Jacquart and Thomasset, pp.61-70, for a discussion of the role of the female sperm. See also Karine van 'T Land, "Sperm and Blood, Form and Food: Late Medieval Medical Notions of Male and Female in the Embryology of *Membra*," in *Blood, Sweat and Tears – The Changing Concepts of Physiology from Antiquity into Early Modern Europe*, ed. by Manfred Horstmanshoff, Helen King and Claus Zittel (Leiden and Boston: Brill, 2012), pp.365-392 (pp.371-372). In her essay, van 'T Land describes the complex philosophy behind the concept of female and male sperm and their respective contributions, which, distilled, so to speak, come down to male spermatic substance generated, distilled, and purified from blood, resulting in a white foamy substance, while the female contribution (the "matter") was considered inferior and remained closer to menstrual blood, and thus served as nourishment for the embryo. See van 'T Land, p.382.

and pluralities of thought, in particular about the role of male and female (seed) in conception – which is related to the notion of vital heat and the theories around how sperm was produced. While Hippocrates and his corpus theorized that conception occurred as a result of mixing the two seeds of the mother and father, the theory of pangenesis, Aristotle's theory was that male and female seed were essentially different in nature. For him, sperm was the "fourth stage of surplus nutriment" – a more purified form of excess blood.[25]

While men and women were considered equal in sharing the biological responsibility of generation, the nature of their contributions, which resulted in the production of a male or a female depending on the nature of the dominate fluid, was considered essentially different. According to Aristotle, the male sperm, in providing the "form," and the female menses, in providing the "matter," vary in degree rather than in kind.[26] With its mixture of *pneuma* and water, moreover, male semen played a special role as "the Soul that would guide and determine embryological development."[27] For Galen, both the male and female provide seed. Both of these principles were supported in various Islamic notions of generation.[28]

Fluidity of Sex, Rigidity of Gender

The one-sex model, as described in Thomas Laqueur's *Making Sex,* viewed the male and female reproductive organs as essentially the same in structure, but differentiated by their location as a result of male and female basic humoral temperament, as noted earlier. This theory of the one-sex model – and all of its cultural and social implications that created it and resulted from it – dominated the landscape of medical thought in Europe until the Enlightenment. Because, according to the one-sex model, men were considered naturally hotter, their reproductive organs were located on the outside of their body. And because females were naturally colder, their reproductive organs were both smaller and internal to the body. This difference was hierarchized, of course: males were considered more perfect (hotter) than the females, whose imperfect (colder) reproductive parts were contained inside their bodies as a result of a failure to develop properly. The one-sex model

25 Gadelrab, p.50.

26 B.F. Musallam, *Sex and Society in Islam: Birth Control before the Nineteenth Century* (Cambridge: Cambridge UP, 1983), pp.39-59 (p.44).

27 B.F. Musallam, p.45.

28 For example, we see this evident in the work of the eleventh-century scholar Ibn Sina, in which he both accepts Galen's theory of male and female seed and Aristotle's notion of the female's contribution to generation as that of "matter" and that of menstrual blood (Musallam, p.48).

allowed for a fluidity of *sex* (but not of what we call *gender*). For example, in this model, it was possible for a female to become hot enough that her internally located reproductive parts could become external; she would then take on the social (and legal) roles of a man.[29]

Documented examples of this transformation exist, both in terms of females and hermaphrodites switching from one sex to another, and in examples of males and females exhibiting flipped characteristics (either physiological or social from a modern perspective). In a notorious account of Montaigne's, a penis and testicles emerged from a girl as she was chasing a pig.[30] In the one-sex model, such a transformation is easily explained through the tenets of humoral theory and pneumatics, which understands sexual difference as a matter of degree – from superior male to inferior version of a male, female. Furthermore, the changes to the body were thought to be fluid, and external or cultural context could have an influence on the form of the body. That is to say, while we might view the body as "real" and culture as "epiphenomenal," the view of the body before the Enlightenment (to over simplify) was more fluid.[31] This fluidity also involved the relationship between the body and the soul. For example, early Christian theologian Origen of Alexandria posited that, as the spirit "warmed" to God, the body would become less "thick" and less "hard."[32] Similarly, Gregory of Nyssa suggested that, prior to the fall, the physical manifestation of sexual differentiation was the result of a "representation in the flesh of the fall from grace."[33]

Though corporal sex was fluid, social sex – what we would consider gender – was not nearly as fluid. Apparent biological sex came with particular privileges, rights, and obligations. The movement from one corporal structure (from female to male, or, in some cases, from hermaphrodite to male or female) was the result of a shift in micro- and macrocosmic elements: hot/cold, active/passive, etc. This shift would also be accompanied by a shift in a legal category. The physical manifestation of these elements was the external sexual organ. That is, "the penis was only the diagnostic sign" of sexual differentiation as defined by these other elements.[34] The case

---

29 Thomas Laqueur, *Making Sex: Body and Gender from the Greeks to Freud* (Cambridge and London: Harvard UP, 1990), pp.43-52.

30 Laqueur, p.129.

31 Laqueur, p.7.

32 Laqueur, p.7.

33 Laqueur, p.7.

34 Laqueur, p.135. See also Paula Sanders, "Gendering the Ungendered Body: Hermaphrodites in Medieval Islamic Law," in *Women in Middle Eastern History: Shifting Boundaries in Sex and Gender* (New Haven: Yale UP, 1992) pp.74-95 (pp.77-78). Sanders' essay gives a fascinating reading of the negotiation of the "biological process of

of those considered hermaphrodites illustrates that the (social) concern was less with the reality of the body, and more with maintaining categories of gender hierarchy.[35] The difficulty was in men or women not acting in concert with their role – it was considered a dysfunction or imbalance of humors, and, if one did not behave according to one's social and juridical role (as male or female), there were potentially legal as well as social and physiological, consequences. The essential qualities (and thereby roles) of male and female were "embedded in both the view of the cosmos and in the social structure."[36]

From Greek to Arabic

The translation of Greek texts into Arabic (some through Syriac, some through Hebrew) began in earnest in the eighth and ninth centuries in large part as a result of the Abbasid dynasty's cultural policies as they came into power. They sponsored the translation of Greek works during a time when there was an active exchange of knowledge between Christian, Jewish, and Muslim communities in centers of learning in the Near and Middle East.[37] In particular, the ninth century Nestorian Christian physician Hunyan ibn Ishaq, under the patronage of the Abbasid Caliphate, translated, interpreted, and added to the "codification and amplification" of pneumatic theory, first into Syriac, then into Arabic. His work is considered to be critical in the early solidifying of the concept of the three sprits, or *pneuma*.[38] In the early eleventh century, Ibn Sina (known in the Christian West as Avicenna) also produced a text (*Al-Qanun*) that synthesized Galenic medical writings, which

---

sex determination" (according to Muslim philosophers and physicians) and how the theory of generation "intersected with Qur'anic doctrines of creation" (p.76). As other scholars have noted, she points out that there was "deep coherence" between the Qur'an (and I would argue also Prophetic Medicine) and scientific literature. She also notes the "insistence on the two sexes of male and female and the articulation of this assumption in the theory of generation" which complicates Laqueur's notion of the "one-sex" model of biological sex (p.76).

35 Sanders, pp.74-95. Sanders discusses in particular the medical, social, cultural, and legal context of hermaphrodites in medieval Islamic law, pp.79-84, which illuminates the complexities around sexual differentiation and gender boundaries and rules.

36 Sanders, p.76.

37 See Sherry Sayed Gadelrab, "Discourses on Sex Differences in Medieval Scholarly Islamic Thought," *Journal of the History of Medicine and Allied Sciences,* 66.1 (January 2011), 40-81. She locates these transactions as occurring in, but certainly not limited to, the areas of the Fertile Crescent, Mesopotamia, Gundishabur, Mosul, North Africa, and Persia – locations where there were significant Arab conquests (pp.53-54).

38 Julius Rocca, "From Doubt to Certainty. Aspects of the Conceptualisation and Interpretation of Galen's Natural Pneuma," in *Blood, Sweat and Tears* (2012), pp.629-659 (pp.647-648).

held that there were two types of *pneuma*, to include three kinds of *pneuma* (as described above). This delineation corresponded to Aristotle's (and Plato's) doctrine of the "tripartite division of the soul."[39] In his discussion, Ibn Sina also reveals the adherence to principles of sex differentiation, conception, and biological sex that have come to be known as the one-sex model of sex and reproduction (also described above).[40] This is significant as it is the one-sex model that emerges from Arabic discourse and is adopted in the Christian European West. Along with this adoption also came the notions of the relationship between vital force, procreation, and the soul.

Insofar as medieval Islamic discourse was concerned, the works of the Greeks, and in particular Galen's medical theory, were compatible in certain respects with the Islamic holistic view of the world in which there is a kind of unity and interdependence between the body and the cosmos and God's will – in Galen terms, the Creator.[41] Nevertheless, while Galen's medical theories were widely influential and authoritative, his philosophical arguments were challenged by Muslim and Jewish practitioners and philosophers alike, most notably the tenth-century scholar Al-Razi, and late-twelfth-century Jewish scholar Maimonides. Likewise, Aristotle's theory of the First Cause – which relied on the premise that all things in nature have a cause, and that these causes reach back to a First Cause or Primary Mover – was also considered compatible in certain respects with Islam, Judaism, and Christianity.[42] Thus we can see the conversation about the connection between the cosmos (First Cause, God, the Creator) and man (*pneuma* or vital force) and soul was ongoing over the centuries. It should also be noted that both Muslim and Jewish scholars became involved in conversations about sex difference and reproduction in relation to the various discourses found in Hippocrates, Aristotle, and Galen. These scholars and theorists sought for these discourses not only to be compatible with contemporary religious practices and ideologies, but also to be compatible with the practical methods of treating diseases and complaints related to sex and sexual desire as they related to theological concerns.[43] In addition, Arabic scholars drew on Indian and Chinese medical texts, treatises, and practices as they synthesized

39 Jacquart and Thomasset, p.49.

40 As carefully explored in Thomas Laqueur's seminal work, *Making Sex*.

41 Gadelrab, p.56.

42 Gadelrab, p.57-58.

43 It should be noted that there were a variety of schools of thought about this subject and a great deal of intellectual debate and controversy, which included "rational" – that is empirical – as well as philosophical, metaphysical and theological debates (Gadelrab, p.59-60).

their own practices and interpretations along with their translations of Greek medical philosophy.[44]

It is has been suggested that by the time of the thirteenth century the philosophical and religious aspects, in conjunction with the medical discourses of the works of Ibn Sina (and others), came to be judged from a theological viewpoint that had an interest in spiritual revelation as much as reason and empirical rationale.[45] This added to the plurality of views on the nature of the soul in relationship to physiological discussions. The nature of the soul was a major topic of interest in Islamic discourse, and various medical and theological philosophers subscribed to the Aristotelian notion of the "one soul connected to the heart as the 'chief organ' of the body."[46] Ibn Sina argued for the immateriality of the soul, which recurred later in Descartes' discussion of the soul. In his commentary on Ibn Sina's anatomy, thirteenth-century scholar Ibn al-Nafis describes how the soul is produced and survives the body. The substance of the soul is called "*ajb al-dhanab,*" and the organs of the body are organized around this core "which is produced through a fine balance of matter at conception, 'from sperm and similar things.'"[47] In this schematic, the spirit is the substance that is generated from air and then introduced into the blood, while the soul connects with the entire bodily system, and is "particularly fine (*raqiq jiddan*)."[48] Aristotle's *De generatione animalium* explains this connection, seeking to answer the question of how the offspring comes to acquire the "faculties of soul."[49] It is not clear how well known or influential Ibn al-Nafis' particular work was upon transmission into the Latin West, but versions of the texts survived into the eighteenth century.[50]

---

44 See, for example, Christopher Beckwith, *Warriors of the Cloisters: The Central Asian Origins of Science in the Medieval World* (Princeton: Princeton UP, 2012), pp.1-10; Frederick S. Starr, *Lost Enlightenment: Central Asia's Golden Age from the Arab Conquest to Tamerlane* (Princeton: Princeton UP, 2013) and M. Shefer-Mossensohn and K. Abou Hershkovitz, "Early Muslim Medicine and the Indian Context: A Reinterpretation," *Medieval Encounters*, 19 (2013): 274-299 (pp.277-282).

45 Rainer Bromer, "The Nature of the Soul and the Passage of Blood through the Lungs: Galen, Ibn al-Nafis, Servetus, Itake, Atar," in *Blood, Sweat, and Tears* (2012), pp.339-360 (pp.343-344).

46 Bos and Ferwerda, pp.19-20.

47 As quoted in Bromer, p.344; see also pp.345-346.

48 Bromer, pp.345-346.

49 See Solmsen, pp.122-123.

50 Bromer, p.350. While Bromer's piece is largely concerned with the transmission of knowledge and theories about the circulation of blood and seeing the relationship between Ibn al-Nafis' work on circulation in the thirteenth century and William Harvey's early seventeenth-century "discovery" of circulation, his study on the ways in which anatomical ideas develop over time and between cultures is instructive for see-

The Qur'an, Prophetic Medicine, and the Role of *Pneuma*

Other contributions to the discourse on *pneuma*, both prior to the initial transmission of Arabic texts into Latin and thereafter, came in the form of Qur'anic verses and through *hadith*, which are collections of reports and sayings attributed to the Prophet Mohammed. In addition to *hadith* that addressed all manner of topics, those *hadith* that were specifically related to diet, hygiene, and the treatment of illnesses – and which also included management and guidelines for the sexual body – were collected and known as Prophetic Medicine, or Medicine of the Prophet. The Qur'an, while not providing an exhaustive account of sexual and reproductive theories and practices, does establish God's overarching "procreative agency" while at the same time drawing on Greek medical theories, in particular Hippocratic and Galenic theories of conception and fetal development.[51] In so doing, it "appropriate[s] and revise[s] Greek medical wisdom."[52] The Qur'an draws on the concept of *pneuma* as it deals with sexual intercourse, conception, fetal development, with particular emphasis on the analogy of *breath*. As an example, Kathryn Kueny describes God's creation of Adam as drawing on the theory of *pneuma* "that animates life from passive, inanimate material."[53] In the Qur'an, when God produces Adam out of clay the act is described using language that "functions analogously to the male semen that transmits the soul in Aristotle's paradigm:" "[…] I have fashioned him and breathed into him from my spirit (*wanafakhtu fihi min ruhi*) […]." [54] In addition to the theory of conception which privileges the male (sperm), this image evokes the source of the soul (from God) as transmitted through air or breath. While the Qur'an "draws on Aristotelian theories of *pneuma* to describe how life comes to be," it also "amends them according to the theological belief

---

ing the cultural implications of such relationships. In addition, Ibn al Nafis, while not entirely rejecting Galen's anatomical understanding of male and female genitalia, did object to his "one-sex model" of sex. Specifics of his objections can be found in Gadelrab, pp.72-75.

51 Kathryn M. Kueny, *Conceiving Identities: Maternity in Medieval Muslim Discourse and Practice* (Albany: SUNY Press, 2013), p.27. Kueny notes that, "While mechanistic processes drawn from Greek philosophy and medical wisdom may play an important role in the production of offspring as they generate heat, friction, or *pneuma* from seminal fluid, the laws of nature become subordinate to the one God who alone possesses the power to engender humans from nonliving matter" (pp.27-28).

52 Kueny, p.29.

53 Kueny, p.31.

54 Sura 38:71-72 and also 32:9, as quoted in Kueny, p.31; see also Kueny, p.262, note 83.

that God alone, as opposed to man or nature, has the power to produce life."[55]

Just as the Qur'an aligned or reconciled medical philosophy with Islamic religious discourse, albeit in limited and metaphorical ways, Prophetic Medicine was actively involved in reconciling and appropriating Greek tradition along with the medicical ideas of Arabia. Similar to the Qur'an, Prophetic Medicine also posited God, articulated through the revelations of the Prophet Muhammad, as the final authority on medicine.[56] A number of treatises on Prophetic Medicine, dating mostly from the tenth through the fifteenth centuries exist (although other treatises date from before and after this period). They existed alongside more empirical, non-religious medical treatises. Prophetic Medicine is also marked by its appeal to religious authority along with reference to Greek-based medical treatments and philosophies. Included are rules, remedies, and treatments for the sexual body. For example, Ibn Qayyim al-Jawziyya's fourteenth-century treatise on Prophetic Medicine includes a section on sexual intercourse and marriage. In it he draws heavily on Galenic tradition for remedies (directly quoting Galen), but he also draws on Galenic theory to support Islamic theology and Prophetic wisdom. In a discussion of the benefits of sexual intercourse, he explains this integration as follows:

> Sexual intercourse was originally created for three matters, which are its primary aims: (1) preservation of progeny and the continuation of the species, so as to complete the total of persons which God has decreed to be in the world; (2) expulsion of the fluid (*ma*) which if restricted and retained would harm the entire body; and (3) fulfillment of desire, attainment of pleasure and enjoyment of God's bounty. This last, alone, is the benefit which will remain in Paradise; for there is not procreation there, nor any bodily retention to be evacuated by the descent of fluid.[57]

The first primary aim – preservation of progeny in light of God's decree – conjoins Galenic philosophy and Qur'anic principles of procreation. The second aim, "the expulsion of fluid" to avoid bodily harm draws on the

55 Kueny, p.30.

56 Emilie Savage-Smith, "Medicine," in *The Encyclopedia of the History of Arabic Science*, vol. 3, ed. by Roshdi Rashed (London and New York: Routledge, 1996), pp.903-962 (p.927).

57 Ibn Qayyim al-Jawziyya, *Medicine of the Prophet*, trans. by Penelope Johnstone (Cambridge: Islamic Texts Society, 1998), p.181. This treatise dates to the early fourteenth century, and tends more towards describing therapies and ailments found in Greek-based medical texts, while also drawing heavily on Prophetic and other religious authorities, while some other treatises were more inclined to therapies that were more reliant on the spiritual than the medical (Savage-Smith, "Medicine," p.928).

theory of humors and has elements of treatment for lovesickness.[58] The third aim, that states intercourse is for the "fulfillment of desire [...] and enjoyment of God's bounty," speaks to the physiological necessity to evacuate bodily retention and God's purpose for humans to procreate, but puts at its center sexual desire.[59] Implicit in these primary aims is the importance of sexual intercourse in preserving spiritual and bodily health and wholeness. Also evident is the ways in which the culture reconciled the theological with the scientific. This reconciliation between the rational Greek-based theories and practices and religious law would be more problematic as these texts made their way to the Christian European West.

From Arabic to Latin

Through the filter of Arabic selection, translation, interpretation, and expansion of Galenic theories and practices a body of medical texts was bequeathed to the Christian European West. This cultural exchange occurred most intensely from the eleventh through the thirteenth centuries via multiple points of entry: Salerno, Sicily, Toledo, and Antioch to name a few major points of contact.[60] This inheritance included a wide range of classical texts in addition to those on medicine and medical philosophy, including texts on philosophy, mathematics, astronomy, and physics, etc.[61] The first medical translation from Arabic into Latin is thought to be a version of Hunyan ibn Ishaq's ninth-century translation and interpretive work, *Medical Questions.* Translated by Constantine the African in the late-eleventh century

58 Lovesickness is addressed at length in al-Jazzar's tenth-century text later translated into Latin by Constantine the African as the *Viaticum.* See Mary Wack, *Lovesickness in the Middle Ages: The* Viaticum *and its Commentaries* (Philadelphia: University of Pennsylvania Press, 1990) for its translation and commentaries.

59 This section on the guidance on sexual intercourse and marriage also indicates a connection between the production of semen and purification: "predominant over the substance of semen are fire and air. Its temperament is hot and moist, because of its being form from pure blood by which the primary organs are nourished" (Ibn Qayyim al-Jawziyya, p.181).

60 Scholars such as Monica Green have noted that the translations coming out of Southern Italy (particularly by Constantine the African) had a greater impact than those from other centers of learning. In a paper given at the Society for the Medieval Mediterranean meeting in 2011 on "The Differential Impact of Medical Translations Coming out of Southern Italy, Spain, and the Crusader Kingdoms in the 11th and 12th Centuries," Monica Green argues that the "Latin medical manuscripts produced between c.1075 and c. 1255 in Western Europe" indicate that the "Antiochene and Toledan translation had only very limited impacts on western medicine" (Monica Green Abstract, academia.edu).

61 Roshdi Rashed, ed., *The Encyclopedia of the History of Arabic Science*, vol. 3 (London and New York: Routledge, 1996).

as the *Isagoge,* it was one of the "five key texts of the twelfth-century medical *Canon"* of the famous school of Salerno.[62]

In addition to the *Isogoge,* Constantine the African's *Pantegni Practicum*[63] and the *Viaticum*[64] had a significant impact on Christian European medical practices. These works opened up further discourses on the nature of *pneuma* in relation to the ways in which the soul operated (through *psychic pneuma*), and regarding medical principles involved in dealing with the sexual body and its diseases such as lovesickness, infertility, impotence, conception, and contraception. The tripartite articulation of *pneuma* found in Constantine's translations follows both the Greek formulation described above as well as the ninth- through the thirteenth-century Arabic interpretations and contributions. Essentially, it codified tripartite pneumatology in theoretical medicine, while linking medical theory and philosophy to questions of functions of the soul.[65]

While the theory of humors in general and the theory of *pneuma* in particular were adopted into the Christian European West through these translations, the theories and practices surrounding the sexual body and its treatment presented a set of potential conflicts. Monica Green notes: "Given the radically different theological positions that Christianity and Islam held on these subjects, the potential for tension between the Islamicized medical science Constantine was translating and the Christian milieu into which he was introducing it was considerable."[66] Muslim theology held a

62 Hunyan's articulation of "natural *pneuma* (*ruh tabi'i*), vital *pneuma* (*ruh hayawani*), and psychic *pneuma* (*ruh nafsani*) would become a mainstay of Arabic medical theory [...] and would prove to be a text of enduring influence in the Western mediaeval medical tradition and beyond" (Rocca, p.648). See also Pormann and Savage-Smith, p.45-46; Emilie Savage-Smith, "Medicine," pp.910-991) on the theories and influence of Arabic science on Western medieval medical tradition.

63 The *Pantegni* is made up of two parts, *Theoretica* and *Practica*, and approximately ten books make up each section. For a thorough and deeply erudite history and context of the *Pantegni* and its significance to the practice of medicine in the Christian European West see Monica Green, "The Re-Creation of *Pantegni, Practica*, Book VIII," in: *Constantine the African and Ali Ibn Al-Abbas Al-Magusi: The Pantegni and Related Texts,* ed. by Charles Burnett and Danielle Jacquart (Leiden: E.J. Brill, 1994), pp.121-160,.

64 Monica Green notes that Chapter 19 of the *Viaticum* appears in Book VIII of the *Practica.*

65 Constantine's translation articulates this relationship as follows: "Also, the first of these moves into the veins which do not have pulse (*non habent pulsum*) and the whole body, the second into the arteries (*arterias*) and the third into the nerves (*nervos*), which seventh of the natural things [meaning spirit] obeys the fifth of the natural things [meaning force], that is, it [meaning force] is recognized or known in [the form of] spirit," as quoted in Rocca, p.649-651; also see his note 74.

66 Monica Green, "Constantinus Africanus and the Conflict Between Religion and Science," in *The Human Embryo: Aristotle and the Arabic and European Traditions,* ed. by

more permissive view of contraception and abortion than medieval Christian theology. In addition, Islamic theology and medical theory supported the notion of a right to pleasure for women as well as men that was not emphasized in medieval Christian discourse. [67] The results of this tension can be seen both in the translations, as noted by Green, and in literary texts that take up the topic of sexual desire, and it is particularly evident in texts such as Andreas Capellanus' twelfth-century work, *The Art of Courtly Love*, and the thirteenth-century *Roman de la Rose* by Guillaume de Lorris and Jean de Meun. This tension is also evident, and came to figure prominently in the European Romance genre from twelfth-century Chrétien de Troyes' Arthurian romances to the late-fourteenth-century Pearl Poet's *Sir Gawain and the Green Knight.* These texts in their representation of courtship and courtly love reflect some of the tensions seen between an Islamicate discourse that recognizes male and female sexual desire (*'ishq*, *eros*) in certain ways and a Christian discourse that seeks to control or deny those desires. In so doing, these texts reflect some of the humoral principles in their description of sexual desire and sex differentiation and highlight what was at stake in medical discourse of the physiology of pleasure. At the same time, these texts can be read to indicate moments when "oriental influences and Christian values interpenetrated."[68]

G.R. Dunstan (Exeter: U of Exeter Press, 1990), pp.42-69 (p.48). Further, Green points out that in Western medicine there was "a conflict between the Latin Christian ethos and a medical science embodying fundamentally different Islamic (and even classical Greek) views of sexuality, life, the nature of the soul, even of the divinity itself. The problem Constantine's translations posed to the West was simple yet fundamental: how does one culture assimilate the science of another without assimilating the ethical principles that underlie it?" (Green, p.50). I would aver that the imaginative literary tradition suggests that in certain respects the West, even if it did not "assimilate" Islamic (and Greek) views, did incorporate and negotiate questions of the sexual body and the nature of the soul using those views as filter.

67 B.F. Musallam, pp.31-34. Musallum discusses a woman's right to sexual pleasure in the contexts of contraception (in particular coitus interruptus in this section); he draws on Qur'anic writings, *hadith,* and opinions of Muslim jurists. In addition, Carmen Caballero-Navas argues that the Kabbalah maintains a "permissive and favourable attitude towards the pleasures of sex within marriage." See Caballero-Navas, *The Book of Women's Love and Jewish Medieval Medical Literature on Women, Sefer Ahavat Hashim* (London: Kegan Paul, 2004), p.67.

68 Jacquart and Thomasset, p.87. Jacquart and Thomasset refer here to the process by which physiology of pleasure "came to be included in the description of bodily mechanisms"; and in their chapter, "Medicine and the Art of Love," they look at the ways in which (Arabic) medical philosophy and practice moved into the Christian European West, suggesting an interplay between repression and the erotic (pp.87-138).

Literary Production: Medieval Arabic and European "Art of Love" Texts

As noted, one area of interpenetration and tension – or concert and conflict – emerged through the production of texts dedicated to the "Art of Love." While it is outside the scope of this essay to do a full review of the numerous Arabic and European texts that emerged on the topic of the art of love and erotic love, I will briefly identify a couple of key texts and the ways in which they developed out of the medical philosophies and principles which governed understandings of sexual desire, sex differentiation, and conception in relationship to the vital spirit. As in the case of the Qur'an and earlier literary texts, the concept of *pneuma* or life force often emerges metaphorically or analogically. Texts that describe and address the conditions of love and lovesickness are useful for inquiring into the extent of cultural transfer. They also offer us a window into how medical theories and practices played out in other documentary forms, such as treatises on erotic love and lovesickness, manuals for courtly love, and imaginative literature.

One such manual is Ibn Hazm's early-eleventh-century Andalusian treatise *The Ring of the Dove.* Ibn Hazm's text takes up the subject of love, its causes, signs and process, its virtues and sins. The work is a blend of the sacred and the profane in that it draws on quotations from the Qur'an and the *hadith*, while also describing erotic pleasure and pains of the flesh.[69] The first half of the work, describes the "root-principles" or "Signs of Love," whereas the second half explores the "praiseworthy and blameworthy attributes" of love.[70] In his description of the relationship between lovers, Ibn Hazm draws on the theory of humors and of *pneuma* to explain how fragmented parts of the soul scattered in the physical universe become unified, asserting that "a union [is] effected within the substance of their original sublime element" so that a "spiritual approbation, a fusion of souls" comes about.[71] The description offers a sense both of the macrocosmic and microcosmic: the physical universe contains the "scattered parts of souls" – whose eventual union in the body, and by extension through sexual intercourse, carry with it their "original sublime element" – that is, God. He further describes how lovers find one another in terms of magnetic attraction. His use of the analogy of how a "magnet draws iron" through "essential force" evokes the idea of the life-force dimension of *pneuma.* On the other hand, Ibn Hazm also addresses the "delightful malady" and "most desirable

---

69 Ibm Hazm, *The Ring of the Dove*, trans. by Anthony Arberry (London: Luzac Oriental, 1994), p.12.

70 Ibn Hazm, p.19.

71 Ibn Hazm, p.25-26.

sickness" that is love. This apparent paradox appears in later European texts that take up the mysteries of love, lovesickness, and courtly love.[72]

Thus, while on the one hand medical and philosophical texts are translated from Arabic into Latin with their sober assessments, on the other hand erotic literature, love poetry, and secular texts similar to Ibn Hazm's *Ring of the Dove* enter the European literary tradition via other avenues. They resulted from the Crusades and, most notably through the dissemination of songs and poetry of the Andalusian troubadours.[73] In particular, scholars such as Maria Rosa Menocal have argued that Arabic literature, particularly that related to the theme of love, had a clear influence on European literature.[74] Similarly, Green notes that the ideas of courtly love are clearly related to "medieval ideas on medicine and psychology, on the soul and its passions" and that the troubadours were instrumental in the spread of love poetry "to all the courts of Europe."[75]

As a result of this inheritance, and seen as perhaps one of the most significant works of the late twelfth century, Andreas Capellanus' *De amore* (The Art of Courtly Love), like its predecessor Ibn Hazm's *The Ring of the Dove*, also takes on defining love and its conditions, providing guidelines for behaviors between lovers in the context of the court and the chivalric code. Capellanus also frames love in terms of illness, and he too is "devoted to the problems of sexuality."[76] *De amores* and the concept of courtly love had a far-reaching impact on the European literary tradition of the Romance, such as *Roman de la Rose*, the stated purpose of which was to instruct court members in the art of love. The text is written in two parts by two separate authors, Guilliaume de Lorris (who wrote the first 4,058 lines in about 1230) and Jean de Meun (who wrote the last 17,724 lines in about 1275). The *Roman de la Rose* is an allegorical dream that details the art of love through a courtship between the dreamer and his object of desire, the Rose. The latter stands variously for the name of his lover, a virginal woman, and generally for female sexuality. The text as a whole renders the protagonist's search for

72 Ibn Hazm, p.31.

73 Menocal, pp.87-88.

74 Maria Rosa Menocal, *The Arabic Role in Medieval Literary History: A Forgotten Heritage* (Philadelphia: U of Pennsylvania Press, 1987), pp.77-78. See also John Jay Parry's Introduction to Andreas Capellanus' *The Art of Courtly Love* (New York: Columbia UP, 1960). Parry compares the contents and missions of these two works (pp.7-12); he also traces, as other later scholars do, the influence of Andalusian troubadours on the troubadours and *trobairitz* of Provencal, p.7.

75 O.H.Green, "Courtly Love in the Spanish Cancioneros," *PMLA*, 64.1 (1949): 247-301 (p.247).

76 Jacquart and Thomasset, p.99.

the attainment of his love – the Rose. Its numerous erotic passages reflect medieval notions of the theory of humors, sexual desire, and conception.

One passage in particular is worth noting here as it reflects the medieval theory of conception and, to a more subtle degree, the role of vital force in the sexual act and of conception. The passage comes at the very end of the text, when the dreamer attains his love:

> I seized the rosebush […] I began very softly, without pricking myself, to shake the bud, since I had wanted it as undisturbed as possible. However, I could not help making the branches stir and shake […] Finally, I scattered a little seed on the bud when I shook it, when I touched it within in order to pore over the petals. […] As a result, I so mixed the seeds that they could hardly be separated; and thus I made the whole tender rosebush widen and lengthen.[77]

In this description of sexual intercourse, ejaculation, and conception, the author clearly draws on the theory of humors and principles of conception, down to a commentary on the role of sperm in conception. Here, the "seed" is actively scattered by the male on the female "bud," which waits passively. Further, the agency seen in the lover who "mixed the seeds that they could hardly be separated; and thus I made the whole tender rosebush widen and lengthen" evokes not only the response of the cold female body to the heat of the male, but it also alludes to a kind of agency we see in the Qur'an's description of God creating Adam. The dreamer has complete control over the entire process. Thus we see the medical and erotic, the sacred and profane, work together to produce a highly charged discourse on sexuality.

Notions of *Pneuma* into the Seventeenth Century … and Beyond

Even as there were certain shifts in medical philosophy and practice in the Age of Science at the end of the sixteenth and into the seventeenth century, the tripartite notion of the function of the *pneuma* or spirits remained commonplace into the Early Modern period and even beyond.[78] In writing about the "spirits" and their functions in his 1621 treatise, *The Anatomy of Melancholy*, Robert Burton connects three kinds of spirits (i.e. dimensions of *pneuma*) with the three principle parts of the animal body: brain, heart, and liver, referring to them as forces "natural, vital, [and] animal."[79] Burton details the movement of these vital spirits through the principle organs of the body and their interconnectedness, concluding: "if the spirits cease, then life

77 Guillaume de Lorris and Jean de Meun, *The Romance of the Rose*, trans. by Charles Dahlberg (Princeton: Princeton UP, 1971), p.353.

78 Rocca, p.630.

79 Robert Burton as quoted in Rocca, p.630.

ceaseth."[80] By 1649, William Harvey would comment on this tripartite schema, noting that "there are many and opposing views" about about the relation of the vital forces to one another and "whether they are separate and distinct from blood and the solid parts, or mixed."[81]

Descartes of course also participated in debates regarding the interactions between the soul and the body. His views are considered anti-Aristotelian and Platonic, as he conceived of the soul as "immaterial thought without any connection with life." He also saw the "seat of the soul" in the pineal gland as the location where soul and body interacted.[82] While the seventeenth century provided a paradigm of physical science that distinguished between the animate and inanimate in ways not distinct in earlier medieval thought, Descartes nonetheless drew on Galenic principles along with Augustine's "neurophysiological concepts" (which in turn were indebted in part to Galenic principles) that "saw the cosmos as permeated by a vivifying principle, the *pneuma*."[83] Thus, his view of the pineal gland as the location where the soul and body interact reflected the new science; he sought to reconcile the notion of "animal spirits" that filled "the pipes and tubes of his neurophysiology" with the "mechanistic interpretation of inanimate matter."[84] In short, even in light of clear advances in the physical and medical sciences in the seventeenth and eighteenth centuries, the paradigm of *pneuma* and the vital spirits associated with it continued to influence intellectual thought through the seventeenth century. In fact, it formed the basis of numerous physiological paradigms and occurred in psychological, cosmological, and theological contexts until the nineteenth century.[85]

While it is not in the scope of this essay to map out other literary examples – other essays in this volume do that – it is perhaps useful to point to the prevelance of literary expressions influenced by the life-force concepts inherent in *pneuma* eighteenth-century England. There the impact of the scientific discourse of "vitalism," with roots in pneumatology, on literary production was quite striking. Denise Gigante and Catherine Packhman among others speak of "Romantic Vitalism." As is well known, the term describes the scientific and philosophical notion of an "animating force of living

80 Rocca, p.630.

81 Rocca, p.630.

82 Moncef Berhouma, "Beyond the Pineal Gland Assumption: A Neuroanatomical Appraisal of Dualism in Descartes' Philosophy," *Clinical Neurology and Neurosurgery*, 115 (2013): 1661-1670 (p.1661).

83 C.U.M. Smith, "Descartes' Visit to the Town Library, or how Augustinian is Descartes' Neurophysiology?," *Journal of the History of Neurosciences: Basic and Clinical Perspectives*, 7.2 (1998): 93-100 (pp.97-98).

84 C.U.M. Smith, p.99.

85 Rocca p.631.

form" and prompted a sweeping inquiry into the phenomena of life as expressed particularly in Romantic literature.[86] Gigante observes how cell theory "demystified" conceptions of the life force in human and animal nature, ultimately influencing medical practice itself.[87] Indeed, the concept of sensibility, grounded in the physiological principles of an animating life force, grew into a veritable literary movement. The essays in this collection explore those traces of what essentially amounts to a relationship between the human and non-human in the formation of mind, soul, and consciousness via ancient concepts inherent in *pneuma.* The past thus clearly resonated in the long eighteenth and ninteenth centuries.

---

86 Diane Gigante, "Zeitgeist," *European Romantic Review*, 18.2 (2007), 265-272 (p.266).

87 Gigante, p.267. See also Catherine Packham, *Eighteenth-Century Vitalism: Bodies, Culture, Politics* (Basingstoke: Palgrave Macmillan, 2012), pp.2-3.

Ordnung des Lebendigen. Naturgeschichtliche Malereien im Kabinett der Franckeschen Stiftungen zu Halle

Ingo Uhlig

Abstract: The Exhibit of Art and Natural Artifacts in Halle a.d. Saale has been restored to its mid-eighteenth-century configuration. It is the last remaining collection of the Baroque era with such broad scope and systematic through-composition. In this essay I explore the question of how this amalgamation of artifacts renders the great chain of life as then conceived. Of special value in this regard are the illustrations on the cabinets themselves that house the art and natural artifacts; they lend a definite vitalism to the static forms. I interpret these allegories of dynamic, living forms against the backdrop of Gottfried Wilhelm Leibniz's *Monodalogy* and theorize that the arrangement of these art works and natural artifacts mirror Leibniz's natural philosophy which presented life as the manifestation of an organizing principle.

In der Mitte der 1730er beginnt in Halle ein reger und historisch neuartiger Transfer von Objekten. Bei den in Bewegung geratenen Gegenständen handelt es sich nicht um Konsum- oder Gebrauchsgüter, es handelt sich nicht um neuestes technisches Gerät, Luxusgüter oder religiöse Gegenstände, sondern um verschiedenartige Objekte des Wissens. Ganze Sammlungen oder umfängliche Sammlungsteile wechseln Orte und Besitzer, um schließlich in den meisten Fällen zu einer wissenschaftlichen Kollektion der Friedrichs Universität zu werden. Die etwa einhundert Jahre ab 1736 stellen sich, auf einige der wesentlichsten Daten beschränkt, wie folgt dar: 1736-41 katalogisiert Gottfried August Gründler das Naturalienkabinett der Glauchaschen Anstalten (der heutigen Franckeschen Stiftungen) und bringt die Objekte in eine neue, in Teilen an das Linnésche System angelegte Ordnung. Bereits im ausgehenden 17. Jahrhundert hatte August Hermann Francke damit begonnen, diese Objekte vor allem dank eines sich über den Globus spannenden Netzes der pietistischen Mission zusammenzutragen, um mit ihnen einen Realienunterricht avant la lettre zu betreiben.

1768 kaufte die Universität eine Sammlung antiker Münzen, die der Mediziner und Historiker Johann Heinrich Schulze angelegt hatte. Schulze hatte sie in seinen Vorlesungen in den 1730er Jahren für altertumswissenschaftliche Demonstrationszwecke eingesetzt, bevor er 1738 die Sammlung nicht mehr exemplarisch verwendet, sondern sie als zentralen Gegenstand eines Kollegs behandelt und damit die Numismatik als Universitätsfach begründete.

Die Fülle der naturgeschichtlichen Objekte gerät ab den 1780er Jahren in Bewegung, auch hier handelt es sich um private Kabinette, die zunächst in den Stadthäusern ihrer Besitzer untergebracht waren, bevor sie Teil des wissenschaftlichen Sammlungsbestands der *fridericiana* wurden. 1787 verkaufte Johann Friedrich Goldhagen sein Naturalienkabinett an die Universität. Er hatte in den vorausgehenden Jahren mehrere private Sammlungen, unter anderem die des Mediziners Friedrich Hoffmann und die des bereits erwähnten Archivars der Kunst- und Naturalienkammer der Franckeschen Anstalten Gottfried August Gründler, aufgekauft und in sein Kabinett eingegliedert. Das Goldhagensche Naturalienkabinett, von einem Zeitzeugen als das „vollständigste, was irgend auf einer Universität in Teutschland anzutreffen ist“, bezeichnet,[1] bildete den Grundstock der im 19. Jahrhundert unter den Direktoren Nitzsch und Burmeister weltweite Berühmtheit erlangenden Halleschen Zoologischen Sammlung. Bis es soweit war, wurde der Goldhagensche Bestand aber mehrfach geteilt: 1805 wurde der mineralogische Sammlungsteil auf Betreiben des Geologen Henrik Steffens ausgegliedert und erhielt eigene Räume. 1815 überführte der „von einer großen Sammelleidenschaft“[2] besessene Mediziner Johann Friedrich Meckel einen großen Sammlungsteil in die anatomischen Sammlungen. Der Grundstock der anatomischen Sammlung, der sogenannten Meckelschen Sammlungen, wurde bereits 1779 von Berlin nach Halle gebracht und ging, nachdem die Kollektion zwischenzeitlich auf einen Bestand von sechzehntausend Präparaten angewachsen war, 1836 in den Besitz der Universität über.

Zu erwähnen ist schließlich die älteste Sammlung der *fridericiana*: der Botanische Garten. Als *hortus medicus* wurde er bereits 1698 gegründet und von Georg Ernst Stahl geleitet. Nachdem die Anlage in der Mitte des Jahrhunderts verwaiste, erfährt sie in den 1780er Jahren eine Renaissance, wird erweitert und grundlegend umstrukturiert. Johann Reinhold Forster war zu dieser Zeit der Direktor des Gartens und bereicherte, wie sein Nachfolger Kurt Sprengel ausführt, insbesondere das Herbarium mit „7-800 der seltensten Pflanzen, vom Kap, von den freundschaftlichen Inseln, von Neu-Kaledonien, Neu-Seeland und Patagonien“.[3]

Besonders in der Phase nach 1780 befinden sich viele dieser Sammlungen, Sammlungsteile und Objekte in einer Art Übergangs- oder Transferbereich. Diese vielfältigen räumlichen Verlagerungen, die Teilungen und Zusammenführungen, der rege Wechsel von Eigentümern und Sammlungsleitern stehen in Zusammenhang mit einer Gemengelage von konkurrieren-

---

1 Vgl. Otto Taschenberg, *Geschichte der Zoologie und der zoologischen Sammlungen an der Universität Halle 1694-1894* (Halle: Max Niemeyer, 1894), p.23.

2 Rolf Gattermann, Volker Neumann, „Die Geschichte der Zoologie in Halle“, *Zoologie. Mitt. DZG*, 2002, p.2-26 (p.3).

3 Kurt Sprengel, ‚Der botanische Garten bey Halle’, *Der neue teutsche Merkur*, 1 (1799), p.35-44 (p.42).

den epistemischen Systematiken, von veraltenden und entstehenden Wissensordnungen, von Autonomie- und Besitzansprüchen neuer Disziplinen und Fachvertreter. Diese Neuordnung der Objekte und Materialen ist damit vor allem Zeugnis einer Passage zwischen zwei Epochen des Wissens: Zumeist entstanden die Kollektionen im Zuge polyhistorischer Interessen, nun werden sie von den sich formierenden Fachdisziplinen übernommen und grundlegend neugeordnet. Eine Institution wie die Friedrichs Universität zeigt so anhand ihres Sammlungsschatzes exemplarisch, dass die Geschichte von Disziplinen, wissenschaftlichen Paradigmen und Konzepten sowie letztlich auch das Auftreten grundlegender wissensgeschichtlicher Umbrüche von einer vielfältigen materialen Dimension umgeben waren. Die Geschichte dieser Objekte – in Form ihrer Neuordnungen, Archivierungen und Präsentation – verläuft im engen Zusammenhang mit der Geschichte von Begriffs- und Disziplinbildungen, sie umgibt die Akteure, Erkenntnisse und Aussagen.

Während dieses Zeitraums und im Zuge seiner epistemologischen Übergänge tritt nicht zuletzt eine historisch neuartige, uns aber heute völlig vertraute Differenz zwischen die Objekte: Das Panorama der *great chain of being* beginnt sich aufzulösen und statt der Kosmologie einer allseitigen Verkettung der Dinge und Wesen wird ihre disziplinäre Zuordnung sichtbar. So werden aus Naturalienkabinetten moderne Spezialsammlungen, die entweder den Natur- oder den Lebenswissenschaften zuzuordnen wären. Exemplarisch zeigt dies vor allem das Kabinett Goldhagens, das nach dem Ankauf durch die Universität 1787 unter erheblichen logistischen Problemen und disziplinären Lagerkämpfen in drei Sammlungen (der Zoologie, Anatomie und der Geologie) aufgeteilt wurde. Das Auseinanderbrechen der alten naturgeschichtlichen Ordnung lässt sich hier in seiner praktischen Umsetzung – räumlich und institutionell – etwa anhand der Vergabe von Zugangs- und Schlüsselrechten zu den neu entstandenen Sammlungen nachvollziehen.[4]

Im Folgenden möchte ich vor diesen Umbruch zurückgehen und mit der Kunst- und Naturalienkammer der Franckeschen Stiftungen eine Kollektion betrachten, die das alte Panorama der Kette der Wesen nicht infrage stellt,

---

4 Vgl. etwa das im Mai 1805 nach der Aufteilung der Sammlung beschlossene *Regulativ über den Gebrauch des mineralogischen Theils des academischen Naturalien-Cabinets*, das die Befugnisse des zoologischen Sammlungsleiters Johann Gottfried Hübner in der dem Geologen Hendrik Steffen gerade unterstellten mineralogischen Sammlung regelt: „Diese Mineralien-Sammlung soll in einem besonderen Zimmer und vor der Hand im Vorsaale des Naturalien-Cabinets, aufgestellt werden. Zu diesem Vorsaale muss Herr p.Hübner ebenfalls einen Schlüssel haben, da durch denselben der Weg ins Cabinet geht. Die Mineralien aber sollen in verschlossenen Schränken aufbewahrt werden, und zu denselben darf H. Professor Steffens allein die Schlüssel haben." Zitiert nach Taschenberg, *Geschichte der Zoologie*, p.151.

sondern dessen Geschlossenheit behauptet. Ich werde vor allem der Frage nachgehen, wie dieses Kabinett eine dem heutigen Betrachter kaum mehr zugängliche Imagination des Lebendigen entstehen lässt.

Die Kunst- und Naturalienkammer der Franckeschen Stiftungen

Rückt man all jene Umbauaktivitäten, die sich im späten 18. und frühen 19. Jahrhundert so rege in diesen Sammlungen entfalten, in die Perspektive der Foucaultschen Epistemologie, so lässt sich davon sprechen, dass sich diese Aktivitäten um eine wissensgeschichtliche Diskontinuität gruppieren. Gemeint ist damit zunächst keineswegs, dass dieser historische Verlauf lückenhaft, also etwa von katastrophischen Einschnitten unterbrochen wäre. Die gemeinte Diskontinuität wirbelt wenig Staub auf und sie ist auf der Ebene der Chronologie nicht ohne weiteres zu entdecken. So erstreckt sich die Entstehungsgeschichte der Halleschen Universitätssammlungen und der Transfer all jener Objekte über mehrere Jahrzehnte, ohne dass es an historischer Zeugenschaft und historiographischem Nachvollzug fehlt. Die Ereignisverläufe lassen sich, auch wenn es zahlreiche Objekte und sogar ganze Kollektionen nicht mehr gibt, dokumentieren und in den wesentlichen Zügen rekonstruieren (dies gilt selbst noch für die Wirren der napoleonischen Kriege). Von einer Diskontinuität gekennzeichnet ist allerdings unsere intellektuelle Annäherung vor allem an die frühen Sammlungsordnungen. Hier verbleibt das Denken im Mittelbaren und entdeckt zunächst eine gewisse Distanz gegenüber der ebenso befremdlichen wie aufwendigen Künstlichkeit solcher Objektensembles, die anderen als den uns heute vertrauten Gesetzen und Systematiken folgten.[5]

Im Bereich der Halleschen Sammlungen ist diese Erfahrung heute am eindringlichsten zu machen, wenn man die auf dem Stand des mittleren achtzehnten Jahrhunderts konservierte Kunst- und Naturalienkammer der

---

[5] Die Problematik dieser Diskontinuität sei nochmals in den Worten Foucaults dargestellt, der hier vor der leichtfertigen Verwendung moderner Begrifflichkeit warnt: Im „klassischen Zeitalter hat sich ein Raum der Empirizitäten herausgebildet, der bis zum Ende der Renaissance nicht existiert hat und seit dem Anfang des 19. Jahrhunderts wieder verschwinden sollte. Es ist für uns heute so schwierig wiederherzustellen, denn er ist durch das System der Positivitäten tief verdeckt worden, zu dem auch unser Wissen gehört, so daß er lange unbemerkt geblieben ist. Man deformiert, man maskiert ihn mit Kategorien oder Zerlegungen, die unsere sind. Man will das wiederherstellen, so scheint es, was im siebzehnten und im achtzehnten Jahrhundert die ‚Lebenswissenschaften', die ‚Naturwissenschaften' und die ‚Wissenschaften vom Menschen' gewesen sind. Dabei vergißt man ganz einfach, daß der Mensch, das Leben und die Natur keine Gebiete sind, die sich spontan und passiv der Neugier des Wissens anbieten." Foucault, *Die Ordnung der Dinge. Eine Archäologie der Humanwissenschaften*, 14nd edn. (Frankfurt am Main: Suhrkamp, 1997), p.107.

Franckeschen Stiftungen betritt. Unter den heute noch erhaltenen barocken Sammlungen besteht die Einmaligkeit dieses Kabinetts darin, dass es nicht nur hinsichtlich seines Objektbestandes, seines Mobiliars und seiner Räumlichkeit kaum verändert erscheint, sondern dass hier auch die konkrete Sammlungsgestalt und -systematik rekonstruiert werden konnten, so dass im Mansardengeschoss des Haupthauses der Franckeschen Stiftungen ein Kabinett auf dem Stand des mittleren 18. Jahrhunderts wieder begehbar ist.[6] Die umfassende Überlieferung – nicht nur der Objekte, sondern die ihrer Ordnung – ist ein wesentlicher Punkt: Man trifft nicht allein auf Gegenstände, die historisch und original sind, sondern man betritt hier einen geschlossenen Kosmos, in dem auch die räumliche Anordnung dieser Gegenstände ebenso wie ihre Präsentation im Interieur und ihre ästhetische Aufbereitung weitgehend den Gegebenheiten des 18. Jahrhunderts entsprechen. Etwa dreitausend Objekte nehmen also die Plätze wieder ein, die sie bereits zu jener Zeit innehatten und sie stehen somit wieder in Nachbarschaften, Beziehungen und letztlich in einem epistemischen wie sinnlichen Gesamtzusammenhang, in welchem sie auch damals schon standen.

Gerade dieser kompakten Erscheinung ist es zu verdanken, dass dieser Ort dem heutigen Besucher keinen einfachen unmittelbaren Zugang gewährt, sondern eine regelrechte Grenze zu errichten scheint. Natürlich kann man diesen Raum wie jeden anderen modernen und öffentlichen Museumsraum betreten, die Dinge bei Licht und aus der Nähe betrachten und doch bleibt der Eindruck, dass man sich auf einem fremden Terrain bewegt. Die Fremdheit dieses Raumes und seines ebenso präsenten wie unnahbaren Objektensembles geht also nur vorderhand von den einzelnen, zum Teil kuriosen oder exotischen Objekten aus, vielmehr sind es das historische Ordnungsmodell und die in ihm entfaltete Kosmologie, die sich dem modernen

---

6 Vgl. Thomas Müller-Bahlke, der im Rückblick auf den Wiederaufbau der Kunst- und Naturalienkammer in den 1990er Jahren festhält: „Die erstaunlichste Feststellung dabei war, dass es sich hier offenbar um die einzige überhaupt erhaltene Wunderkammer aus der Barockzeit handelt, deren originale Grundkomponenten aus dem 18. Jahrhundert noch alle vorhanden sind und nun wieder zu einer Einheit zusammengefügt werden konnten. Mit diesen Komponenten sind gemeint: die Sammlungsgegenstände, das Mobiliar der Kammer, der Sammlungsraum, das museumstheoretische Konzept." Müller-Bahlke, *Die Wunderkammer der Franckeschen Stiftungen*, 2nd edn (Halle: Verlag der Franckeschen Stiftungen, 2012), p.11. Zu ergänzen ist hier, dass eine zweite Kammer aus der Barockepoche weitgehend erhalten ist: Die von Peter dem Großen gegründete Kunstkammer in Petersburg. Hier allerdings handelt es sich um eine fürstlich-dynastische Sammlung, während man in Halle ein frühes bürgerliches Museum vorfindet.

Denken sperren und ihm in einer ebenso beeindruckenden wie irritierenden Künstlichkeit gegenüberstehen.[7]

Die heutige Gestalt und die Ordnung dieser Kollektion geht zurück auf Gottfried August Gründler, der zwischen 1736 und 1741 den Versuch unternahm, die in Folge einer globalen, vier Jahrzehnte währenden Sammeltätigkeit im Waisenhaus zu Halle angelangten Objekte in ein geschlossenes Ensemble zu überführen. Diese Objekte waren bis zu diesem Zeitpunkt nicht in einem gemeinsamen Raum untergebracht, sondern auf verschiedene Räume der Anstalten verteilt. Gründler übernahm nun die Aufgabe, den gesamten Objektbestand zu katalogisieren und ihn im ehemaligen Knaben-Schlafsaal des Waisenhauses auszustellen. Hier stand ihm ein Raum von etwa 24 mal 6 Metern zur Verfügung, in dem der damalige Bestand der über viertausendsechshundert Objekte in einer Ordnung erscheinen sollten.

Gründlers Leistung bei der Umsetzung dieses Plans besteht zunächst in einem strengen systematisierenden Vorgehen. Zunächst teilt er die Sammlung, wie zeitgenössisch üblich, in die zwei Bereiche der natürlichen und künstlichen Gegenstände. An der Südseite des Raumes sind in einer Gruppe von sechs Schränken die Objekte der *res naturales* zu sehen, an der Nordseite in einer Gruppe von acht Schränken die der *res artificiales.* Für die Übersicht über die Mannigfaltigkeit der natürlichen Objekte orientiert sich Gründler an der aristotelischen Dreireiche-Lehre, von der die naturgeschichtlichen Ordnungsvorstellungen bis ins 18. Jahrhundert geprägt wurden. So finden sich die drei Basisgruppen der *naturalia*: Mineralien, Pflanzen- und Tierpräparate. Anders als bei Naturaliensammlungen älterer und zeitgenössischer fürstlicher Kammern fehlt hier aber weitgehend die Akzentuierung des Außergewöhnlichen. Die zentrale Regel „Quanta rariora tanta meliora" scheint nicht mehr in Kraft und das Hauptaugenmerk liegt folglich nicht auf dem bizarren und exotischen Objekt. Natürlich enthält die Sammlung zahlreiche *kuriosa*, aber sie werden nicht als Attraktionen hervorgehoben und keineswegs ins Zentrum der Aufmerksamkeit gerückt. Der Linné-Übersetzer

---

7 Ein markantes Dokument dieser Diskontinuität stammt von Max Sauerlandt, einem der maßgeblichen deutschen Museumsgestalter und -theoretiker des frühen 20. Jahrhunderts. Sauerlandt war von 1908-1919 Leiter des Halleschen Museums für Kunst und Kunstgewerbe (heute das Kunstmuseum Moritzburg) und veröffentlichte 1911 einen Aufsatz über Franckes Kabinett. Er bemerkt angesichts einiger Objekte: „Das moderne Leben hat diese Symbole des universelleren Erkenntnisstrebens jener merkwürdigen Zeit beiseite geschoben. So sind auch die ganzen einst so lebensvollen Sammlungen des Naturalienkabinetts der Franckeschen Stiftungen seit Jahrzehnten tot und der Vergessenheit anheim gefallen. Nur noch als ein merkwürdiges Ausstellungsobjekt in dem Gesamtorganismus eines modernen Museums würde das Kabinett eine würdige Rolle spielen." Sauerlandt, „Die Naturalienkammer des Halleschen Waisenhauses", *Museumskunde* VII (1911), p.133-146 (p.145).

Gründler scheint weniger an der Wirkung der Staunen erregenden Einzelobjekte, als an der instruktiven Vermittlung von Ordnungen und Taxinomien interessiert: es gilt, im Format der Sammlung und eines beigeordneten Kataloges, den universellen Zusammenhang der Schöpfung zu repräsentieren. So erhält der Blick auf die Mannigfaltigkeit der natürlichen Dinge seine Ordnung durch die Anwendung zeitgenössischer naturgeschichtlicher Klassifikationsraster, den Schrank mit den Tierpräparaten systematisiert Gründler bereits nach den Klassenbegriffen des Linnéschen Systems.[8] Im Vordergrund der Kollektion der natürlichen Gegenstände steht also die Einbindung des Besonderen: Das Nahe und Ferne, das Vertraute wie das Exotische finden so ihren Platz in der Klassifikation der Wesen und Dinge.[9] Auch das kuriose und wunderliche Ding lässt sich systematisch zuweisen und wird schon allein dadurch – also durch seine Platzierung in der naturgeschichtlichen Ordnung – aus dem Bereich der Legende und des Aberglaubens herausgelöst. Nun steht es in Nachbarschaftsbeziehungen, die es – und sei es als Ausnahme von der Regel – als ein Element im Plan der Schöpfung explizieren.[10]

Von einem systematischen Anspruch zeugt auch der Bereich der *artificialia*, auch hier rückt das exotische Objekt in einen Kanon ein, dessen Anliegen es ist – im strengen Zusammenhang mit den Maßgaben der pietistischen Programmatik, vor allem des Missionsprojektes – systematisches Wissen über fremde Kulturen zu vermitteln. Das Beispiel des Indien-Schrankes zeigt dabei, dass sich das theologisch-missionarische Bestreben und ein hohes Maß an völkerkundlicher Genauigkeit nicht ausschließen.[11] Auch wenn im Vis-à-vis vieler Objekte der Kammer die exotistische Schaulust sicherlich die spontane Rezeptionsform gewesen sein dürfte, besteht der von Gründler intendierte Rezeptionsmodus auch im Bereich der *artificialia* nicht mehr in jenem Staunen, das sich beim Durchwandern fürstlicher Kabinetten einstellte, zugrunde liegt nun vielmehr eine pädagogische Idee.

---

8 Die Aufstellungssystematik des Kabinetts ist bislang nicht grundsätzlich erschlossen. Wichtige Hinweise gibt allerdings Annelore Rieke-Müller, „Die außereuropäische Welt und die Ordnung der Dinge in Kunst- und Naturalienkammern des 18. Jahrhunderts – das Beispiel der Naturalienkammer der Franckeschen Stiftungen in Halle," *Das Europa der Aufklärung und die außereuropäische Welt*, ed. by Hans-Jürgen Lüsebrink (Göttingen: Wallstein Verlag, 2006), pp.51-73.

9 Vgl. im Hinblick auf den Gesteinsschrank, Müller-Bahlke, *Die Wunderkammer*, p.68.

10 Horst Bredekamp sieht in der Systematisierung des Abnormen und folglich in der Zurückdrängung des Aberglaubens eine Funktion der Kunst- und Naturalienkammer Peters des Großen. Bredekamp, „Leibniz' Idee eines ‚Theaters der Natur und Kunst' für Russland," in *Palast des Wissens. Die Kunst und Wunderkammer Zar Peters des Großen*, 2, ed. by Brigitte Buberl, Michael Dückershoff (München: Hirmer Verlag, 2003) p.116-123 (p.122).

11 Vgl. Müller-Bahlke, *Die Wunderkammer*, p.124-155.

Dies führt zur eigentlichen Intention dieses Sammlungsprojekts: Im Zentrum des museologischen Konzeptes der Franckeschen Anstalten und der Kunst- und Naturalienkammer steht die Vermittlung von Wissen, die hier gruppierten Objekte haben sachlich-instruktiven Charakter. Diese Pädagogik der Realien geht zurück auf den Sammlungsgründer August Hermann Francke. Francke hatte in den letzten Jahren des 17. Jahrhunderts begonnen, Objekte für den Lehrgebrauch zu beschaffen. In einem sammlungsgeschichtlich berühmten Schreiben von 1698 erbat er sich einige Dubletten aus dem Berliner Kabinett Friedrichs III. Er erhielt daraufhin 47 Objekte für seine Hallesche Kollektion. Franckes pädagogische Intention legitimierte seine Bitte um Überführung der Objekte aus dem dynastischen in den bürgerlichen Kontext. Vor allem im *Pädagogium* des Waisenhauses, also in jener Schule der Franckeschen Institution, in welcher adlige und betuchte bürgerliche Zöglinge aufgenommen wurden, sollte Wissen auch durch die physische Präsenz und die Anschauung der Dinge vermittelt werden. Der Ortswechsel von den Schulräumen zu den Objekten gehörte früh auf systematische Weise zum Unterrichtskonzept August Hermann Franckes, auch wenn dafür zunächst kein eigenes Kabinett zur Verfügung stand.[12] Etwa vierzig Jahre später, nach Gründlers 1741 abgeschlossener Neuordnung, erscheint die Sammlung schließlich auch als ein räumlich definierter Mikrokosmos, der von dem systematischen pädagogischen Anspruch durchdrungen wird, die empirische Breite und zugleich die Einheit der Schöpfung zu vermitteln.

Das ästhetische Programm der Kammer I: Gründlers Kompositportraits nach Arcimboldo

Gründler überführte die Sammlung aber nicht nur in einen systematischen Wissenszusammenhang, er machte daraus zudem einen ästhetischen Gegenstand. Dies zeigen vor allem die Bemalungen der Sammlungsschränke. Zwölf dieser Schränke bestehen, betrachtet man sie hinsichtlich ihrer Verzierungen, aus zwei Teilen: einem hölzernen bemalten Korpus und einer ebenfalls hölzernen und bemalten Bekrönungsplatte.

Der mit einer Glastür versehene Schrankkorpus ist an seinen sichtbaren Außenseiten in einer blau-grauen Marmorierung gehalten. Den vier jeweils über Eck nahtlos aneinander stehenden Schrankgruppen (die beiden Hauptgrup-

---

12 Vgl. Dorothea Hornemann, Claus Veltmann, "'Zu Erziehung der Jugend'. Die Naturalienkammer August Hermann Franckes in der Tradition der frühneuzeitlichen Bildungskultur," in *Die Welt verändern. August Hermann Francke – Ein Lebenswerk um 1700*, ed. by Holger Zaunstöck, Thomas Müller-Bahlke, Claus Veltmann (Halle: Verlag der Franckeschen Stiftungen, 2013), p.129-143; Stefan Laube, *Von der Reliquie zum Ding. Heiliger Ort – Wunderkammer – Museum* (Berlin: Akademie Verlag, 2011), p.307-385.

pen der *naturalia* und *artificialia* sind jeweils durch ein Fenster unterbrochen) wird dadurch ein ausgesprochen kompakter, beinah baulicher Charakter verliehen. Durch ihre Frontbemalungen, ihre Gruppenanordnung und die an den Nahtstellen aufgesetzten Säulenimitate überwinden sie den Eindruck eines hölzernen Möbels und erscheinen eher als architektonische Gruppen, sie stehen dem Betrachter als feste Blöcke mit einer steinernen Anmutung gegenüber. Konterkariert wird das monolithische Bild aber durch die Bekrönungsmalereien; die Kabinettmöbel, die in ihrer Tiefe die Objekte bergen, werden von aufwendigen Verzierungen überragt. Den Objekten werden bildliche Kommentare zugeordnet, die sich in ihrer Auffälligkeit als ein drittes Element in die Relation von Subjekt und Objekt drängen. Der so entstehende Bühnencharakter der Schränke ist verschiedentlich bemerkt worden.[13] Dem Kabinett werden Eigenschaften eines Theaterraums verliehen, verstärkt und thematisch angeleitet wird diese Imagination durch die auf den Schrankgiebeln angebrachten Malereien.

Zunächst ist offensichtlich, dass die Malereien Hinweisschildern gleich Auskunft über die im Schrank befindlichen Objekte geben. Weiterhin beziehen sich einige der Motive aufeinander und verbinden die Schränke und Objekte thematisch: So findet sich das Palmblatt des Indien-Schrankes auf dem Schriftenschrank wieder. Derartige Themenzusammenhänge und Kontexte ergänzen und umspielen die Systematik der Aufstellung und tragen sicherlich dazu bei, dass die Schränke und ihre Objekte einen „geschlossenen Reigen“ bilden,[14] dessen Effekt wiederum die Sichtbarkeit des Gesamtzusammenhangs der Schöpfung ist.

Aber die bildlichen Kommentierungen gehen – formal wie inhaltlich – noch einen Schritt weiter: Es handelt sich um Allegorien, welche den kommentierten Objekten zu weiteren Bedeutungsdimensionen verhelfen. Besonders bei den Verzierungen der Naturalien-Schränke (auf die ich mich im Folgenden beschränken möchte) fällt auf, dass sie jene Objektklassen, auf die sie Bezug nehmen und deren materiale Vertreter sich im Inneren der Schränke befinden, in ebenso aufwendige wie kapriziöse Anordnungen überführen. Dieses formale Kriterium gilt für die Darstellungen auf dem Pflanzen- und Conchylien-Schrank, hier ergeben einzelne Früchte oder Muscheln, die nach dem Vorbild der Kompositportraits Guiseppe Arcimboldos angeordnet sind, die lebendigen Züge eines männlichen Gesichts, das frontal auf den Betrachter blickt. Eine ähnliche Anordnung, die allerdings nicht in ein explizit anthropomorphes Bildprogramm mündet, liegt bei den Mineralien vor. Einzelne Steine und Kristalle treten hier zu einer Girlande und schließlich zu einer wappenartigen Struktur zusammen. Ebenfalls nicht

---

13 Vgl. Laube, *Von der Reliquie zum Ding*, p.379f.

14 Müller-Bahlke, *Die Wunderkammer*, p.46.

anthropomorph, aber nach heraldischem Modell konzipiert, zeigt sich die Bekrönungsmalerei des Tierschranks, die heute eines der bekanntesten Motive aus der Geschichte der Franckeschen Stiftungen darstellt. Hier handelt es sich um eine Szene aus der Tierphantastik: In spielerischer Beweglichkeit und verschiedenen *Trompe-l'œils* gruppieren sich reale wie phantastische Wesen (Schlangen, Schmetterlinge, Reptilien, Spinnen, wiesel- und sogar koboldartiges Getier) um das frontal dargestellte Gesicht eines lachenden Leoparden. Motivisch angedeutet werden die Themen der Verwandlung, der Jagd, des Fressens und Gefressenwerdens.

Eine aufwendige, durchaus manieristische Ordnung erscheint zunächst als das gemeinsame formale Merkmal dieser Malereien. Hinzu kommt der inhaltliche Aspekt, dass—angefangen beim Kristallwachstum der Mineralien[15] über die komponierten Figuren aus Muscheln und Pflanzen bis hin zur Tiergruppe—diese Ordnungen eine Verlebendigung vollführen. Als zentrales Thema erscheint geradezu das Lebendige, und zwar das Lebendige als Phänomen des Ordnens oder Anordnens. (Siehe Abb. 1-4, die vier Bekrönungsmalereien der Naturalienschränke im Anhang. Halle, Franckesche Stiftungen, Kunst- und Naturalienkammer. Fotos: Klaus E. Göltz.)

Auch wenn diese Bilder, würde man sie außerhalb ihres pietistischen Kontextes betrachten, weit weniger außergewöhnlich erschienen (zum Beispiel fanden die nach Arcimboldo gefertigten Kompositportraits im 16. und 17. Jahrhundert eine weite Verbreitung an den mitteleuropäischen Höfen), ist ihre Verwendung in einem der Kernräume des Halleschen Pietismus ein bemerkenswerter, nicht leicht zu erklärender Sachverhalt. Wie ist angesichts des Misstrauens, das der Hallesche Pietismus dem Bild entgegenbrachte, eine derart verspielte, in Teilen phantastische und in die Simulation übergehende Bildhaftigkeit zu erklären? Wie lässt sich erklären, dass bei einem zentralen Projekt der Franckeschen Anstalten, wie es die Kunst- und Naturalienkammer darstellte, mit einem nicht unerheblichen konzeptuellen und künstlerischen Aufwand auf die Tradition manieristischer Malerei zurückgegriffen wurde? Müssten diese Bilder nicht für sich betrachtet als Adiaphora gelten? Sie wären opulente und etwas anrüchige Spielereien, ‚Mitteldinge', die von den wesentlichen Aspekten pietistischer Frömmigkeit ablenken. Ich möchte diesen großen Fragekomplex nach dem berühmten Bildervorbehalt des Hallenser Pietismus hier ausklammern, um lediglich in einigen Aspekten die Funktion zu beschreiben, die diese Bilder im Sammlungskonzept der Kunst- und Naturalienkammer übernommen haben könnten.[16] Ich möchte

---

15 In der zeitgenössischen Naturgeschichte war es durchaus üblich, die Mineralien dem Bereich der lebenden Wesen zuzuordnen. Vgl. Foucault, *Ordnung der Dinge*, p.207f.

16 Sicherlich ließe sich ohnehin erst nach der Beantwortung dieser Frage der Funktion die theologische Problematik dieser Bilder weiterführend erörtern. In jedem Fall wird

für diese Frage eine Antwort vorschlagen, die bei der dargestellten und simulierten Vitalität ansetzt. Das heißt es geht mir darum, das vitale surplus im Erscheinungsbild der Sammlung näher zu bestimmen. Stützen kann man sich dabei auf die Beobachtung, dass diese Bekrönungsmalereien auf einen Punkt anzuspielen scheinen, an dem Ordnung und Leben konvergieren. Dieses Zusammentreffen von Harmonie und Belebung, von Repräsentation und *vis activa* führt direkt zu einem barocken Philosophen, der im Umfeld Franckes sowie in der Sammlungstheorie der Zeit um 1700 eine wichtige Rolle gespielt hat: Gottfried Wilhelm Leibniz.[17] Die Leibnizsche Metaphysik macht das Lebendige grundsätzlich zu einer Frage der Ordnung: Lediglich die göttliche Kunst ist demnach in der Lage, die Dinge so zu ordnen, dass sie – als Natur – lebendig werden. Er verdeutlicht dies in der *Monadologie* an der Unterscheidung von natürlichen und von Menschenhand gefertigten Maschinen. Bei allen natürlichen Maschinen greifen die Elemente mit unendlicher Feinheit ineinander, während die aus menschlicher Kunst hervorgegangenen Apparate nur bis zu einem bestimmten Punkt geordnet sind:

> So ist jeder organische Körper eines Lebewesens eine Art göttlicher Maschine oder so etwas wie ein natürlicher Automat, der alle künstlichen Automaten unendlich übertrifft, weil eine durch die Kunst des Menschen geschaffene Maschine nicht in jedem ihrer Teile Maschine ist. Zum Beispiel: Der Zahn eines Messingrades hat Teile oder Bruchstücke, die für uns nichts künstliches mehr sind und nichts an sich haben, was in Hinblick auf den Gebrauch, zu dem das Messingrad bestimmt war, die Maschine auszeichnet. Die Maschinen der Natur aber, das heißt die lebendigen Körper, sind doch noch im kleinsten ihrer Teile bis ins Unendliche Maschinen. Das macht den Unterschied zwischen Natur und Kunst, das heißt zwischen der göttlichen Kunst und unserer aus.[18]

Leben zeigt sich dieser barocken Harmonie zufolge als ein Problem der Ordnung: Die Ordnung muss ins unendlich Kleine vorgedrungen sein, um

---

man ihre ausgesprochen starke Kontextgebundenheit in Rechnung stellen müssen: Diese Malereien können nicht isoliert betrachtet werden, sondern für die Bestimmung ihres Status muss ihr inniger Zusammenhang mit den Objekten der Kammer Berücksichtigung finden.

17 Die Beziehungen Franckes zu Leibniz sind bislang vor allem hinsichtlich der Gemeinsamkeiten und schließlichen Divergenzen der jeweiligen Institutionsprojekte (Anstalts- bzw. Akademieprojekt) untersucht worden. Vgl. Erich Beyreuther, August Hermann Francke und die Anfänge der ökumenischen Bewegung (Leipzig: Köhler & Amelang, 1957) sowie Hornemann/Veltmann, „Zu Erziehung der Jungend". Zu Leibniz' Interesse Sammlungen vgl. Horst Bredekamp, *Die Fenster der Monade. Gottfried Wilhelm Leibniz' Theater der Natur und der Kunst* (Berlin: Akademie Verlag 2004).

18 Gottfried Wilhelm Leibniz, „Monadologie," in *Kleine Schriften zur Metaphysik*, ed. by Hans Heinz Holz (Frankfurt am Main: Suhrkamp, 1996), p.439-483 (p.469).

hier jenen „Maschinen der Natur" zur Existenz zu verhelfen.[19] Erst diese Vollendung der Repräsentation als einer restlosen Einheit in der Vielheit und im Chaos der empirischen Phänomene vermag auch das Leben zu erklären. Ich möchte diese Leibnizsche Idee aufgreifen und im Folgenden die Bekrönungsmalereien der Naturalienschränke von der Fragestellung ausgehend interpretieren, inwiefern hier das Leben als ein Thema des Ordnens respektive der Ordnungsverfahren erscheint. Die Bilder wären dann weit mehr als nur Kulisse und theatrales Beiwerk; sie wären pointiertes Lebenszeichen und womöglich zentraler Signifikant der Kunst- und Naturalienkammer der Franckeschen Anstalten.

Zunächst zu den Motiven in Arcimboldescer Manier: Guiseppe Arcimboldo (1526-1593) verbrachte zwischen 1562 und 1580 achtzehn Jahre an den Wiener und Prager Höfen der Habsburger Kaiser, zunächst bei Ferdinand I., dann bei Maximilian II. (dies ist jener Sammler, von dem die oben zitierte Regel „Quanta rariora tanta meliora" überliefert ist) und schließlich Rudolph II. Besonders die Wissenschafts- und Alchemiefaszination Rudolfs sowie seine Sympathien für die Künste sollten in die Geschichte der frühen Neuzeit eingehen: Kepler und Brahe waren für ihn tätig, der Rabbi Löw und der Mythos von der Erschaffung des Prager Golems gehören in seinen Umkreis. Rudolf legte eine der größten Kunst- und Gemäldesammlung an und förderte eine Reihe bildender Künstler, die schließlich unter der Bezeichnung Rudolfinische Kunst bzw. Rudolfinischer Manierismus zusammengefasst werden sollten.

Ab 1563 begann Arcimboldo die an den Höfen bald so beliebten Kompositportaits zu malen, aber er erfüllte bei Hofe eine Reihe weiterer Aufgaben: Er war Erfinder, Ratgeber, *Maitre de Plaisir* und er war an der Einrichtung von Kunstkammern beteiligt. Seine Kompositportraits werden häufig in den Zusammenhang mit den Kunstkammern der Habsburger Kaiser gebracht und möglicherweise waren sie tatsächlich im Prager und Wiener Kabinett zu sehen. Der Zusammenhang der Bilder und der Sammlungen erklärt sich insofern, als einige der Portraits ebenso wie die Sammlungen das prachtvolle Gefüge der Monarchie ersichtlich machen. So konnten die Portraits als politische Allegorien entworfen werden, welche etwa eines der Ele-

---

19 Eine solche Lebendigkeit fiele folglich auch nicht unter die pietistische Ablehnung der zeitgenössischen Automaten. Im Gegenteil: Die pietistische Ablehnung künstlicher Wesen lässt sich treffend mit der Kritik des künstlichen Automaten durch Leibniz vereinbaren. (In der Tat finden sich in der Sammlung Franckes weder mechanische Wesen noch ihre Vorläuferfiguren, die antiken Statuen. Zu diesem historischen Zusammenhang vgl. Bredekamp, *Antikensehnsucht und Maschinenglaube. Die Geschichte der Kunstkammer und die Zukunft der Kunstgeschichte* (Berlin: Verlag Klaus Wagenbach, 1993.)

mente oder eine Jahreszeit darstellen und zugleich eine heraldische Anspielung oder sogar das Konterfei von Rudolf II. ergeben. Der doppelte Sinngehalt lässt sich als Ausdruck absolutistischer Souveränität verstehen: Der Makrokosmos der Natur spiegelt sich thematisch variiert und wohlgeordnet im Bild, die Habsburgisch regierte Welt und die Natur fügen sich in Eins.[20] (Siehe Abb. 5, Rudolf II. als Vertumnus [1590], Skoklosters Slott, Stockolm.)

Dabei erscheint es wichtig, jenen Referenzwechsel oder Umschlag festzuhalten, den diese Bildgattung zwischen Einzelobjekten und ihrer Gruppierung initiiert. Je nach Blick erkenne ich konkrete Einzelobjekte in fein ausgearbeiteter Naturtreue oder ich realisiere den Gesamtzusammenhang: das komponierte Portrait. In aufsteigender Richtung passiert der Blick eine Schwelle, auf der durch den Einsatz manieristischer Techniken und die merkliche Betonung des Artifiziellen eine neue, wiederum natürliche und lebendige Gestalt sichtbar wird. Insofern ist der Manierismus generativ: Durch die ebenso geschickte wie explizite Kombination vorgefundener Objekte wird ein neues Objekt geschaffen. Nur eine minimale Differenz des Blicks wird vorausgesetzt, um diese zahlreichen Objekte zu einem Bild zusammentreten zu lassen, in dem der Souverän sichtbar wird.[21]

Durch diesen stets präsenten und leicht wiederholbaren Moment ihrer Synthese (ich kann diese Schwelle immer wieder in beide Richtungen überqueren) scheinen die Einzelobjekte der Kompositportraits auf die Objekte der fürstlichen Sammlung zu verweisen: Für beide Objektklassen gilt (auch wenn bei Aufstellung realer Objekte das mimetische Prinzip nicht in Kraft ist), dass durch ihre treffliche Auswahl und Anordnung die Herrlichkeit des Habsburger Kaisers und die Harmonie der von ihm regierten Welt sichtbar wird.

Gründlers Arcimboldeske Darstellungen auf dem Pflanzen- und Conchylien-Schrank übernehmen das Vexierspiel zwischen Portraitfigur und natürlichem Objekt. Das Gesicht jenes Wesens, das uns auf diesen Bildern entgegentritt, ist Ergebnis einer Artikulation, es entsteht durch ordnende Eingriffe der Kunst in den Bereich natürlicher Objekte. Gerade diese Künst-

---

20 Vgl. Franz Kirchweger, „Zwischen Kunst und Natur: Arcimboldo und die Welt der Kunstkammern," in *Arcimboldo 1526-1593*, ed. by Sylvia Ferino-Pagden (Wien: Kunsthistorisches Museum, 2008), p.189-194; Sven Alfons, „The Museum as Image of the World," in *The Arcimboldo Effect. Transformations of the Face from the Sixteenth to the Twentieth Century*, ed. by Pontus Hulten (London: Thames and Hudson, 1987), p.67-88.

21 Roland Barthes fasst dies folgendermaßen zusammen: „Arcimboldo erzeugt Phantastik mit sehr Bekanntem: Die Summe hat eine andere Wirkung als die Addition der Teile. Man könnte meinen, sie ist deren Rest." Barthes, „Arcimboldo oder Rhetoriqueur und Magier," *Der entgegenkommende und der stumpfe Sinn* (Frankfurt am Main: Suhrkamp, 1990), p.136-154 (p.147).

lichkeit erzeugt eine neue, "jüngere" und lebendige Figur zweiter Ordnung; sie lässt sie aus den Einzelelementen hervortreten. Wobei jene synthetische, supra-natürliche Gestalt die Spuren ihrer Entstehung offensichtlich macht und den Sachverhalt des künstlichen Anordnens dokumentiert. Das Zusammentreten der einzelnen Elemente zur Figur vollzieht sich also nicht in einem Schwung, gerade ihr Manierismus hält ihre Entstehung präsent. Es sind ‚komplizierte' Bilder, auf eine ebenso geschickte wie aufwendige Weise ineinandergefaltet.[22]

Auch für Gründers Kompositportraits gilt damit, dass sie nicht allein Lebewesen abbilden, sie zeigen auch und vor allem die Ordnung beziehungsweise das Geordnetsein der Wesen. Es sind regelrechte Allegorien des Sammelns und Ordnens. Im Unterschied zu den Bildern aus den Kaiserlichen Kammern entsteht hier aber nicht mehr das erkennbare Bild eines absoluten Monarchen, sondern lediglich eine anonyme, bis auf die männliche Gestalt kaum weiter bestimmbare Figur. In diesem Zurückdrängen der Individualität und ihrer personalen Merkmale gewinnen diese Portraits zum einen ihre seltsame Maskenartigkeit zum anderen wird durch dieses Übergewicht des formalen Kriteriums die Repräsentation nochmals in gesteigerter Weise wahrnehmbar: Es erscheint der reine Sachverhalt der Ordnung. Gründlers Portraits sind, mit Foucault gesprochen, nicht mehr Figuren der Ähnlichkeit (dies würde für Arcimboldos magische Bildsprache noch gelten), sondern reine Repräsentationen.[23] Aus den alchemistischen Gestalten des sechzehnten Jahrhunderts wurden Symbole für die Kette der Wesen. Noch das Wappenmotiv auf dem Mineralienschrank folgt diesem Prinzip, indem die heraldische Anordnung – weitgehend inhaltsleer – nur prinzipielle Ordnungsstrukturen anzeigt und frontal dem Betrachter darbietet.

Die Aussage, die solche Portraits über die Sammlung und ihre Objekte treffen, lautet wohl zunächst, dass es die Ordnung ist, von der alles abhängt. Ohne das Artifizielle, ohne den systematisierenden Eingriff, bliebe das Natürliche völlig im Bereich des Verworrenen. Erst durch die Anordnung gewinnen die Objekte einen Zeichencharakter und werden physikotheologisch relevant.

Zudem zeigen die Portraits, dass dieser Ordnung ein poietisches Moment zugehört: Durch das Zusammentreten der Einzelwesen entsteht eine neue, wiederum lebendige Figur. Die Portraits nehmen sich, insofern ihre Ordnung ein Moment der Verlebendigung enthält, wie ein naturphilosophischer, monadologischer Kommentar aus: Entweder ich sehe viele Einzelwesen oder ich sehe die eine, wesentlich höherstehende Monade. Entweder ich

22 Barthes, *Arcimboldo*, p.152.
23 Vgl. Foucault, *Ordnung der Dinge*, p.45.

sehe Fische oder ich sehe den Teich, der aus lauter Fischen besteht.[24] Es könnte sich um diese monadologische Szene handeln, die hier die naturgeschichtliche Sammlung umspielt und die das naturphilosophische Thema einer universellen Belebtheit in die naturgeschichtliche Sammlung einbringt: „toute la Nature est pleine de vie“, heißt es bei Leibniz, und dieses Leben organisiert sich in singulären Einheiten, den Monaden oder Entelechien: „[D]ie Glieder des lebendigen Körpers sind voll von anderen Lebewesen, Pflanzen, Tieren, deren jedes wiederum seine beherrschende Entelechie besitzt.“[25] Möglich wird dieses Leben – und hier unterscheidet es sich wesentlich von den Automaten und mechanischen Wesen der Epoche – erst durch eine unendlich fein abgestimmte Ordnung, die Komposibilität aller Elemente. Damit stellt sich natürlich die Frage, ob für Gottfried August Gründler und seine Zeitgenossen diese Sammlung als eine derart poietische Ordnung funktioniert haben könnte, die Leben zu generieren oder zu simulieren sucht?

Bevor ich dieser Frage anhand der Bemalung des Tierschranks weiter nachgehe, lohnt es gerade hinsichtlich der vitalen Komponente der Kompositportaits auf die kunstgeschichtlichen Vorläufer Arcimboldos zu verweisen. Thomas Da Costa Kaufmann ging diesen Herkünften nach und fand sie vor allem in der Renaissance, hauptsächlich bei Leonardo, und in antiken Satyrdarstellungen.[26] So liegt es durchaus nahe, in Arcimboldos Kompositwesen einen Rückgriff auf Leonardos Hinweise zur Erschaffung eines fiktiven Wesens (*animal finto*) zu sehen, die sich im *Traktat über die Malerei* finden. Zudem gibt Vasari eine Episode wieder, wonach Leonardo aus verschiedenen Einzelwesen den Kopf einer Medusa zusammensetzte. Das entstandene Bild war auf einem Schild angebracht und erwirkte in der Dunkelheit den Eindruck eines furchtbaren schrecklichen Ungeheuers.[27] Die antiken Herkunftslinien der Kompositköpfe führen zu den sogenannten Grilli, dabei handelt es sich um zusammengesetzte Bilder menschlicher und monströser Formen auf Gemmen oder Kameen. In Betracht zu ziehen sind zu-

---

24 Leibniz, „Monadologie,“ p.471.

25 Leibniz, „Monadologie,“ p.471.

26 Da Costa Kaufmann, „Arcimboldos Kompositköpfe: Ursprünge und Inventionen,“ in *Arcimboldo 1526-1593*, p.97-101.

27 Da Costa Kaufmann, „Arcimboldos Kompositköpfe,“ p.100. Vgl. die Beschreibung bei Ernst Uehli, *Leonardo, Raffael und die geistig-kulturellen Grundlagen der italienischen Renaissance* (Dornach: Philosophisch-Anthroposophischer Verlag am Goetheanum, 1967), p.120: „Vasari berichtet über eine Holzscheibe, die er [Leonardo] bemalen sollte. Er wählte als Motiv eine Medusa. In dieser Absicht trug er allerlei Gezücht wie Eidechsen, Schlangen, Molche, Heuschrecken, Fledermäuse und Schmetterlinge zusammen, aus deren Verknäulung er ein Ungeheuer malte, das bei gedämpften Licht den schreckhaften Eindruck des Lebendigen machte.“

dem die Mischwesen aus dem Umkreis von Dionysos, etwa aus Phalloi zusammengesetzte Satyren auf Majoliken oder Medaillen.

So umreißt die Herkunft des Kompositportraits ein Milieu, in dem eine Belebung zu beobachten ist, die sich in den Begriffen des Natürlichen oder des Organischen nicht erschließt. Es handelt sich um künstliche, widernatürliche, aber dennoch ausgesprochen agile Wesen. Ein solches Zwischenreich entdeckt auch Barthes bei Arcimboldo („ein Leben im Larvenstadium," eine „Verknäulung von vegetativen Wesen") und schlägt von hier aus direkt den Bogen zu den Wunderkammern der Habsburger, denen etwas von den „Laboratorien Fausts und Caligaris" anhaftete. In der Zusammenkunft der Reiche (der Tiere, Pflanzen, Menschen) erlebten die Dinge Verwandlungen – dies zeigt sich auch im Prinzip des Arcimboldesken Monsters.[28]

Das ästhetische Programm der Kunst- und Naturalienkammer II: Die Bekrönung des Tierpräparateschranks

Das offensivste Motiv unter den Schränken der *res naturales* trägt der Tierschrank, auch hier findet sich die wappenartige Struktur, die sich aus der Anordnung von etwa zwanzig einzelnen Tierdarstellungen ergibt (allerdings ohne ein anthropomorphes Kompositportrait darzustellen). Die meisten dieser Tiere werden in ausgesprochen lebendigen und aktiven Haltungen gezeigt. Dominiert wird dieser Reigen von einem breit lachenden Leopardengesicht, das die Szene aus ihrem Zentrum heraus erhellt. Wir finden das Lebendige hier nicht nur in der motiv- und anspielungsreichen Versammlung der Tiere, die sich in vielfältiger Bewegung befinden: sie springen, fliegen, kriechen, greifen oder seilen sich, wie die beiden Spinnen, an einem Faden ab. Die Szene wird nicht nur durch dieses Geschehen belebt, sondern durch eine kaum erklärliche Fremdreferenz, die im dezenten Licht des Museumsraums regelrecht aufblitzt: Mit der lachenden Katze beginnt sich die Sammlung gleichsam mitzuteilen. Der lachende Blick ist unmissverständlich und offensiv an den Betrachter gerichtet, er kommuniziert unverhohlen und durchbricht damit den geschlossenen Rahmen des Gesamtensembles Kunst- und Naturalienkammer. Der Kreis öffnet sich auf den Betrachter hin mit einer Geste – einer Art spontaner Kontaktaufnahme –, die nicht nur einen lebendigen, sondern vor allem einen performativen Charakter hat und damit jeden Adressaten unmissverständlich zu einem Glied der in der Sammlung repräsentierten Kette der Wesen erklärt. Der frontale Anblick, der bereits von den Kompositfiguren ausging, wird hier noch einmal in aller Deutlichkeit aufgenommen.

---

28 Barthes, *Arcimboldo*, p.153f.

So sehr dabei dieses Lachen seinen Träger – das Tiergesicht – überragt und als Lebensäußerung wie eine äußerliche und kapriziöse Geste wirkt; so sehr die Vitalisierung also wie ein Überschuss und ein merkwürdiges Zuviel erscheint, so gelingt es doch, dass sich dieses Lachen adressieren kann. Die Geste integriert: Auch das Betrachter- oder Erkenntnissubjekt findet sich inmitten der Sammlung und unter den natürlichen Objekten und Wesen wieder. Und dieser Einbezug bleibt, sobald ihn dieser Signifikant einmal organisiert hat, auch dann noch in Kraft, wenn sich der Betrachter vom Tierschrank und dem Vis-à-vis des Leoparden abgewandt hat – immer noch weiß er das vereinnahmende Lachen im Rücken, die schelmische Geste behält das letzte Wort.

Dieser Leopard und die um ihn gruppierten Tiere bilden im Zusammenhang mit den Arcimboldo nachempfundenen Darstellungen auf dem Conchylien- und dem Pflanzenschrank eine Art Klimax. Die Vielfalt der Einzelwesen, die aus dem Bereich der Pflanzen und Conchylien stammt, ergibt jeweils eine lebendige Gestalt. Das vitale Element, das die Bilder auf dem Schrank den Objekten im Schrank hinzufügen (indem sie es aus der geordneten Vielzahl hervortreten lassen), trägt anthropomorphe Züge. Diese anthropomorphe Struktur weist die Malerei auf dem Tierschrank nicht auf, aber dennoch erscheint hier die lebendigste aller Szenen und (bezogen auf den Leoparden und insofern man die Kammer als monadologische Komposition begreift) die "wachste" Perzeption.[29] Während in den komponierten Portraits der Manierismus die Ordnung und die sich einstellende Verlebendigung hervortreten ließ, zeigt er sich bei der Tierdarstellung in einer Vitalisierung, die sich physiognomisch ausdrückt.

Zur Verstärkung dieses vitalen Illusionsmoments setzt Gründler hier mehrfach *Trompe-l'œil*-Effekte ein. Er gibt den Tiergestalten durch die Darstellung der von ihnen geworfenen Schatten Tiefe und passt diesen optischen Effekt an den tatsächlichen Lichteinfall an, wie er von dem ins Kabinett einfallenden Tageslicht erzeugt wurde.[30] Diese Tageslichtszene und vor allem der Lauf der Sonne um den nach Süden weisenden Raum erschließt sich dem heutigen Betrachter leider nicht mehr. Die Kammer wird seit ihrer Restaurierung von gleichmäßigem künstlichem Licht dezent erhellt und so

29 Unter den Tieren ist nur der Leopardenkopf am anthropomorphen Bildprogramm angelehnt, weist dabei aber eine bemerkenswerte Ununterscheidbarkeit oder Doppeldeutigkeit auf: Man könnte einerseits davon sprechen, dass hier eine anthropomorphe Tiergestalt erscheint, andererseits aber wäre mit gleichem Recht festzuhalten, dass man es mit einer dem Tier angenäherten menschlichen Geste – einem zoomorphen Lachen – zu tun hat. Das hervorstechenste Motiv der Kammer erhöht also zum einen das Tier durch den menschlichen Zeichengebrauch, zum anderen wird die menschliche Geste dem Tierkörper anverwandelt.

30 Müller-Bahlke, *Die Wunderkammer*, p.48.

stellen sich jene Impressionen, die das variierende natürliche Tageslicht erzeugt haben muss, nicht mehr ein. Man kann nur vermuten, dass hier zwischen Sonnenaufgang und Untergang eine sehr komplexe Szene ans Licht trat, die nicht nur theologische Relevanz hatte, sondern auch die Illusionskraft der Kammer unterstützt hat.

Zusammenfassung

Es liegt auf der Hand, dass das barocke Ensemble der Kunst- und Naturalienkammer der Franckeschen Anstalten eine systematische Darstellung des Schöpfungszusammenhangs ergeben soll. Mit dem Blick auf die Naturaliensammlung des Kabinetts ging es im Vorausgehenden darum, zu zeigen, dass sich zu dieser naturgeschichtlichen Systematik ein naturphilosophisches Thema gesellt, das in den Bekrönungsmalereien der Sammlungsschränke entwickelt wird. Auf verschiedenen Ebenen wird hier versucht, Verlebendigungseffekte zu erzielen: 1. formalästhetisch, durch die Ordnung der Kompositportraits und *Trompe-l'œil*-Effekte. 2. inhaltlich, indem in diesen Ordnungen das Leben erscheint, 3. performativ, durch den frontal an den Beschauer gerichteten Blick und das an ihn adressierte Lachen.

Man hat es hier noch nicht mit einem Leben zu tun, das sich dem modernen biologischen Verständnis folgend als Zusammenhang von Organfunktionen und einem ökologischen Milieu entwickelt und das man etwa nach den Regeln der vergleichenden Anatomie zu einer Sammlung anordnen würde. Die Ideen der Organisation oder der Lebenskraft sucht man vergebens. Leben wird wesentlich als Fülle und geordnete Kontinuität verstanden und folglich als eine durch Blick und Sprache, durch Objekt und Katalog klassifizierte naturgeschichtliche Sammlung repräsentiert: „Für die klassische Erfahrung war das Lebendige ein Feld oder eine Folge von Feldern in der universellen *taxinomia* des Seins.“[31]

Die durch die zahlreichen Einzelheiten simulierte Lebendigkeit folgt dabei nicht der Idee eines künstlichen Automaten, sondern scheint – hier wäre durchaus eine Parallele zur Leibnizschen Monadologie zu bemerken – an der Idee orientiert, dass nur die göttliche Ordnung eine lebendige sein kann.

---

31 Foucault, *Ordnung der Dinge*, p.207.

Die vier Bekrönungsmalereien der Naturalienschränke, Halle, Franckesche Stiftungen, Kunst- und Naturalienkammer. Fotos: Klaus E. Göltz.

Abb. 1. Bekrönungsmalerei, Schrank 1

Abb. 1. Bekrönungsmalerei

Abb. 2. Bekrönungsmalerei, Schrank 2

Abb. 2. Bekrönungsmalerei

Abb. 3. Bekrönungsmalerei, Schrank 3

Abb. 3. Bekrönungsmalerei

Abb. 4. Bekrönungsmalerei, Schrank 4

Abb. 4. Bekrönungsmalerei

Abb. 5. Rudolf II als Vertumnus (1590) (Skoklosters Slott, Stockolm)

Abb. 5. Rudolf II. als Vertumnus (1590), Skoklosters Slott, Stockholm

Haller, Unzer, and Science as Process

Brian I. McInnis

Abstract: Imagining human beings as "ontological centaurs" Howard Tuttle argues in *Human Life is Radical Reality* that humans have a biological nature but that their fundamental essence lies in the activity of the mind. Eighteenth-century physiologists Albrecht von Haller and Johann August Unzer viewed the balance of the dualistic human existence differently. Haller employed experiment and hypotheses to describe more closely the three forces of contractility, irritability, and sensibility that regulate muscle movement. These forces highlight the ability of the body to react without a soul or consciousness. In his challenging synthesis of physiology and Baumgartian empirical psychology, Unzer redefined traditional terminology to emphasize the biological sources of sensation and thought. He thus conceived of thought as embodied. Even if neither physician pursued a theory of life force (*Lebenskraft*), the value of their work in the present context lies in their linking of biological nature to reason and volition.

In chapter ten, "The Directions of Human Life," of *Human Life is Radical Reality,* Howard Tuttle strongly differentiates humans' biological nature from their existence as self-activating rational subjects. He argues with José Ortega y Gasset that while biological nature is a necessary condition of embodied life, it is not a sufficient condition to explain the life of a self-creating subject in history. Tuttle clearly finds the essence of the human in the activities of the mind. Nonetheless, humans are inescapably both physical and conceptual, and in recognition of this dualistic existence, he imagines human beings as "ontological centaurs."[1]

This metaphor prioritizes the intellectual and obscures the indebtedness of thought and volition to biological processes. In their mixed being, half horse and half man, centaurs were generally known for their violence and the prominence of their animal drives.[2] Chiron was the exception, a centaur known for intelligence and wisdom.[3] Tuttle's metaphor implies a common biological and instinctual basis for centaurs and humans while leaving open their potential to develop wisdom. The metaphor creates a certain tension. It intimates, like Tuttle claims, that humans are either defined more by their biological being or by their mind. Tuttle's vision of humankind, his preference for the reflective human being, and his use of the centaur metaphor

1 Howard N. Tuttle, *Human Life Is Radical Reality: An Idea Developed from the Conceptions of Dilthey, Heidegger, and Ortega Y Gasset* (New York: Peter Lang, 2005), pp.141.

2 Edith Hamilton, *Mythology*. (Boston: Back Bay Books, 1998), pp.43, 155.

3 Edward Tripp, *Crowell's Handbook of Classical Mythology* (New York: Crowell, 1970), pp.154.

obscure the complex relationship between biological urge and self-aware contemplation in the human body.

The debate over the character of the relationship of mind and body extends back to antiquity.[4] In the early modern period, many of the faculties of reason and volition we now attribute to the mind were largely attributed to the soul. Eighteenth-century writers transmitted existing philosophical and theological models that prioritized the soul above the body. They also transformed these discourses. In his didactic poem "Über den Ursprung des Übels" (1734), Albrecht von Haller (1708-1777), like Tuttle, highlights the dualistic nature of human beings. Humans occupy a middle link in the famed chain of being between angels and animals as what Haller calls citizens of the heavens and of nothingness.[5] They have one foot in eternity and the other in mortal decay. As such Haller classifies the human being as a "Zweideutig Mittelding von Engeln und von Vieh, / Es überlebt sich selbst, es stirbt und stirbet nie" (Ambiguous hybrid of angels and of beasts, / It outlives itself, it perishes yet never dies).[6] At the beginning of the eighteenth century, human beings were seen through a theological lens as endowed with a mortal body and an immortal soul. Haller's poem reflects the common belief that while the body dies, the soul survives for eternity. Haller further juxtaposes the body and the soul in the history of salvation and the moral fate of humankind. In the poem's Christian anthropology, he notes that persons who veer from moral virtue sacrifice happiness and are prone to corporeal vice (*Laster*) in temporary bodily indulgence. The "truly" human is rooted in the spirit (*Geist*) and eschews the purely pleasure-driven experience of this world.[7] But the body is far from being depraved. The body's senses are central in the perception of truth.[8] In this poem, Haller's accentuation of sensation as the genesis of thought creates a more nuanced relation-

---

4 John P. Wright and Paul Potter, *Psyche and Soma: Physicians and Metaphysicians on the Mind-Body Problem from Antiquity to Enlightenment* (New York: Oxford, 2000).

5 Arthur Lovejoy argues that this eighteenth-century position contested with egoism, humility, and a kind of tragi-comic internal disharmony to define the place of humans in the cosmic order. Arthur O. Lovejoy, *The Great Chain of Being: A Study of the History of an Idea* (Cambridge, Mass.: Harvard UP, 1936), pp.186-200.

6 Albrecht von Haller, "Über den Ursprung des Übels," in *Die Alpen und andere Gedichte*, ed. by Adalbert Elschenbroich (Stuttgart: Reclam, 1965), pp.53-74 (p.62-63; book II, lines 103-108).

7 Haller, 'Über den Ursprung,' p.70 (book III, lines 81-88).

8 "Vielleicht empfangen wir, bei trüber Dämmerung Klarheit, / Nur durch fünf Öffnungen den schwachen Strahl der Wahrheit" (Perhaps we receive clarity amid the twilight haze, / Only through five senses the weak ray of truth). Haller, 'Über den Ursprung,' p.62 (book II, lines 81-82).

ship between body, soul, and mind than Tuttle's mapping of the biologically determined body and the aspiring mind.

Of concern here is how Haller and fellow physician Johann August Unzer (1727-1799) advance the science of physiology and how their evolving knowledge of the human nervous system shapes their conception of the brain and the soul.[9] Haller's *Erster Umriß der Geschäfte des Körperlichen Lebens* (Initial Outline of Bodily Functions of Life, 1770), which is an updated version of his *Primae lineae physiologiae* (First Lines of Physiology, 1747), and Unzer's *Erste Gründe einer Physiologie der eigentlichen thierischen Natur thierischer Körper* (First Principles of a Physiology of the Actual Animal Nature of Animal Bodies, 1771) each address the connection of the body and the nerves to voluntary and involuntary muscle movement, volition, and reason. The nerves are an important site of interaction between the body and the mind and thereby of Tuttle's biological and reflective human dimensions. My analysis evaluates Haller's and Unzer's medical training and philosophy of medicine, the stated goals in their physiology, and their application of the scientific method as contexts that shape their perspectives on how body, soul, and nerves interact.

In their scientific work, Haller and Unzer participated in a physiological line of inquiry that underscores the body's influence on the formation of the self-creating subject. While Tuttle links the activities of the mind to the essence of the human, Haller – and especially Unzer – argue for a biological contribution to mental function. Indeed, for Unzer it is the physiological structure of the brain and nerves that enable the higher human faculties and thereby make human subjectivity possible. One of Unzer's important contributions to physiology involves his investigation of the biological basis for Alexander Gottlieb Baumgarten's (1714-1762) empirical psychology. Unzer's uncommonly persistent pursuit of the interplay of neurology and psychology meant that he pushed at the borders of developing disciplines. This and his redefinition of scientific terms may have complicated the reception of his ideas. Haller's detailed experimentation on irritability and sensibility, his extensive incorporation of new knowledge into existing physiology, and his

9 This essay is not concerned with where the soul is located (head, heart, stomach, etc.) or according to which theory it interacts with the body (Cartesianism, Occasionalism, pre-established harmony, physical influx, etc.). The soul is examined primarily according to its naturalistic function in human life. By the end of the eighteenth century, developments in psychology and physiology begin to necessitate replacement of the naturalistic soul by the mind.

far-reaching reviews of scientific and literary publications have garnered Haller much more critical attention than Unzer.[10]

Haller's laboratory testing of the causes of muscle motion and Unzer's investigation of the neurology of emotions and movement contributed to a tradition of explanations about the source of muscle movement or *Lebenskraft.* Since antiquity, physicians had identified neural spirits or neural fluids as sources of human motion.[11] In the early modern period, some writers proposed a unified explanation of the forces that shaped human life: Johann Baptista von Helmont (1577-1644) saw a gas shaping the *Archäus*; Franz Sylvius de la Böe (1614-1672) thought fermentation and the blood produced motion, and Georg Ernst Stahl (1659-1734) postulated the *anima* as vital force regulating life.[12] Though Francis Glisson (1597-1677) had differentiated irritability and sensibility, Haller utilized his extensive lab results

---

10 Recent publications on Haller have addressed his research on physiology and neurology. The collection of essays *Albrecht von Haller: Leben, Werk, Epoche* (Göttingen: Wallstein, 2009) edited by Hubert Steinke, Urs Boschung, and Wolfgang Proß contains important surveys of Haller's contributions to anatomy and physiology, his views on religion, his role as researcher and scholar, his research program, his literary production, and his place in the Enlightenment. In *Irritating Experiments: Haller's Concept and the European Controversy on Irritability and Sensibility, 1750-90* (Amsterdam, New York: Rodopi, 2005), Hubert Steinke evaluates the European reception of Haller's dissertation on irritability and sensibility. Brigitte Lohff's older diachronic study, "Die Entwicklung des Experiments im Bereich der Nervenphysiologie," *Sudhoffs Archiv: Zeitschrift für Wissenschaftsgeschichte,* 64 (1980), 105-29, notes two general tendencies in the reception of Haller's 1753 dissertation: one to experimentally validate Haller's findings, another to disprove them. She places Haller at the beginning of a scientific tradition that shapes the *Lebenskraft* debate, but mentions Unzer only in passing. More generally, Unzer receives less attention than Haller in the history of science. In Zelle's edited volume *"Vernünftige Ärzte." Hallesche Psychomediziner und die Anfänge der Anthropologie in der deutschsprachigen Frühaufklärung* (Tübingen: Niemeyer, 2001), Matthias Reiber's monograph *Anatomie eines Bestsellers: Johann August Unzers Wochenschrift "Der Arzt" (1759-1764), Achtzehnte Jahrhundert, Supplementa 8* (Göttingen: Wallstein, 1999) and Stefan Bilger's study *Üble Verdauung und Unarten des Herzens: Hypochondrie bei Johann August Unzer, 1727-1799* (Würzburg: Könighausen und Neumann, 1990), Unzer's *Erste Gründe* (1771) receives scant attention. Tanja van Hoorn and Yvonne Wübben's edited volume *"Allerhand nützliche Versuche" Empirische Wissenskultur in Halle und Göttingen (1720-1750)* (Hannover: Wehrhahn, 2009), provides a first comparison of empirical investigations in Halle and Göttingen, including Carsten Zelle's contribution, "'Ey was hat der Arzt mit der Seele zu thun?' Physiologie und Psychologie bei Albrecht von Haller und Johann Gottlob Krüger," that compares physiology and psychology according to Haller and Unzer's mentor Johann Gottlob Krüger (pp.21-40).

11 Brigitte Lohff, "Die Entwicklung des Experiments im Bereich der Nervenphysiologie," *Sudhoffs Archiv: Zeitschrift für Wissenschaftsgeschichte,* 64 (1980), 105-29 (pp.108-09).

12 Lohff, "Die Entwicklung," p.109.

to argue in *De partibus corporis humani sensibilibus et irritabilibus* (1753, trans. *A Dissertation on the Sensible and Irritable Parts of Animals,* 1755) that all muscle fibers were irritable and all nervous fibers sensible.[13] This enumeration of forces and other scientists' interest in the unified causes of human life led physicians and philosophers to engage Haller's results in their own postulation of theories of *Lebenskraft.* The focus here is on the process of science that leads to the shifts of thought by Haller and Unzer.

Haller's philosophy of medicine was initially shaped by his training in Tübingen and Leiden, then evolved through his scientific work in conjunction with his adherence to religious tradition. In Leiden Haller studied with the physiologist Hermann Boerhaave (1668-1738) and the anatomist Bernhard Siegfried Albinus (1697-1770). Boerhaave was acknowledged as one of the leading doctors in Europe and had taught numerous physicians including the Habsburg Imperial Court physician Gerhard van Swieten (1700-1772), the British physician and proponent of seawater therapy Richard Russell (1687-1759), the British physician Robert Whytt (1714-1766), and the materialist Julien Offray de La Mettrie (1709-1751). Boerhaave was an adherent of Newton, argued for the necessity of an empirical basis for the natural sciences, and explained human anatomy and physiology mechanically as a result of hydraulic forces.[14]

Haller understood his inquiry into human physiology and his explanation of its function as the product of experiment and hypotheses. As a student, he grew skeptical of empirically unverified hypotheses in his training with Elias Camerarius (1673-1734) in Tübingen.[15] This skepticism was likely instilled and extended to Cartesianism during his study from 1725-1727 in Leiden with Boerhaave. It is likely that Boerhaave gained exposure to Cartesianism in the late 1680's through his professors of logic and natural philosophy Wolferd Senguerd (1646-1724) and Burchard de Volder (1643-1709). De Volder had visited England in 1674, learned about the new experimental philosophy, and returned to Leiden where he established a laboratory and added experimental physics to his repertoire.[16] De Volder's eventual

---

13 Roy Porter, *The Greatest Benefit to Mankind. A Medical History of Humanity from Antiquity to the Present* (London: HarperCollins, 1997) p.250. Albrecht von Haller, *A Dissertation on the Sensible and Irritable Parts of Animals.* trans. Samuel Auguste André David Tissot (London: J Nourse, 1755).

14 Hubert Steinke, "Anatomie und Physiologie," in *Albrecht Von Haller: Leben-Werk-Epoche*, ed. by Hubert Steinke, Urs Boschung and Wolfgang Pross (Göttingen: Wallstein, 2008), pp.226-54 (p.228).

15 Urs Boschung, "Lebenslauf [Albrecht von Haller]," in *Albrecht von Haller: Leben - Werk - Epoche*, ed. by Hubert Steinke, Urs Boschung and Wolfgang Pross (Göttingen: Wallstein, 2009), pp.15-82 (p.21).

16 Rina Knoeff, *Herman Boerhaave (1668-1738): Calvinist Chemist and Physician* (Amster-

criticisms of Cartesianism have been attributed to the study of Nicolas Boyle and Isaac Newton.[17] This educational milieu, the publication of Newton's *Principia* (1687), and Boerhaave's 1692 stint as inspector of a newly acquired collection for the Leiden University library helped shape his empirical stance toward medicine.[18]

Boerhaave was the primary proponent of experimental natural philosophy at the University of Leiden between the retirement of his teacher de Volder in 1705 and the appointment of Willem s'Gravesande (1668-1742) in 1717. He reiterated his commitment to the experimental method and trial and observation in his rectorial addresses, his clinical teaching, and his chemistry textbook. In his first rectorial address of 1715, Boerhaave distanced himself and his natural philosophy from Cartesianism. Though scientists might speculate on basic principles in the natural sciences, hypotheses should be confirmed by observation and experiment.[19] In his second rectorial address of 1731, Boerhaave called for physicians to observe the body carefully in order to discover how it works within itself and also upon other bodies.[20] He assumed some clinical teaching duties in 1714 and continued them through 1738, that is, during Haller's time in Leiden, 1725-1727. Boerhaave's clinical lectures took place twice a week, following a regular pattern. These practical sessions involved surveying all patients in St. Caecilia hospital and evaluating the case of one patient in detail, during which students took notes on his discussion of the patient's general appearance, state of nutrition, constitution, posture, pulse and respiration, temperature, swelling, and the condition of the lungs, as well as the prognosis and proposed treatment. On some days the hospital rounds included impromptu mini-examinations, in which experienced students were asked to evaluate a patient's signs and symptoms and offer a prognosis and plan of treatment.[21] In this way, Boerhaave's clinical teaching incorporated empirical observation; his later revaluation of a patient's case and treatment plan constituted experimental feedback on the therapy. Boerhaave employed new technologies in his clinical practice, including a pocket lens for clinical and

---

dam: Koninklijke Nederlandse Akademie van Wetenschappen, 2002), p.237 (p.23-24).

17 C. de Pater, "Experimental Physics," in *Leiden University in the Seventeenth Century: An Exchange of Learning*, ed. by Thomas H. Lunsingh Scheurleer and G. H. M. Posthumus Meyjes (Leiden: Universitaire Leiden, Brill, 1975), pp.309-27 (pp.314-18).

18 Gerrit Arie Lindeboom, *Herman Boerhaave: The Man and His Work* (London: Methuen, 1968), p.25-27.

19 Lindeboom, *Boerhaave*, p.100.

20 Lindeboom, *Boerhaave*, p.174.

21 Lindeboom, *Boerhaave*, pp.287-292.

postmortum diagnosis and thermometers to monitor body temperature.[22] We can assume that Haller was exposed to these experimental techniques during his studies in Leiden.

Haller experienced Boerhaave as a professor whose empirical knowledge stemmed not only from his clinical teaching but also from laboratory experiments. In particular, Boerhaave modeled his understanding of the scientific method through proficiency in experimental chemistry. This was his personal interest and became an area of professional expertise.[23] His textbook *Elementa Chemiae* (Chemical Elements, 1732) outlines his experimental method, which establishes general procedures for investigations throughout natural philosophy. The last of these steps indicates that he viewed chemistry as a sister science with the potential to serve medicine. In Boerhaave's view, the chemist should 1) examine the material to understand its properties; 2) precisely describe the experimental procedure; 3) relate the effects of the test; 4) inspect what remains and compare it with the original materials to identify observable changes; 5) describe the experimental apparatus; 6) observe and record the physical state of the trial's products; and 7) identify possible applications of the observed results for medicine, whether beneficial, innocuous, or hazardous. Boerhaave moreover writes that he strives to serve the scientific community and not to deceive.[24] Drawing on long experience in a laboratory, this protocol established an empirical procedure for use with the textbook experiments and a methodological foundation for chemistry as a natural science. His rigorous investigation, for example with mercury, where one experiment ran for fifteen-and-a-half years and another was repeated 877 times, set a scientific expectation for deriving conclusions from extensive testing and careful reflection.[25] These were standards that Haller strove to emulate.

Haller's Empirical Predisposition and Use of Hypotheses

Haller's scientific approach involved a review of existing research, detailed experimentation with specific structures or stages of an organism's devel-

22 Lindeboom, *Boerhaave*, p.293-297.

23 During his initial studies, Boerhaave began working in the laboratory with his brother, later delivering private lectures in chemistry from 1702 to 1718 and winning the interest of numerous students. In 1718, he was named professor of chemistry and began public lectures on the subject at the university. See Lindeboom, *Boerhaave*, p.59-61.

24 Herman Boerhaave, and Peter Shaw, *A New Method of Chemistry: Including the History, Theory, and Practice of the Art*. trans. Peter Shaw. 2 vols. Vol. 1 (London: T. Longman, 1741) p.7-8.

25 Lindeboom, *Boerhaave*, p.338.

opment, and the production of new hypotheses.[26] The latter were based on the comparison of existing published knowledge and experimentally observed phenomena. As a student, Haller refuted anatomist Johann Georg Coschwitz's labeling of a vein as a saliva duct.[27] His focus on detailed studies led to his differentiation of irritability from sensibility and to his new explanation of the cardiac cycle without reference to the nerves. Moreover, his studies of development of chicken embryos led to a shift in his generative theory. Hence, through sustained experimentation, Haller gained greater insight into physiology. His empirical practice of experimentation and observation superseded Cartesian rationalism.[28] The compilation of observations aided his review of scientific literature and together with his lab work and extensive knowledge of developments in medical research provided a foundation for his own hypotheses.

The public face of science concerned Haller. Like some contemporaries, Haller went out of his way to deny materialism and reconcile causation with the notion of divine direction. A touchstone for the nuanced connection between science and belief was Haller's position in the debates on generation and materialism. His training under Boerhaave led him to espouse preformation theory, the doctrine that living beings grow from miniatures of themselves. The opposing theory, epigenesis, argued that new organisms developed from seeds, spores, or eggs and subsequent cell differentiation. Epigenesis rose to the fore in the wake of Abraham Trembley's (1710-1784) experiments on the regeneration of fresh water polyps. In light of the new evidence, Haller switched to support epigenesis. However, several attempts to distance his views from materialists who espoused epigenesist theories led him back to preformationism. In a review of Julien Ouffray de la Mettrie's *Institutions de médecine de Mr. Herman Boerhaave* (Herman Boerhaave's Principles of Medicine, 1740), Haller accused La Mettrie of plagiarizing his commentated edition of Boerhaave lectures, *Praelectiones academicae in proprias institutiones rei medicae* (Academic Findings on the Proper Instruction in Medical Matters, 1739-1744), and distanced himself from the French materialist.[29]

---

26 Steinke, "Anatomie," pp.237.

27 Steinke, "Anatomie," pp.228.

28 Gerrit Arie Lindeboom, *Descartes and Medicine* (Amsterdam: Rodopi, 1979), p.57.

29 In his review, Haller states that La Mettrie acknowledges integrating Haller's commentary into his work, but without Haller's extensive integration of secondary literature (p.377). He confronts La Mettrie for misrepresenting Haller's commentaries as his own, assuming credit for Haller's observations, improper attribution in footnotes, using multiple pseudonyms, and repetition of the errata that Haller had acknowledged (p.378). See Albrecht von Haller, "Rezension Von M. de la Mettrie *Institutions de médecine de M. de Boerhaave*," *Göttingsche Zeitungen von gelehrten Sachen* (1745), pp.377-78.

In response, La Mettrie dedicated *L'homme machine* (Machine Man, 1747) to Haller and feigned friendship with him. In the public eye, this aligned Haller closely to the controversial author. Fearing charges of materialism against his own work and under pressure from his closest correspondent Charles Bonnet, Haller reverted to preformationism. He initially made this shift based on his scientific observations, yet when this position was superseded by Caspar Friedrich Wolff's (1733-1794) account of epigenesis in *Theoria generationis* (1759), he defended his adherence to preformation theory on a theological basis rather than a scientific one, thus betraying the findings of the scientific community.

In contrast to his mentor Albinus, Haller prioritized empirical observation in his scientific representation of the human body. Albinus refined knowledge of human anatomy by idealizing the represented body. In his anatomical drawings, *Tabulae sceleti et musculorum corporis humani* (Tables of the Skeleton and Muscles of the Human Body, 1747), artists improved illustrations of the most representative cadavers through a process of aggregation of bones, muscles, and organs according to ideals of proportion, exclusion of unique features, and enhanced depiction of parallel anatomical structures to make them appear symmetrical.[30] Haller approached the challenge of physiological representation differently, choosing instead for his anatomical atlas, *Icones anatomicae* (Images of Anatomy, 1743-1756), to dissect many cadavers and then have artists depict a representative example of a particular physiological structure. The inevitable anomalies were retained and explained in the commentary to the illustration.[31]

Haller's goals for physiology identified the sphere in which his procedure of literature review, experiment, and new hypotheses should be applied. Shirley Roe cites Haller's preface to the first volume of *Elementa physiologiae* (Elements of Physiology), in which Haller presents the goals for his multivolume physiology. These goals served in a larger sense for the establishment of physiology as a separate field of practice and study. Haller claims: "Whoever writes a physiology must explain the inner movements of the animal body, the functions of the organs, the changes of the fluids, and the forces through which life is sustained." He then goes on to affirm that the physiologist

> must explain the forces 'through which the forms of things received by the senses are presented to the soul; through which the muscles, which are governed by the commands of the mind, in turn have strength; the forces through which food is changed into such different kinds of juices; and through which, finally, from

30 Steinke, "Anatomie," pp.233.
31 Steinke, "Anatomie," pp.232-34.

> these liquids both our bodies are preserved and the loss of human generations is replaced by new offspring.' Sensation, motion, digestion, assimilation, growth, reproduction–these are the functions of the living organism the physiologist must explain.[32]

Clearly, Haller proposed a functional explication of the human body based on an analysis of relevant forces.

His experimental protocol for the investigation of sensibility and irritability deserves special consideration in terms of the biological contributions to mental function and in terms of the *Lebenskraft* debate. Oswei Temkin notes that Greco-Roman physician Galen of Pergamon (CE 129-ca. 200/ca.216) had already developed a doctrine of sensibility in antiquity.[33] In addition, in the seventeenth century, Francis Glisson had identified a property of irritability in works such as his *Tractatus de ventriculo et intestines* (Treatise on the Ventricle and Intestines, 1677).[34] While Haller's conclusions were not completely new, his reliance on repeated, systematic experimentation, rigorous analysis of results, the incorporation of other scientists' experimental findings, and comparison with previous scientific studies created an enduring interest in his process as a general model for scientific research. Haller wanted to demonstrate the necessity of experimentation as the basis for physiological knowledge instead of mere analogical comparisons in anatomical structure and physiological function proposed by many scientists and philosophers.[35]

Haller also disagreed with a tendency of some writers to appropriate sensibility as the basis of a system of physiology describing all corporal mobility. This included Frederik Winter (1712-1760), Johannes Lups, De Magni, Guillaume Mauquest de La Motte (1655-1737), Johann Gottlob Krüger (1715-1759), Ernst Anton Nicolai (1722-1802), Robert Whytt (1714-1766), and Heinrich Friedrich Delius (1720-1791).[36] The experimental method used in his dissertation on sensibility and irritability enabled Haller to differentiate

---

32 Haller, Albrecht von. *Elementa physiologiae,* 8 vols., (Lausanne: Sumptibus Marci-Michael. Bousquet, 1757) I, p.i. Translated and cited in Shirley A. Roe, "Anatomia animata. The Newtonian Physiology of Albrecht von Haller," in *Transformation and Tradition in the Sciences*, ed. by Everett Mendelsohn (Cambridge: Cambridge UP, 1984), pp.273-300 (p.276).

33 Oswei Temkin, "Introduction [to A Dissertation on the Sensible and Irritable Parts of Animals]," *Bulletin of the History of Medicine,* 4 (1936), 651-55 (p.651).

34 Peter J. Koehler, "Neuroscience in the Work of Boerhaave and Haller," in *Brain, Mind and Medicine: Essays in Eighteenth-Century Neuroscience*, ed. by Harry A. Whitaker, C. U. M. Smith and Stanley Finger (New York: Springer, 2007), pp.213-32 (p.225).

35 Albrecht von Haller and Samuel Auguste André David Tissot, *A Dissertation on the Sensible and Irritable Parts of Animals* (London: J Nourse, 1755) p.658.

36 Haller and Tissot, *A Dissertation,* p.658.

the cause of voluntary and involuntary muscle movement according to two factors. There he defines irritability as muscle or other tissue stimulation that produces contraction without activation of the nervous system (p.658). Sensibility does include activation of the nervous system. In humans, it is a facility by which body parts transmit an impression to the soul. In other animals, similar impressions might only cause signs of pain or discomfort. Haller insists on experiment as the means to differentiate these qualities (p.659). Starting in 1746, his work featured trials on one-hundred-ninety animals (e.g., dogs, goats, rats, cats, rabbits, frogs). Through animal vivisection, Haller stimulated an animal's body part with air, heat, a scalpel, or chemicals to determine the tissue's sensitivity and recorded all results (p.660). Haller's vivisections identified connections between the physical body and the brain via the nervous system in animals, and he transferred the knowledge to humans via comparative zoology.[37] Like many predecessors, he continued to postulate a body/soul divide. Haller rejected Whytt's theory of the dispersion of the soul throughout the body in favor of localization of the soul in the brain, but provided little evidence for the nature of the soul.[38]

The experimental method in specific studies also shaped Haller's approach to surveys of human physiology. In the preface to volume one of the *Elementa physiologiae*, Haller qualifies himself for the multivolume work by noting his accomplishments, among them the dissection and observation that provided the basis for his 1747 handbook *Primae lineae physiologiae.*[39] This handbook also incorporated new knowledge from the work of Giovanni Battista Morgagni (1682-1771), Jakob Benignus Winslow (1669-1760), Bernhard Albinus, and Douglas.[40] The *Erster Umriß der Geschäfte des körperlichen Lebens* (1770) was, as previously mentioned, a vernacular revision of *Primae lineae physiologiae* that updated Haller's text based on new discoveries.[41]

---

37 Haller states in his introduction that one result of the animal experiments is to better understand the human body (p.657). He discusses the results of cutting a dog's Achilles tendon and compares that to the similar insensitivity of human tendons in a finger and the forearm (pp.662-663). Drawing on his animal experiments, Haller notes that the dura matter in the skull of animals is resistant to pain by physical or chemical stimulation. He compares this to a similar insensibility in humans as observed by Mr. Zinn and Mr. Mekel (p.668).

38 On Whytt's theory of the location of the soul see Julius Rocca, "William Cullen (1710-1790) and Robert Whytt (1714-1766) on the Nervous System," in *Brain, Mind and Medicine* (2007), pp.85-98 (p.89).

39 Shirley A. Roe, "Introduction," in *The Natural Philosophy of Albrecht von Haller* (New York: Arno Press, 1981), 2r.

40 Peter J. Koehler, "Neuroscience in the Work of Boerhaave and Haller," in *Brain, Mind and Medicine* (2007), pp.213-32 (p.222).

41 Albrecht von Haller, *Erster Umriß der Geschäfte des körperlichen Lebens für die Vorlesungen eingerichtet. Aus dem Lateinischen unter der Aufsicht des Verfassers übersetzt.* trans. Franz

In *Erster Umriß,* Haller uses observation and experiment to analyze the location of the soul and the operation of the nerves. His understanding of sensation and muscle movement is similar to our contemporary understanding. Sensation involves transmission of information from a sensory nerve via the nervous system to the soul for processing. Muscle movement can be directed by the soul via the nerves. Drawing on observations of twitching in cases of damaged brains, he surmises that the soul lies not on the surface of the brain, but in the neural tissue.[42] Experiments provide a means to evaluate this hypothesis (p.199). Similarly, laboratory observations prompt questions about existing hypotheses. Popular theories at the time proposed that nerves function like strings of an instrument, vibrating with sensations to and from the control center. Yet Haller's laboratory observations indicated that the nerves were structured less like a piano or violin string than a web. Some sensatory nerves, including the auditory and olfactory nerves, did not appear to be under tension. Moreover, he noted that for nerves stimulating motion, not all muscles adjacent to the nerve networks in the body move upon the actuation of the nerves (pp.201-202).

Since empirical evidence makes tensioned nerves seem improbable, Haller evaluates other models of neurological function. One approach proposes that the nerves are activated by vapors that dissipated into surrounding tissues (p.200). Another theory suggests that the nerves are hollow to accommodate a fluid that is elastic, rigid, or electric. Though the electrical hypothesis seems promising, Haller doubts that the body could transmit electrical impulses without agitating other bodily tissues, which equally conduct electricity. Observations of nerve stimulation seem to indicate that nerve function is isolated. One indication was that binding a nerve halts sensation and movement, apparently restricting transmission of neurological signals. By comparison, other electrical impulses cannot be restricted in this fashion and upon introduction disperse throughout the body (p.203). Haller hypothesizes that the localization of neurological function points to a hydraulic nerve system. Though popular, this proposition is also far from certain, for the hollow structure of the nerves is at this point unverifiable even with magnification (p.202).

Haller differentiates three sources of muscle movement: contractility ("die tote Kraft," "vis mortua"), irritability ("die angeborene Kraft," "vis

---

Ludwig Tribolet (Berlin: Haude & Spener, 1770). In the preface, Haller states that this version is based on the 1765 edition of his *Primae lineae physiologiae.* He notes that he translated one-third of the text and authorized the other two-thirds translated by Tissot.

42 Haller, *Erster Umriß*, p.200.

[viva] insita"), and sensibility ("die bewegende Kraft," "vis nervosa").[43] Contractility ("die tote Kraft") describes the general facility of nearly all tissues to draw together. Physical responses to chemical exposure and physical pressure, and the rhythmic movement of vital organs such as the heart and the intestines occur via irritability ("die angeborene Kraft") (p.212). This motion can be activated and increased through irritation, and is not directed by the soul (pp.216, 218). Sensibility ("die bewegende Kraft") describes the neural stimulation of the muscles (pp.213-14).

Haller saw these powers as important qualities of human tissues, but did not interpret them as the basis for a system, as some subsequent physicians and philosophers did. Members of the European scientific community put his hypotheses to the test, validating or replacing his conclusions.[44] For Haller, the study of physiological forces did not lead him to deduce an overarching life force.[45] For others, a life-defining force could be derived from Haller's work on irritability and sensibility. In Göttingen, Johann Friedrich Blumenbach (1752-1840) modified Haller's work in the creation of the *Bildungtrieb,* a principle that determined processes of growth and regeneration.[46] In Mainz, Samuel Thomas Soemmering (1755-1830) translated Haller's *Elementa physiologiae* in such a way as to collapse irritability and sensibility as variations of a single *Lebenskraft.*[47] The conclusions in *A Dissertation on the Sensible and Irritable Parts of Animals* proved fruitful for writers in diverse ways.

Haller's investigation of anatomy and physiology included detailed analyses of organ structure, organ function, and the flow of fluids, and favored conclusions based on observable phenomena. This concentration on details and empirical evidence helped make Haller a respected physiologist. In the

---

43 Shirley A. Roe provides the Latin terms from the fourth volume of Haller's *Elementa physiologiae* (1762) in "Anatomia Animata. The Newtonian Physiology of Albrecht von Haller," in *Transformation and Tradition in the Sciences*, ed. by Everett Mendelsohn (Cambridge: Cambridge UP, 1984), pp.273-300 (p.279). The German terms and all other information are cited from Haller's *Erster Umriß* (1770), p.212.

44 Lohff, "Die Entwicklung des Experiments," p.113.

45 In chapter six, paragraph 181 of *Erster Umriß,* Haller appears to question the viability of the concept of *Lebenskraft.* When discussing arteries, temperature, muscle movement, and movement of the blood, he argues that the pumping of the blood warms it and writes: "Daß die Wärme aus einem Anfange der Fäule entstehe, ist leicht zu wiederlegen [sic]. Wann die Fäulung [sic] gewiß [sic] ist, so wird das Blut kalt. Soll man sich auf die Lebenskraft beruffen [sic], und ist dieses nicht ein blosses [sic] Wort?" (It is easy to disprove that the warmth derives from the beginning of decay. When decay has set in, the blood becomes cold. Should one refer to *Lebenkraft* [life force], and is this not just a mere word?) See Haller, *Erster Umriß,* p.94.

46 Roy Porter, *The Greatest Benefit,* p.251.

47 Lohff, "Die Entwicklung des Experiments," p.114.

case of the nerves, however, these methods limited his development of a compelling explanation of the sources of sensation and motion.[48] The laboratory observation of muscle movement aided his understanding of nerve function, yet numerous questions remained for him. While binding a nerve limited voluntary muscular function, removing the restriction on the nerve restored that function. Nerves irritated during vivisection prompted the muscles they controlled to contract.[49] Haller posited the now antiquated hypothesis that muscles contract where more neural fluid flows.[50] Johann Gottlob Krüger also argued for the proportionality of nerve and muscle function. Haller's experimental predisposition is evident in his scientific evidence about the neural fluid: he cannot confirm its conductivity and presumes it has no electrical properties (p.217).

Haller's observation of the nerves and muscles also reveals the limits of his knowledge about neural function. While acknowledging that the cause of some muscle movement is unknown, he finds it so difficult to describe the manner in which nerves influence muscle movement that he concludes it is nearly hopeless to understand the exact means of connection (p.216). The nerves move the body's extremities in response to individual mental prompts, which he presumes are the purview of the soul. He believes something in the neural fluid activates the muscles, but is unable to identify that substance (p.217). He also cannot identify the physiological reason for the speedy response of the nerves to the will and does not offer a hypothetical physiological explanation (p.220). Moreover, involuntary muscle movement stems from a different source (pp.212, 218).

When thinking at the limits of medical knowledge on neurology, therefore, and despite insistence on experimental data, Haller nonetheless uses inference and analogy in formulating hypotheses about nervous system functions. In his preface to the first volume of the German translation of Georges-Louis Leclerc, Comte de Buffon's (1707-1788) *Histoire naturelle, générale et particulière* (Natural History, General and Particular, 1749-1788), Haller insists on empirical study, yet argues for the benefits of hypotheses.[51] They do not replace the factual evidence as truth, instead they lead research-

---

48 Cf. Fielding H. Garrison, and Lawrence C. McHenry, *History of Neurology* (Springfield, Illinois: Thomas, 1969), p.109.

49 Haller, *Erster Umriß*, p.214,

50 Because he cannot verify the electrical nature of the nerves, Haller hypothesizes a hydraulic nervous system (*Erster Umriß*, p.217).

51 Albrecht von Haller, "Vorrede" in George-Louis Leclerc, Comte du Buffon, *Allgemeine Historie der Natur nach allen ihren besondern Theilen abgehandelt, nebst einer Beschreibung der Naturalienkammer Seigneur Majestät des Königes von Frankreich* (Hamburg: Grund; Leipzig: Holle, 1750).

ers in a promising direction of discovering the truth.[52] This is but one, representative example of how Haller views science not as a definitive set of evidence-based laws, but as a process. He illustrates the incremental change in scientific theory with the example of botany's development by scientists Andrea Cesalpino (1519-1603), Joseph Pitton de Tournefort (1656-1708), and Carl von Linné (1707-1778) (p.xvi). He argues for a sociology of science that anticipates by nearly two hundred years some of Ludwig Fleck's conclusions in *Entstehung und Entwicklung einer wissenschaftlichen Tatsache* (Genesis and Development of a Scientific Fact, 1935).

Additionally, Haller claims that scientists profit from making their ideas public. The iteration of hypotheses contributes to individual and professional advancement with benefits for society at large. (This phenomenon is still in evidence today in the Citation Index.) When scientists must consider the accuracy of a hypothesis, Haller argues, they become willing to risk everything: their existence, wealth, and life of ease. They become absorbed in demonstrating the accuracy of their own hypotheses or disproving those of others (p.xvii). The hypothesis approach prompts personal investment in the promotion of one's own ideas. Moreover, the formulation of a hypothesis gives rise to the search for empirical evidence that otherwise would not have occurred. The creation and dissemination of hypotheses thus leads to more empirical, verifiable research as well as to the investigation of particular questions. The act of postulating or disproving an hypothesis inspires a new series of questions. In sum, a hypothesis focuses the researcher's attention in the processing of empirical data. (p.xvii). In this way, theory aids the advancement of science by supporting the identification of relationships, promoting a synthesis of increasing information that is seen as a contribution to scientific query as a whole (p.xviii). Clearly, Haller believed that the production and verification of hypotheses benefits the advancement of the sciences (p.xx).

All the above highlights just how intensely Albrecht von Haller's medical investigations included both experiment and hypothesis. He thereby offered a revised model for scientific investigation. Boerhaave had provided a precendent of a rigorous researcher and surely shaped Haller's expectations for seeking detailed, verifiable results across large data samples. Like Boerhaave Haller also achieved success as a laboratory experimenter and clinician, combining theoretical and applied science. In the preface to the German translation of Buffon's *Histoire naturelle,* Haller again acknowledges that hypotheses have an important function in encouraging empirical re-

52 Haller, "Vorrede," p.xiv.

search. Yet he embraces them less openly than his contemporary, physiologist Johann August Unzer.

Unzer's Philosophical Synthesis of Neurology and Psychology

Johann August Unzer was born and studied medicine in Halle an der Saale. After completing medical studies, he moved first to Hamburg, then settled in 1750 in Altona, where he profiled himself as a physician and author. In a publication career shaped by monographs, journal articles, and the dissemination of science, the journal *Der Arzt: eine medicinische Wochenschrift* (The Doctor: A Medical Weekly, 1759-1764) proved the most popular. [53] His work frequently investigated neurology as a basis to understand human physiology and psychology. As a writer, physician, and intellectual, he lamented the limited engagement between physicians and philosophers.

Unzer's training at the Friedrichs Universität in Halle was shaped by the philosopher Georg Friedrich Meier (1718-1777) and the physician Johann Gottlob Krüger (1715-1759). Meier studied theology with Sigmund Jacob Baumgarten and philosophy with Alexander Gottlieb Baumgarten, then achieved prominence as a transmitter of the latter's aesthetics in German. In Meier's work, Unzer experienced the empirical investigation of the soul through psychology and the development of a popular weekly magazine. Unzer's other primary mentor, Krüger, had studied with the mechanist physician Friedrich Hoffmann in Halle and achieved success with his *Naturlehre* (1740-1750, 3 vols.).[54] Krüger's theories about the importance of the nerves in the *Physiologie* and his 1744 experiments on electrical impulses in the human body provided a context for Unzer's 1746 study of the nerves in the emotions, the *Neue Lehre von den Gemüthsbewegungen* (New Doctrine of the Emotions).[55] *The Neue Lehre* responds to Meier's philosophy in *Theoretische Lehre von den Gemüthsbewegungen überhaupt* (Theoretical Doctrine of the Emotions in General, 1744), which analyzed the use of the passions for self-advancement in the Baumgartian/Wolffian tradition.[56] As a physician, Unzer in the *Neue Lehre* shifts the discussion to include the body in the organiza-

[53] Johann August Unzer, *Der Arzt: eine medicinische Wochenschrift*, 12 v. in 6 vols. (Hamburg: G.C. Grunds Erben, 1759-1764).

[54] Krüger, Johann Gottlob. *Naturlehre* (Halle: Hemmerde, 1740); *Physiologie* (Halle: Hemmerde, 1743); *Pathologie* (Halle, Hemmerde, 1750).

[55] Susan Splinter, "Die Anfänge der 'Elektrifikation.' Untersuchungen zur Wirkung der Elektrizität auf den menschlichen Körper 1744 an der Friedrichs-Universität zu Halle," in *"Allerhand nützliche Versuche." Empirische Wissenskultur in Halle und Göttingen* (2009), pp.89-103. Johann August Unzer, *Neue Lehre von den Gemüthsbewegungen (*Halle: Hemmerde, 1746).

[56] Georg Friedrich Meier, *Theoretische Lehre von den Gemüthsbewegungen überhaupt* (Halle: Hemmerde, 1744).

tion of the emotions and psychology, and revises the prevailing system of humoral pathology to a system based on the nerves.

Unzer pursued a sustained goal to integrate more closely physiological and psychological knowledge. This was an aim of *Gedanken vom Einfluß der Seele in ihren Körper* (Thoughts on the Influence of the Soul on the Body, 1746) and *Philosophische Betrachtung des menschlichen Körpers überhaupt* (Philosophical Observations on the Human Body in General, 1750). His *Erste Gründe einer Physiologie der eigentlichen thierischen Natur thierischer Körper* (1771) is an innovative attempt to further synthesize psychological and physiological knowledge. The latter conveys his palpable frustration with the development of physiology. He avers that physiologists have explained the physical and mechanical means of human motion with little consideration of the "animal forces" ("tierische Kräfte"). For Unzer, such forces mean the faculties of the understanding and the will that coordinate sensation and muscle movement for nearly any bodily functions.[57] This includes involuntary processes such as the cardiac cycle and pulse regulation as well as irritability. He approaches the term historically. *Animal forces* as understood by Unzer for the period prior to 1740 primarily included ideas. Haller broadened the scope of these forces to include irritability, but Unzer judges this expanded account also to be incomplete (p.a4r). Seeking to overcome the confusion surrounding the concept (p.a4r), he asks:

- How does the imagination affect the body?
- Which conditions improve the transmission of nerve signals from the sense organs to the soul?
- Which conditions transform sensation into movement?
- Which principles of imagination, prophecy, reflection, desires, drives, aversions, emotions, and volitions produce bodily movement?
- And why did God create humans as a hybrid of body and soul (p.a4v)?

Since the animal forces work so closely with the cognitive faculties and the nervous system, Unzer refers the reader to the fourth volume of Haller's *Elementa physiologiae* for an anatomical description of the brain and the nerves (p.a7r).[58]

57 Johann August Unzer, *Erste Gründe einer Physiologie der eigentlichen thierischen Natur thierischer Körper* (Leipzig: Weidmann, 1771), p.a4r. This page annotation locates the front matter by signature (printer's grouping of pages), the page number, and the front (recto) or back (verso) side of the page.

58 Unzer references other standard works including Haller's *Primae lineae physiologiae* (1747, updated and translated as *Erster Umriß der Geschäfte des körperlichen Lebens*, Berlin: Haude & Spener, 1770), Haller's *Opera minora* (Lausanne: Grasset, 1763), Alexander Gottlieb Baumgarten's *Metaphysik* (trans. Georg Friedrich Meier, Halle: Hemmerde, 1766).

The common contemporary use of the term "tierisch" denotes being or acting like an animal or at the level of an animal. It can imply an inferior bearing of rationality and morality compared to humankind. In Johann Georg Walch's *Philosophisches Lexicon* (1740^2, 1776^4), "Thier" in a narrower sense referred to "Vieh" (livestock in general). The broader definition refers to an animated body, which can grow and reproduce and possesses faculties of sensation and motion (Walch, vol. 2, col. 1144). Zedler's *Universal-Lexicon* draws similar distinctions.[59]

Unzer criticizes the narrow perspectives on animal forces iterated by the representatives of many of the major physiological theories. He disapproves of Georg Ernst Stahl's physiological perspective for limiting such faculties to the imagination (p.a5r). Adherents of the mechanist model also fell short because they sought to explain even biological processes according to physical and mechanical principles. They thereby denied the influence of other animal forces or of the imagination on those biological principles (p.a5r-v). Moreover, some physiologists insisted on explaining biological processes solely on the basis of physiological structure. Functions they could not explain they relegated to the soul. Yet too often they assigned functions without understanding how they govern the human body (p.a5v). Unzer's *Erste Gründe* thus criticizes those like Stahl who took a narrow view of the influence of the soul on the body. He disapproves the position of the mechanist physicians who deny any influence of the mind on the physical body. He rejects physicians who lack intellectual engagement and explain only what they can mechanistically and what they cannot by writing if off to the soul (p.a5v).

The book title *Erste Gründe einer Physiologie der eigentlichen thierischen Natur thierischer Körper* and the book's contents focus on animal forces. Unzer's use of animal ("tierisch") does not infer the pejorative meaning of "beastly." Instead, he steers *tierisch* away from brutality and rawness as pejorative qualities that the individual should renounce. In Walch's *Philosophisches Lexikon*, an animal is defined in a broad sense as "ein belebter Körper, welcher sich nicht nur ernähret, wächset, und sein Geschlecht fortpflanzen kann; sondern auch mit der Kraft zu empfinden und sich zu bewegen begabet ist" (an animated body that not only nourishes itself, grows, and can reproduce its species, but also possesses the power of sensation and motion).[60] For Walch, an animal possesses mobility and the lower cognitive faculty of sensation. Haller argued in *Über den Ursprung des Übels* that sensation is a source of truth for

59 "Thier" in Johann Heinrich Zedler, ed., *Grosses vollständiges Universal-Lexicon aller Wissenschaften und Künste.* 64 vols (Leipzig, Halle: 1732-1750), vol. 43, col. 1333-1334.

60 Johann Georg Walch, and Justus Christian Hennings, *Philosophisches Lexicon.* 4 edn (Leipzig: Gleditschens Buchhandlung, 1775) II, column 1144.

human beings.[61] The faculties of sensation and motion also define *tierisch* for Unzer, who uses the word to describe the origin of faculties that cannot be explained merely through physiological or mechanical powers. Included among these other powers are the faculties of reason and volition. The sources of these "animal" faculties are "animal machines," the brain and the nerves (Unzer, *Erste Gründe*, pp.5-6, 11). In the human domain, *tierisch* refers to the mechanisms for cognitive and volitional faculties.

In *Erste Gründe* Unzer conjoins Haller's physiology with the psychology from the third portion of Baumgarten's *Metaphysik*.[62] The latter work emphasizes empirical psychology over rational psychology, and Unzer makes a similar choice. Baumgarten's empirical psychology defines the science of the soul based on experience, i.e. through observation and experiment, whereas rational psychology derives from a long series of syllogisms.[63] Unzer's *Erste Gründe* features three sections that examine the physiology of thought and volition in the mind, the physiology of thought and volition in the nerves, and animal nature generally over the human life cycle. The first two parts are relevant here. Unzer's impetus for the study is that medical scientists have not provided sufficient analysis of the functions and contributions of the brain and the nerves in human physiology. He argues that physicians must understand the physiology of the *tierische Kräfte* (which now include thoughts and desires, and the reciprocal interplay between the body and the soul) in order to develop a pathology of mental and neurological disease.[64]

Erste Gründe exhibits strong structural parallels to Baumgarten's description of empirical psychology. The headings from the first section, second chapter, third subchapter on the "animal" faculties of the nerves in their relationship to the "tierische Seelenkräfte des Gehirns" (animal powers of the soul in the brain) follow relatively closely Baumgarten's explication of em-

---

61 Haller, *Über den Ursprung*, section II, lines 81-82.

62 Unzer had begun working with Baumgarten's *Metaphysica* (Halle: Hemmerde, 1739) at least twenty years earlier. In a series of letters to his wife, the poet Johanna Charlotte Ziegler, Unzer had informally translated and commentated the text. Upon the urging of her uncle, Unzer's mentor Johann Gottlob Krüger, Ziegler published an edited version of the exchange of philosophical letters with her husband as *Grundriß einer Weltweißheit für das Frauenzimmer* (Halle: Hemmerde, 1751, 1767[2]). Georg Friedrich Meier prepared the German translation of Baumgarten's work as *Metaphysik*, trans. Georg Friedrich Meier (Halle: Hemmerde, 1766). Baumgarten's *Metaphysica* is a systematic work in the tradition of Wolff's *Vernünftige Gedancken von Gott, der Welt und der Seele des Menschen, auch alle Dinge überhaupt [Deutsche Metaphysik]* (Frankfurt am Main: Andrae, 1719). Baumgarten's psychology diverges from the Wolffian model by combining the empirical and rational psychology into one chapter and significantly curtailing the discussion of the rational side.

63 Baumgarten, *Metaphysik*, p.158.

64 Unzer, *Erste Gründe*, pp.a3v, a4r, a5v, a7v.

pirical psychology. The order of the discourse in this subchapter of the *Erste Gründe* moves from sensation and the senses to free will, paralleling the structure in Baumgarten's *Metaphysik*. Unzer's book covers in §31-33 external impressions and §34-54 sensations (cf. Baumgarten's *Metaphysik* part 3, chapter 1, section 3). Unzer then discusses in §55-64a the five senses (cf. Baumgarten, section 3) and in §67-72 the imagination (cf. Baumgarten, section 4). *Erste Gründe* analyzes the physiology of current and past sensations in §73-75 on sensual prophecy and forecasts (cf. Baumgarten, sections 8, 10) and reason in §76-79 (cf. Baumgarten, section 12). Paragraph 80 covers physical sensation as a source of enjoyment and discontentment (cf. Baumgarten, section 15). Unzer outlines the craving and loathing of the higher faculties of volition in §81-89 (cf. Baumgarten, section 18) and the drives and passions of the lower faculties of volition in §90-95 (cf. Baumgarten, section 17). Free will is the topic of §96 (cf. Baumgarten, section 20-21), and the relationship of the body to the soul is treated in §345-352 (cf. Baumgarten, section 22). Paragraphs 31-96 parallel Baumgarten's progression from a consideration of the senses to a discussion of free will, and in paragraphs 181-344 Unzer revisits how the faculties in §31-96 pair with the nerves to effect change in the physical structures of the body. In this context, the first half of *Erste Gründe* deals with the physiological origins of thought, emotion and volition. He accords special consideration to the function of the entire nervous system, including the brain. What Unzer offers, then, are insights into what we today call neurological functions. His intent is not to render a complete image of human physiology as Haller did. Unzer used Baumgarten's Wolffian-inspired psychology to organize existing empirical knowledge describing how sensations lead to thought and emotion along the pathway of the nerves into the brain.

To emphasize the embodied nature of human existence, Unzer repurposes traditional terminology.[65] Thus "animal" (*tierisch*) now connotes the biological origins of sensation and thought. Similarly, in contrast to Aristotle, he redefined the term "material ideas" employing the term to describe changes in the brain's white matter (*Gehirnmark*) when an idea takes form.

65 In his review of *Erste Gründe* in the *Allgemeine deutsche Bibliothek*, Philipp Gabriel Hensler (1733-1805) comments on the difficulty of Unzer's neologisms, encouraging Unzer to complete a pathology of animal nature so that physicians can better understand his new system. See "Rezension von Unzer, Johann August. *Erste Gründe einer Physiologie der eigentlichen thierischen Natur thierischer Körper*," *Allgemeine deutsche Bibliothek* (1772) 16:2, pp.502-513 (p.513). In a separate review attributed to Haller, the reviewer criticizes Unzer for ascribing cardiac irritability to sensibility, which the reviewer claims is experimentally unprovable. [Albrecht von Haller], "Rezension: *Erste Gründe einer Physiologie der eigentlichen thierischer Natur thierischer Körper*," *Göttingsche Anzeigen von gelehrten Sachen* (1772), vol.1, pp.548-553 (p.551).

Material ideas are, however, not simplified images of perceived objects, nor remnants of sensation in the brain, nor unknown qualities that enable analysis of the powers of the soul (pp.42-43, §25). Unzer underscores the influence of the biological systems when he concludes that since material ideas are movements in the brain, their effects on the body must be movements (p.113, §118).

Part two of *Erste Gründe* underscores the important role of the nerves in human life. Unzer acknowledges that the nerves frequently communicate with the soul, especially in the processes of sensation and motion.[66] In addition to normal functioning in combination with the brain, nerves continue to transmit signals, even if the path to the brain is temporarily impeded or permanently severed. Since nerve signals follow established pathways, they may be blocked on their way to the brain, but still arrive in other parts of the body. Unzer considers the possibility that nerves might transmit along pathways of motor and sensory nerves and partially evoke sensation, motion, or thought (pp.345-349). Thus, Unzer proposes that the nerves can execute some actions that normally occur in combination with the mind without contact to that control center. This observation underscores the importance of the body in conscious processes, including what Tuttle describes as self-actualization.

In accord with his earlier publications, Unzer again argues in *Erste Gründe* for the interaction of body and the soul. Similar to Baumgarten in his *Metaphysik,* he too argued for their reciprocal association. Unzer thus summarizes his findings from the first half of *Erste Gründe* and notes that the body transmits impressions that shape the mind. He also lists the ways that the mind and will transmit sensations that link to and influence the body. Though it is possible for individuals to influence each other in their physical and intellectual/moral being, the discussion of interaction aims not to describe intersubjectivity but the more intimate link of physical body and mind in one person. Unzer is critical of those who fail to acknowledge the full ramifications of the mind/body interaction: Stahlians, mechanist physicians, those who only acknowledge external sources of sensation, those who only acknowledge muscle movement that originates in the will, and those who only interpret the actions of the soul on the basis of physical and mechanical laws (pp.336-340).

Unzer employed different research methods than Haller did. This change contributed to differing trajectories of thought and conclusions regarding their research. Haller developed a scientific process based on controlled experimentation and verified results through repetition. Moreover, he pointed-

66 Unzer, *Erste Gründe*, p.343.

ly narrowed his physiological investigations to test animal nerve function and embryonic development of chicken eggs. His extensive knowledge of secondary literature shaped his experimental design and influenced his conclusions. By contrast, Unzer primarily synthesized existing knowledge, integrating his own observations with secondary literature in an innovative fashion. He favored observation over systematization of unsubstantiated hypotheses in the search for the laws of nature (p.b1v).

Haller's and Unzer's choice of scientific methods helped shape their views on the role of the brain and the nerves in human life. Haller sought new information or confirmation of a hypothesis primarily by experimental means, which required painstaking preparation and replication. This perspective shaped what one could ask, what one could learn and how quickly one could learn it. Haller's goal of verifiability led him to address research questions at a detailed level. His pursuit of accurate and repeatable experimental data in the laboratory helped him achieve greater certainty and reliability in his findings. In some instances, his narrow focus may have prevented him from asking broader questions such as the nature of the soul or how his three sources of muscle movement (contractility, irritability, and sensibility) relate. The studies of irritability and sensibility affirmed the contractility of muscle tissue on its own. They inadvertently underscored the fact that the body could react without participation of the soul or consciousness.[67]

Unzer's horizon was determined more by how what one has observed shapes what one can think. For Unzer, the realm of the observable was more encompassing than the realm of the experiment; therefore, one could engage more topics intellectually or emotionally. In hindsight the overlay of philosophical and physiological knowledge in *Erste Gründe* may seem an obvious step in challenging and, consequently, advancing scientific knowledge. His description of the role that the nerves and physiological structures play in the operation of the mental faculties did prove to be innovative. Moreover, the repurposing of traditional terminology such as *tierisch* to underscore the biological provenance of sensation and thought occasioned an important shift in thinking. A drawback of the semantic maneuvering is the confusion that it can prompt. A second new idea is the conception of thought as movement in the brain that subsequently prompts movement in the body. According to this model, then, thought is inherently embodied and material.

Neither Haller nor Unzer were drawn to the idea of a single, all-encompassing cause of an organism's movement, i.e. to a theory of *Lebenskraft*. Haller's three forces of contractility, irritability, and sensibility reflected the specificity of forces in the animal body but lacked the unity of a

---

67 Philipp Sarasin, *Reizbare Maschinen: eine Geschichte des Körpers 1765-1914*, (Frankfurt am Main: Suhrkamp, 2001), p.56.

single life force. In *Erster Umriß* Haller seemed merely to scoff at the idea of such a concept. Unzer also imagined that forces in the soul and in the nerves significantly contributed to human motion, willing, and thought. However, these forces were separate, and *Lebenskraft* was not a component of his neurological theory.

Haller and Unzer also shared the idea of science as process. This common ground was primarily methodological. For Haller, this meant that his vivisections of animals in the study of irritability or his studies of poultry embryology produced evidence that could be summarized in a hypothesis and could be tested to elicit new and more accurate knowledge. For Unzer, the development of hypotheses stemmed from several sources: observation, a synthesis of Haller's physiology, and adoption of Baumgarten's psychology. Via this path he was able to advance the field of neurology. Like contemporary scientists, the physicians conceived of their scientific contributions as stages in the development of knowledge and the means of inquiry.

In Tuttle's claim that human beings are "ontological centaurs," he argues that the thinking self takes precedence over the biological organism. Yet drawing on Haller's experimental results, Unzer makes a case that the activities of the soul are biologically determined to a greater degree than had previously been acknowledged. Indeed, if the process of thought includes movement in the brain, Tuttle's self-creating subjects achieve self-knowledge and a concept of history in their biological being. They cannot be imagined without it. Haller's lab experiments and Unzer's thought experiments underscore the role of the nerves, and thereby the necessary participation of the body, in the complex functioning of the brain.

Blood, Nerves, Resonance

Lebenskraft, the Body, and Will Power: The Life Force in German Musical Aesthetics

James Kennaway

Abstract: This chapter considers the development of theories that have proved influential in German aesthetics and medicine since the Enlightenment that link music to various notions of a vital principle or life force (*Lebenskraft*). First it considers the role of the eighteenth-century physiological conceptions of life force in the musical aesthetics of writers such as Johann Georg Sulzer, who portrayed music as a kind of nervous stimulant. It then looks at how this physiological model of *Lebenskraft* was recast around 1800 in terms of a mental or spiritual model of life force understood in terms of will power, creating the basis of a new metaphysics of music. Despite this shift away from medicine, it is striking that the body and medical concerns about sanity, health and masculinity continued to play an important role as music became incorporated into the debate on *Lebenskraft* and "diseases of the will," in the work of Friedrich Nietzsche, Max Nordau, and Eduard Hanslick. In conclusion, the chapter examines how this discussion of musical *Lebenskraft* and will power and its physical effects took on a more sinister aspect in twentieth-century German music criticism and aesthetics.

Medical Vitalism and Music in the Enlightenment

A belief in music's power to communicate directly with the self, unmediated on the whole by concepts or language, has long been central to its importance in Western culture. But this conception of music raises crucial questions about what kind of self is involved. Does music speak to a transcendental soul or a material brain? How might the two be related? Since the early modern period, thinking on the subject of music's impact has been profoundly affected by changing views of this issue, and indeed by the whole mind-body problem. In the wake of Descartes and his contemporaries such as the English physician Thomas Willis, many eighteenth-century observers argued that the self could be understood in essentially material, neurological terms, often in relation to a supposed animating *Lebenskraft* or life force. In that context, some observers argued that music too could be understood in essentially neurological terms, as a way to stimulate that life force via the nerves. As we shall see, thinking on music and the life force was radically reimagined during the early nineteenth century in terms not of the nervous system but in the context of a mental conception of *Lebenskraft* that focused on will power. Since today's spectacular rise of "neuromusic", or neurological explanations for how music works, is in some ways taking us back to this Enlightenment view of music as a form of nervous stimulation, a look back on this development could not be more timely.

The concept of life force, or *Lebenskraft,* has a long and complex history in Western medicine. In particular, it came to play a central role in medical debates on the nature of the nervous system and its relationship with the mind in the wake of the Scientific Revolution. Cartesian and Newtonian mechanical conceptions of the body provoked questions about the difference between living and non-living matter, which many physicians attempted to answer in terms of an unknown extra vital force, drawing on older Christian roots and ideas related to Aristotle's concept of the "vegetative soul." To explain the difference between living and dead tissue, a variety of vitalist and animist theories were put forward by Georg Ernst Stahl in Berlin and Théophile de Bordeu and Francois Boissier de Sauvages in Montpellier that posited various animating principles linking body and soul. These theories were at once a response to the mechanistic medicine of the period and a reaction against its materialist implications. The experimental work on the distinction between sensibility and irritability of tissue in 1750s by the Swiss polymath Albrecht von Haller that showed the ability of living tissue to respond to stimulation without the patient being conscious of it reignited a prolonged international discussion on the subject. That debate often drew on the work of the Edinburgh physician Robert Whytt on the sympathetic nervous system and reflexes, which he argued required an unconscious yet sentient principle for explanation. The idea of a life force as a kind of battery that could be drained or recharged with nervous stimulation that emerged from these debates proved enormously influential.

This Enlightenment medical debate on the self, the nerves and a life force came to have great relevance for thinking on music. Since the seventeenth century, the emerging fields of experimental acoustics and neurology had challenged older cosmological traditions of music theory, and by the eighteenth century, music was widely understood to be a matter of nervous stimulation. As Penelope Gouk has put it, after Newton "the paradigm became 'music is natural' and its effect on human nature became amenable to medical and scientific experiment."[1] Works on music theory from the first half of the eighteenth century, such as Jean-Philippe Rameau's *Traité de l'harmonie* (Treatise on Harmony, 1722) and Johann Mattheson's *Der vollkommene Capellmeister* (The Complete Conductor, 1739), generally assumed a mechanistic explanation of music's effects, and the rhetoric of the "harmony of the spheres" gradually became a mere poetic trope.[2] Drawing on

1 Penelope Gouk, "Raising Spirits and Restoring Souls: Early Modern Medical Explanations for Music's Effects," in *Hearing Cultures: Essays on Sound, Listening, and Modernity* ed. By Veit Erlmann (Oxford: Berg, 2004), pp.87-105, (p.104).

2 Johann Mattheson, *Der vollkommene Capellmeister* (Hamburg: 1739), p.16, Jean-Philippe Rameau, *Traité de l'harmonie* (Paris: Ballard, 1722).

John Locke's sensationalist philosophy, Mattheson explicitly rejected the cosmological view of music as a question of pure arithmetical order in favor of a model based on the analogy with rhetoric and the physical character of music.[3] The emphasis on the physical impact of sound waves on the ear meant that the Enlightenment debate on the nerves and the life force had direct relevance to understandings of how it functioned.

The role of the nerves in music became more explicit later in the eighteenth century in the context of the cult of sensibility, which, as George Rousseau has argued, was based on a view of the nervous system.[4] As far as music is concerned, instead of the baroque *Affektenlehre* (theory of emotions) that posited abstract correspondences of feeling and aspects of music, a more physical and subjective idea of *Empfindung* (sensibility) came to dominate the late eighteenth-century aesthetics of the likes of C.P.E. Bach. Aestheticians took a keen interest in the potential of what Johann Gottfried Herder labelled the "Physiologie der menschlichen Seele" (physiology of the human soul) to explain music as a form of stimulation.[5] In the work of people such as the Swiss aesthetician Johann Georg Sulzer, the music theorist and first biographer of Johann Sebastian Bach Johann Nikolas Forkel, the writer Johann Jacob Engel and the physicians Johann Christian Reil and Johann Gottlob Krüger, musical aesthetics was put into the context of a fully-fledged stimulation theory of what might be called psycho-acoustics.

In his *Allgemeine Theorie der schönen Künste* (General Theory of the Fine Arts, 1771-4), Johann Georg Sulzer depicted musical aesthetics as a matter of "Stöße" (blows) delivered to the body, which "ihre Wirkung auf das ganze System aller Nerven verbreiten" (spread their effect throughout the nervous system).[6] The direct impact of music on the sympathetic nervous system, its power to make the body resonate, was central. The effect on the mind was of secondary importance. As Sulzer put it, "Es brauchte weder Überlegung, noch lange Erfahrung, um diese Kraft in dem Ton zu entdekken. Der unachtsame Mensch erfährt sie" (It did not require much consideration or much experience to discover the power of music. The inattentive

---

3 Johann Mattheson, *Der vollkommene Capellmeister* (Hamburg: Herold, 1739), foreword, section 6.

4 See George Rousseau, *Nervous Acts: Essays on Literature, Culture and Sensibility* (Basingstoke: Palgrave, 2004), Anne C. Vila, *Enlightenment and Pathology: Sensibility in the Literature and Medicine of Eighteenth-Century France* (Baltimore: The Johns Hopkins UP, 1998).

5 Johann Gottfried Herder, *Werke*, ed. by Wolfgang Pross (Vienna: Wissenschaftliche Buchgesellschaft, 1987), 2: p.146. Unless otherwise noted, all translations are my own.

6 Johann Georg Sulzer, *Allgemeine Theorie der schönen Künste* (Leipzig: Weidmann, 1793) III, p.422.

person experiences it).[7] He defended the prestige of music on the grounds of its quasi-automatic nervous power, writing that, "Unter allen Künsten aber scheinet die Musik […] die größte Kraft zu haben, weil sie das körperliche Gefühl und das System der Nerven am starksten angreift" (Of all the arts, music has the greatest power because it attacks physical sensation and the nervous system).[8] This argument about the impact of music drew on the contemporary debate on the nature of the sympathetic nervous system. Such was the direct impact of music on the nervous system for Sulzer, for instance, that he suggested that certain kinds of music make us want to dance because the body's motor nerves are directly stimulated.

Drawing on Sulzer, Johann Nikolaus Forkel, in his 1788 *Allgemeine Geschichte der Musik* (General History of Music) set out an explanation of music's effects that also focused on the nerves. Musicality in general was a question of having nerves, he suggested.[9] He argued that music had a direct effect on the nerves akin to the power of the mind itself, and that this effect is enough to explain its emotional impact, without particular recourse to the imagination. He asserted:

> Leidenschaftlichen Vorstellungen der Seele sind nemlich mit gewissen Bewegungen im Nervensystem, oder in den feineren Theilen des Körpers, welche man Lebensgeister nennen kann, unzertrennlich verbunden, und werden durch Wahrnehmung dieser Bewegung unterhalten und verstärkt. Diese entsprechenden Nervenerschütterungen entstehen im Körper, wenn vorher in der Seele eine leidenschaftliche Vorstellung erweckt war, so wie umgekehrt in der Seele die leidenschaftlichen Vorstellungen entstehen, wenn vorher im Körper die verwandten Erschütterungen erregt worden sind. Die Wirkung ist gegenseitig. Eben der Weg, der aus der Seele in den Körper führt, führt zurück in die Seele. Da nun die Erschütterungen des Nervensystems durch nichts so mächtig bewirkt werden können, als durch Töne, so erklärt sich aus dem gegenseitigen Verhältnis der Luft- und Nerven-Bewegungen die Kraft und Gewalt hinlänglich, welche schon einzelne Töne auf das Herz des Menschen haben können.
>
> (The passionate impressions of the soul are thus inseparably linked to certain movements in the nervous system or in the finer parts of the body that one can call the animal spirits, and are maintained and strengthened by the perception of this movement. The corresponding nervous convulsions arise in the body when a passionate impression has previously been awoken in the body, just as the other way round, when passionate impressions are created when related convulsions have been stimulated in the body. The effect is reciprocal. The path that leads from the soul to the body, leads back to the soul. Since the convulsions of

7 Sulzer, *Allgemeine Theorie der schönen Künste*, III. p.422.

8 Sulzer, *Allgemeine Theorie der schönen Künste* III. pp.225-26.

9 Johann Nikolas Forkel, *Allgemeine Geschichte der Musik*, 2 vols (Leipzig: Schwickert, 1788), II, p.8.

> the nervous system are affected by nothing more than by music, the reciprocal relationship of the movements in the air and in the nerves are sufficient to explain the power and force that even individual sounds can have on people's hearts.)[10]

Views like this seemed to provide a justification for taking music seriously that did not rely on text or older cosmological theories, making it a symbol of vigorous physical *Lebenskraft* rather than mathematical rationality. It also seemed to offer a way out of an aesthetics dominated by mimetic theory and rhetoric, linking music to the body and making the sensitive soul more fit for the *Sturm und Drang* era.

The idea of music as a stimulant placed it directly in the debate on vitalism and *Lebenskraft*, leading to both a wave of speculative music therapy and to a sudden and extensive medical critique of the art. Music, it was argued, could boost physical *Lebenskraft* directly via the nerves, in a way more like drugs or electricity than, say, a novel, and could thereby lead either to health or sickness. In contrast to works on music and medicine from only a generation before, such as Michael Ernst Ettmüller's *Musurgia universalis* (Universal Musurgia, 1714) or Adam Brendel's *Dissertationem medicam de curatione morborum per carmina et cantus musicos* (Medical Dissertation on the Curing of Disease Through Song and Music, 1706) that still spoke in terms of cosmic harmony, Ernst Anton Nicolai's 1745 *Die Verbindung der Musik mit der Arzneygelahrheit* (The Nexus of Music and Medical Knowlege) explicitly dealt with music as a source of stimulation. In a model of the body that drew on his Halle colleague Friedrich Hoffmann that focused on the fibres (i.e. nerves, muscles, etc.), Nicolai described the nerves as tightened strings to be struck by sounds.[11] Metaphors of the nerves as musical instruments go back to Antiquity, but Nicolai and many of his contemporaries meant it literally.[12] Musical stimulation could restore health via the nerves.

By the 1790s thinking on *Lebenskraft* was incorporating more speculative ideas linked to *Naturphilosophie,* linking it to Johann Friedrich Blumenbach's teleological principle of *Bildungstrieb* (developmental drive), for instance. More importantly, much of the medical discussion about music's effect on the nerves in relation to a physiological life force drew on the theories and

10 Johann Nikolas Forkel, *Allgemeine Geschichte der Musik,* 2 vols. (Leipzig: Schwickert, 1788) II, p.10.

11 Ernst Anton Nicolai, *Die Verbindung der Musik mit der Arzneygelahrheit* (Halle: Carl Hermann Hemmerde, 1745), pp.viii-ix.

12 Nicolai, *Die Verbindung der Musik mit der Arzneygelahrheit*, p.xiii.

terminology of the controversial Scottish physician John Brown.[13] Brown mixed up Haller's categories of irritability and sensibility to create a general concept of physical stimulation, the level of which determined all sickness and health and amounted to a universal life force, which could be reinforced, he said, by opium and whiskey. Brown's theory and its variants provided an ideal simplified model of *Lebenskraft* that struck a chord with many contemporaries, including many well beyond medicine. The overlap between his ideas and other contemporary theories such as galvanism, *Naturphilosophie* and vitalism led to a wave of popularity for "Brunonianism," particularly in Germany and Italy. Such was the impact, at least in the short term, of Brunonianism's combination of vitalism with a material conception of the soul as a matter of stimulation, that the physician Andreas Röschlaub suggested that Brown's work was one of the most important tendencies of his time along with the French Revolution and the work of Fichte and Goethe.[14]

In this context, music was often recruited as a tonic for the psychophysical *Lebenskraft*, to be measured on a Brunonian "barometer." For instance, Peter Lichtenthal's 1807 *Der musikalische Arzt* (*The Musical Doctor*) discussed at length the need to make musical stimulation proportionate to the physical capacity of the listener in Brunonian terms.[15] In order to stimulate a patient with a "Stumpfheit der Lebenskräfte" (stupor of life force), loud music should be prescribed, he suggested.[16] Likewise, *Erläuterungen der Erregungstheorie* (Comments on Stimulation Theory, 1803) by the Viennese physician Joseph Frank set out a Brunonian view of music as a stimulant for the life force. Indeed, he repeatedly used the terms "Erregung" (stimulation) and *Lebenskraft* as being analogous.[17] Drawing on thinking about the sympathetic nervous system, Frank was convinced that music would boost a patient's life force via the nerves, stating: "Die Wirkung des Schalls schränkt sich nicht auf das Gehörorgan ein, sondern breitet sich über den ganzen Organismus" (the effect of sound is not limited to the organ of hearing, but spreads over the whole organism).[18] For Frank, a physician employing music for medical

---

13 John Brown, *Elements of Medicine* (Philadelphia: Dobson, 1814; 1st Latin edn, 1780). See W. F. Bynum and Roy Porter, eds., *Brunonianism in Britain and Europe* (London: Wellcome Institute, 1988).

14 Andreas Röschlaub, *John Browns Leben beschrieben von dessen Sohne Dr. William Cullen Brown* (Frankfurt/Main: Andreä 1807), p.xx.

15 Peter Lichtenthal, *Der musikalische Arzt* (Vienna: Christian Friedrich Wappler und Beck, 1807), pp.168-172.

16 Lichtenthal, *Der musikalische Arzt*, p.168-172.

17 Joseph Frank, *Erläuterungen der Erregungstheorie* (Heilbronn: Johann Daniel Class, 1803), p.194.

18 Frank, *Erläuterungen der Erregungstheorie*, p.339.

treatment must remember that music functions in accordance with the receptivity of the patient's *Lebenskraft*. It was obvious, he suggested, that the life force of women and members of the *bon ton* would be more sensitive.[19]

Electricity was another common context for discussing the energizing potential of music on the body and *Lebenskraft*, especially among German Romantics, who often appeared to regard life force as analogous to the electricity stored in a battery. Until Emil du Bois-Reymond conclusively demonstrated the electrical character of the nerves in the mid-nineteenth century, a wide variety of theories about how the nervous system operated jockeyed for influence.[20] However, especially after Luigi Galvani's work in the 1780s, electricity increasingly became seen as a semi-mystical life force with hidden influence on many spheres of life. The invisible apparently semi-magical effects of both music and electricity were regularly related to the idea of *Lebenskraft,* especially among those interested in *Naturphilosophie*, and not only at the level of metaphor.[21] The rhetoric of the "galvanic fluid of harmony" became commonplace and was often associated with supposed medical dangers and benefits.[22] The Brunonian physician Wilhelm Schlesinger, for example, described music as "life electricity."[23] Likewise, the influential Saxon physician and speculative philosopher Gotthilf Heinrich von Schubert thought of all sound as a form of electricity. Reflecting the extent to which conceptions of music as a form of stimulation for the listener's *Lebenskraft* had become mainstream by 1800, Novalis included "acoustic" in lists of forms of galvanism,. However, as we shall see, this attempt to reconcile the physical and the mental in a quasi-mystical conception of physiological *Lebenskraft* was challenged by a recasting of both life force and music in terms of a transcendental model of the mind.

Music, *Lebenskraft* and Will Power in German Idealist Aesthetics

In the early years of the nineteenth century, this medical conception of music as a matter of physio-psychological life force was turned upside down and recast in terms of a more mental idea of a *Lebenskraft* of will power, selfhood, and imagination. At a time when, as Thomas Dixon has argued, emotions were reconfigured as essentially mental states, a sense of music as

19 Frank, *Erläuterungen der Erregungstheorie*, pp.140, 340, 198.

20 Emil Du Bois-Reymond, *Untersuchungen über thierische Electricität,* 2 vols (Berlin: Reimer, 1848-9).

21 James Kennaway, 'Stimulating Music: The Pleasures and Dangers of 'Electric Music' 1750-1900', *Configurations* (2011), 191-211.

22 James Johnson, *The Economy of Health* (London: S. Highley, 1837), pp.32-4.

23 Guglielmo Schlesinger, *Dell' influenz fisica e morale della music sull' uomo* (Pavia: Fusi e comp., 1842), p.6.

the expression of the subject was an important part of an immensely influential new metaphysics of music.[24] Rejecting the passive, physiological model of listening, many commentators instead looked to an active, irrational, semi-conscious life force, akin to the Christian soul, which comprehended music not with the nerves, but in an abstract, mathematical and quasi-disembodied way. The idea that music was a matter of a metaphysical kind of *Lebenskraft* expressing the spiritual and mental life force of the composer and demanding the same from the serious listener came to dominate understandings of the subject. For many late eighteenth-century observers, *all* music had been a matter of stimulating the nerves, but in the Idealist paradigm, the language of nerves was increasingly reserved for music regarded as sensual, effeminate and sickly.

In a period when writers such as Ludwig Tieck, Wilhelm Heinrich Wackenroder and E.T.A. Hoffmann were passionately arguing that music was the supreme art form, the model of music's effects based on nervous stimulation and a physiological view of *Lebenskraft* proved inadequate to explain why music was the sublime.[25] If music was to be understood in entirely sensual and physical terms, it seemed to be in danger of being reduced to the level of cookery or perfume. Jean-Jacques Rousseau prefigured the Romantic rejection of this model, arguing that the physiological approach and the way it had been used as a justification for purely instrumental music had led to a moral crisis for music. As he put it:

> Voilà comment le chant devint, par degrés, un art entièrement séparé de la parole dont il tire son origine; comment les harmoniques des sons firent oublier les inflexions de la voix ; et comment enfin, bornée à l'effet purement physique du concours des vibrations, la musique se trouva privée des effets moraux qu'elle avait produits quand elle était doublement la voix de la nature.[26]

> (This is how song has by degrees become wholly separated from words, its source of origin. This is how harmonics of sound have led us to forget vocal inflection and how, its effect being limited to the purely physical concurrence of vibrations, music has found itself shorn of those moral effects that it had produced when it was in two senses the voice of nature.)[27]

---

24 Thomas Dixon, *From Passions to Emotions* (Cambridge: Cambridge UP, 2006).

25 Wilhelm Heinrich Wackenroder and Ludwig Tieck, *Phantasien über die Kunst für Freunde der Kunst* (Hamburg: Perthes, 1799).

26 Jean-Jacques Rousseau, *Essai sur l'origine des langues* (Paris: Berlin, 1817), 65.

27 Translation of Jean-Jacques Rousseau, 'Essai sur l'origine des langues', cited in *Music and Aesthetics in the Eighteenth and Early Nineteenth Centuries* ed. by Peter le Huray and James Day (Cambridge: Cambridge UP, 1981), pp.66-82, (p.82).

If music was mere tingling of nerves, then why should it be taken seriously? Critiques of the physiological view of music's effects went back to studies like Johann Joseph Kausch's 1782 *Psychologische Abhandlung über den Einfluß der Töne* (Psychological Treatise on the Influence of Tone) but it was Immanuel Kant who provided much of the basis for music to be understood in terms of a mental model of life force. Although Kant himself tended to view music as stimulation, saying that it "merely plays with sensations," the way that he put the quasi-disembodied transcendental subject rather than the nerves at the heart of aesthetics was to have a crucial role on future understandings of music. [28] Already around 1800 the essayist Christian Friedrich Michaelis was beginning to incorporate a sublime view of music into Kant's Idealist aesthetic in a series of articles and books.[29] For Michaelis, music could not be reduced to mere "Erschütterungen der Luft und der Nerven"" (vibrations of air and nerves).[30] Michaelis sharply distinguished attentive musical listening from simple nervous stimulation:

> Diese Töne afficieren zunächst die Sinnlichkeit. Das Nervensystem wird von ihnen mehr oder weniger gereizt [...]. Dies ist ihre bloß mechanische oder physikalische Wirkung. Aber durch Einbildungskraft und inneren Sinn werden die Gehörsempfindungen auch als innere Veränderungen aufgefaßt.
>
> (These sounds initially affect the senses. The nervous system is stimulated by them to a greater or lesser extent [...]. This is only its mechanical or physical impact. But through imagination and the internal mind these aural impressions are grasped as internal transformations).[31]

True listening was an active attempt mentally to grasp the whole form of the piece of music, not just passively receiving stimulation.

Thereafter this conception of musical emotions as essentially mental was often associated with a similarly mental conception of life force. Crucially, the self that this music addressed was not understood as a passive abstract receptacle of sound: It was a dynamic, often explicitly masculine life force in its own right. Commentators often borrowed the older language of physiological *Lebenskraft* and also that of will power to express the relationship of

---

28 Immanuel Kant, *Critique of Judgement* (New York: Cosimo, 2007), p.130.

29 Christian Friedrich Michaelis, 'Ein Versuch, das innere Wesen der Tonkunst zu entwickeln' in *Allgemeine musikalische Zeitung* 8 (1806), 673-683, 691-696; Christian Friedrich Michaelis, 'Über den Rang der Tonkunst unter den schönen Künsten' in *Allgemeine musikalische Zeitung* 2 (1799), 183-6.

30 Christian Friedrich Michaelis, *Über den Geist der Tonkunst, mit Rucksicht auf Kants Kritik der aesthetischen Urtheilskraft* (Leipzig, Schaefer, 1800), p.4.

31 Christian Friedrich Michaelis, 'Nachtrag zu den Ideen über die ästhetische Natur der Musik,' *Eunomia* 1.1 (1801), 343-348 (pp.344-345).

the imagination with music. One of the founding fathers of musicology, the influential Berlin critic Adolf Bernhard Marx, often emphasized the role of "ernstliche[n] Wille[ns]" and "ausdauernde[r] Kraft" (serious will, persistent strength) in music.[32] He was explicit about the necessity of "Vergeistigung" (making mental or spiritual), as opposed to physical.[33] In particular this was related to rhythm, the essence of which was "Selbstbestimmung" and "Wille" (self-determination, will).[34] Beethoven, the vital exemplar of titanic musical life force, featured prominently in this context. Marx described the predominance of personal "Nervenkraft, Geist und Wille" ("nerve power, spirit, and will power") in the composer's work, reflecting the mixture of physiological and metaphysical language used in the period.[35]

In particular, large-scale musical form came to be understood as a reflection of the composer's own healthy and potent *Lebenskraft.* It became commonly understood that music represented will or subjectivity in terms of "structural listening," that is, the very Western idea of music focusing on harmonic and formal structures.[36] Adolf Bernhard Marx was clear about the dominant role of will in composition, insisting: "Der Geist baut die Formen, oder er findet und beseelt sie, und modelt sie nach seinem Bedürfnis und Willen" (The spirit constructs the forms, or finds and animates them, and models them according to its needs and will).[37] The dichotomy between healthy music related to structural listening and music that stimulated the nerves created a surprisingly extensive debate throughout the nineteenth century about the medical dangers of music that might undermine will power and leave a disordered mind.

The high point of this trend towards associating music's effects with will power and *Lebenskraft* (which Marx regarded as synonymous) is reached in the work of Arthur Schopenhauer. His view of the will proved extremely influential in placing not just the mind or the imagination but an existential life force at the heart of thinking about the nature of music. In his *Die Welt als Wille und Vorstellung* (The World as Will and Representation, 1819-44), he portrayed the will as the central fact of existence, the thing-in-itself, whose essence is desire, which leads to suffering. Sensory experience was downplayed as mere "Vorstellung" (representation)." However, music is a crucial

---

32 Adolf Bernhard Marx, *Die Musik des neunzehnten Jahrhunderts* (Leipzig: Breitkopf und Härtel, 1855), pp.233, 255, 282.

33 Marx, *Die Musik des neunzehnten Jahrhunderts,* p.297; Adolf Bernhard Marx, *Ludwig van Beethoven,* 2 vols. (Berlin: Otto Janke, 1901), I, pp.286-7.

34 Marx, *Die Musik des neunzehnten Jahrhunderts,* p.356.

35 Marx, *Ludwig van Beethoven,* I, p.35.

36 Rose Rosengard Subotnik, *Deconstructive Variations: Music and Reason in Western Society* (Minneapolis: U of Minnesota Press, 1996), pp.148-176.

37 Marx, *Beethoven,* I, p.107.

exception to this, relating not to the physical world, but directly to the will. He argued: "Die Musik ist also keineswegs, gleich den andern Künsten, das Abbild der Ideen; sondern *Abbild des Willens selbst*, dessen Objektität auch die Ideen sind (Music is thus by no means like the other arts, the copy of the Ideas, but the *copy of the will itself*, whose objectivity the Ideas are).[38] Music is thus the greatest of arts because it relates directly to the will in the sense of an all-embracing vital principle. As late as Kant, music's lack of mimetic power meant that its prestige was in question, but the Romantics, and Schopenhauer in particular, led to a revival of musical metaphysics, exchanging the music of the spheres for a view of music as a vibrant manifestation of mental *Lebenskraft*.[39] The physiological model that had laid the groundwork for this development by making music concept-free, was not disproved so much as turned into the mundane basis for a transcendental model of aesthetics.

However, even this association of music with the "will itself" did not remove the body and the eighteenth-century physiological conception of life force from the equation.[40] For Schopenhauer, the body was the will experienced externally, so to speak, and the will was the body experienced internally. And in *Die Welt als Wille und Vorstellung* he made the unity of body and will very clear, affirming: "Eine höhere Stufe der Objektifikation des Willens ist der Organismus. In ihm wirken physikalische und chemische Kräfte, aber was diese zusammenhält und lenkt, ist die Lebenskraft, welche ihre Wirkung modifiziert. Diese Lebenskraft ist nun an sich Wille (A higher level of the objectification of the will is the organism. Physical and chemical forces have their effect, but what holds it together and directs it all is *Lebenskraft*, which modifies those effects. This *Lebenskraft* is the will in itself).[41] Because of this, Schopenhauer took perhaps a surprising amount of interest in medical and scientific views of *Lebenskraft* and the will, drawing much from the physician Joachim Dietrich Brandis's *Versuch über die Lebenskraft* (Essay on the Life Force) in his *Über den Willen in der Natur* (On the Will in Nature).[42]

---

38 Arthur Schopenhauer, *Die Welt als Wille und Vorstellung*, 2 vols (Munich: Georg Müller, 1902), I, p.310.

39 See Edward Lippman, *A History of Western Musical Aesthetics* (Lincoln, NB: University of Nebraska Press, 1992), pp.59-82.

40 John Atwell, *Schopenhauer on the Character of the World: The Metaphysics of the Will* (Berkeley: University of California Press, 1995), p.81.

41 Arthur Schopenhauer, 'Über den Willen in der Natur,' in *Kleinere Schriften I*, ed. by Arthur Hübscher (Zurich: Diogenes, 1977), 181-340 (p.209). Joachim Dietrich Brandis, *Versuch über die Lebenskraft* (Hanover: Hahn, 1795).

42 Arthur Schopenhauer, *On the Will in Nature*, trans by E.F.J. Payne (New York: Berg, 1992), Joachim Dietrich Brandis, *Versuch über die Lebenskraft* (Hanover: Hahn, 1795).

As the complex relationship between the body and will power in Schopenhauer reflects, the move of musical *Lebenskraft* to the transcendent sphere was never entirely complete. In what could be seen as a case of the return of the repressed, the body and issues of health were never far away. Theories of music's effects may have generally turned away from direct neurological explanations, but they still often drew heavily on embodied conceptions of the inner self, the will and the imagination. Although German critics such as E.T.A. Hoffmann and Adolf Bernhard Marx did sometimes describe their own extreme physical responses, those effects were mediated by the mind. They reserved their fiercest barbs for the low ambitions of Italian operas that merely "titillate the ear."[43] Crucially, the medical dimension remained, as the self that was at the heart of new mental conceptions of *Lebenskraft* increasingly itself became the subject of medical scrutiny.

Music and Will Power 1850-1900

As the life force that music supposedly addressed or expressed was increasingly regarded in mental terms, debates on aesthetics became entwined with emerging medical concerns about the nature of the self. While it is true that the "passions of the mind" were one of the six non-naturals that had been the basis of medical regimen since antiquity, interest in selfhood among physicians gained in importance in the emerging field of psychological medicine that challenged the church's traditional role as spiritual caretaker (i.e., caring for the self). From the 1780s, doctors took an increasing interest in the will and its maladies, as evidenced in the pioneering *Magazin zur Erfahrungsseelenkunde* (Magazine for Experiential Psychology, 1787-88).[44] In the following generation, physicians such as Ernst von Feuchtersleben developed a variety of theories of disorders of the will, understood as a fundamental vital principle in mental more than physiological terms.[45] These disorders, which dealt primarily with less extreme mental conditions, were soon systematized in early French psychiatry under the rubric of *manie sans délire*. Although the vitalism debate continued within physiology with figures such as Johannes Peter Müller, the focus in medicine shifted to mental models of will power and self-control, which became crucial markers of health and sanity.

The highpoint of anxiety about the effects of music on *Lebenskraft* and will power was reached in the second half of the nineteenth century. Discussion of music's dangers managed to combine the pathologized sensualism

43 *E.T.A. Hoffmann's Musical Writings*, ed. by David Charlton, trans. by Martyn Clarke (Cambridge: Cambridge UP, 2004), p.441.

44 German Berrios, *The History of Mental Symptoms* (Cambridge: Cambridge UP), p.357.

45 Ernst von Feuchtersleben, *Zur Diätik der Seele* (Vienna: Armbruster, 1838).

inherited from the eighteenth century with fears about will power, reflecting broader developments in medicine. By the 1850s the psychological medicine of the period around 1800 came to be regarded as insufficiently scientific, and a more somatic approach to mental illness became dominant, but it is striking that even among those who aimed to recast madness as mere brain disease, emphasis on strength of will persisted. For instance, German psychiatrist Wilhelm Griesinger defined mental health at mid-century as self-control. His interest in philosophical conceptions of the will extended to quoting Schopenhauer in his *Pathologie und Therapie der psychischen Krankheiten* (Pathology and Therapy of Mental Diseases, 1845).[46] Will power and its consequences for health were reframed in many ways in the context of new ideas, but still maintained a complex relationship with both the nerves and the rational mind. In the decades around 1900 a whole genre emerged, popularising medical theories of the necessity of building up strength of will, particularly among schoolchildren.[47]

Discussions of music reflected the medical concern about will power and enervated *Lebenskraft*. For example, it has an important role in the aesthetics of music set out by the critic Eduard Hanslick in his *Vom Musikalisch-Schönen* (On the Musically Beautiful, 1854) widely regarded as the most influential nineteenth-century work on musical aesthetics.[48] In it he sharply rejected the popular aesthetic of feeling and expression, instead proposing an aesthetic that suggested that "Tönend bewegte Formen sind einzig und allein Inhalt der Musik" (Sonorous forms in motion are the single and sole content of music).[49] Like Michaelis, he argued that real music should appeal only to a transcendental self, not just the ears. Although he was at pains to exclude the physical effect of music from consideration, he strikingly turned to explicitly medical rhetoric, setting out a division between privileged "aesthetic" appreciation of form and "pathological" listening that merely stimulated the nerves.[50] Music should relate to the will in terms of "freier Selbstbestimmung" and "befördernder Nervenreize" (free self-determination, nervous stimulation), he wrote.[51] Hanslick thus matched a mentalist understanding of serious listening to a neuro-stimulation model of the dangers of dubious music.

---

46 Wilhelm Griesinger, *Pathologie und Therapie der psychischen Krankheiten* (Stuttgart: A. Krabbe, 1861), p.43.

47 See Michael Cowan, *Cult of the Will: Nervousness and German Modernity* (University Park, PA: The Pennsylvania State UP, 2008).

48 Eduard Hanslick, *Vom Musikalisch-Schönen* (Leipzig: Breitkopf und Härtel, 1922).

49 Hanslick, *Vom Musikalisch-Schönen,* p.59

50 Hanslick, *Vom Musikalisch-Schönen,* pp.120-140.

51 Hanslick, *Vom Musikalisch-Schönen,* p.129.

Nonetheless, the role of will power and *Lebenskraft* in thinking about music was never more fully developed than in the work of Friedrich Nietzsche. Although his model of the will owed much to Schopenhauer, he radically inverted its meaning to create a very different positive conception of the "will to power." His view of the will is a life force in every sense, affirming life and a profoundly physical form of health, a kind of vital principle that had elements in common with both Enlightenment and Romantic conceptions of *Lebenskraft.* To give, at least rhetorically, a physiological basis in weak nerves and bad heredity for his critique of degenerate will power in music, he drew extensively on contemporary psychiatry. Although some of his terminology of music and strength of will is superficially similar to that of Hanslick, full of allusions to nervous over-stimulation, Nietzsche's discussion of music, *Lebenskraft,* and sickness is in fact very different. Whereas Hanslick has a basically mentalist conception of music and discussed nervous stimulation only in pathological terms, Nietzsche sometimes wrote approvingly of music's physical effects, but had a broader agenda about music (especially Wagner's), the nerves and will power.

For Nietzsche, the nervous system was the means by which Wagner damaged the *Lebenskraft* and will power of his listeners. Nietzsche gave extensive medical arguments for his view that Wagner's music was both a symptom and a cause of pathologically weak will power.[52] In the composer, himself there was a kind of "Krankheit" (disease), he claimed, "[u]nsre Ärzte und Physiologen haben in Wagner ihren interessantesten Fall" (our doctors and physiologists have in Wagner their most interesting case).[53] The effect of his music was like the "fortgesezter Gebrauch von Alkohol" (sustained consumption of alcohol) or an "Infektion" (infection).[54] Although Nietzsche begins *Der Fall Wagner* with a paean to the sensuality of *Carmen*, he condemned Wagner's appeal to the nerves. Of the Wagner's instruments, he wrote: "Einige von ihnen überreden selbst noch die Eingeweide [...] andre bezaubern das Rückenmark" (some convince the bowels [. . .] others enchant the spine).[55] The real problem with these physiological effects was how they undermine the will to power, the *Lebenskraft,* leading to the "Disgregation des Willens" (disintegration of the will).[56] Without a strong will, listeners were left as degenerate moderns, incapable of living a real life. Drawing on

52 Kennaway, *Bad Vibrations*, pp.84-141.

53 Friedrich Nietzsche, "Der Fall Wagner," in *Nietzsche Kritische Gesamtausgabe* ed. by Giorgio Colli and Mazzino Montinari, 30 vols. (Berlin: De Gruyter, 1967ff.), VI.3, pp.15, 17.

54 Nietzsche, "Der Fall Wagner," VI.3, 38.

55 Nietzsche, "Der Fall Wagner," VI.3, p.21.

56 Nietzsche, "Der Fall Wagner," VI.3, p.21.

the work of Charles Féré, Nietzsche used metaphors of hypnosis about Wagner many times to express the sense of that music could be an insidious threat to autonomy – a fear of what would be later be called brainwashing.[57]

Drawing on similar theories of bad heredity and psychiatric views of will power, the physician and critic Max Nordau, who penned the bestseller *Entartung* (Degeneration) in the early 1890s, also linked the wrong kind of music and the wrong kind of listening to disorders of the will. Nordau set out a view of music in many ways dominated by the idea of will power as the basis for all real composition and listening. Indeed, he portrays all the cultural pathologies he encyclopaedically lists as sub-categories of "[d]ie Krankheit des Jahrhunderts" (the malady of the century): "Willensschwäche" (weakness of will).[58] Wagner, as the chief villain in Nordau's litany of degenerate artists, displays this acquired and hereditary weakness in various forms of the inability to pay attention: that is, a lack of strength of will. This weakness accounts for Wagner's supposed inability to create large-scale forms. Nordau depicts the leitmotifs that in many ways dominate the musical form of the composer's mature work as "die Form, in welcher die Unfähigkeit zur Aufmerksamkeit sich in der Musik äußert" (the form that the incapacity for attention takes in music."[59] He describes Wagner's "unendliche Melodie" (endless melody), his use of dramatic recitative, as a feeble throwback to a more primitive evolutionary era caused by a lack of attention reducing listeners to the level of the mollusc, at a crude sensory level and entirely without will.

Wagner's supposed lack of musical will power was often connected to his sexuality, with insinuations made about his relationships with Nietzsche and Ludwig II. More broadly, the association of music and a medicalized conception of will power linked the idea of *Lebenskraft* in music to masculinity and to anxieties about the increasingly medicalized concept of homosexuality. Based on the combination of the Idealist disinterested appreciation of form, many critics portrayed musical form as masculine as well as "deep" and the nervous power of sound as mere feminine and "surface." The ability to manage large-scale forms was a matter of the expression of will power; it was a serious measure of musical manliness, one which composers such as

57 Nietzsche, 'Der Fall Wagner,' VI.3, p.17. H.C. von Herrmann, 'Induction psychomotrice', in *Electric Laokoon*, ed. by Michael Franz, Wolfgang Schäffner, Bernhard Siegert and Robert Stockhammer (Berlin: Akademie Verlag, 2007), pp.82-96. See James Kennaway, "Musical Hypnosis: Sound and Self from Mesmer to Brainwashing," *Social History of Medicine* 25.2 (2012), pp.271-289.

58 Christoph Schulte, *Psychopathologie der Fin de Siecle – Der Kulturkritiker, Arzt und Zionist Max Nordau* (Frankfurt: Fischer, 1997), p.179.

59 Max Nordau, *Entartung*, 2 vols. (Berlin: Carl Duncker, 1892-1893) I, p.353

Schubert and Tchaikovsky were sometimes seen as failing.[60] In *Richard Wagner und die Homosexualität* (Wagner and Homosexuality, 1902), Hanns Fuchs explicitly dealt with homosexuality conceived as a medical condition and a mental weakness. There could never be a great homosexual or female composer, he wrote, because music is "die vollendetste Offenbarung des Willens" (the most complete revelation of the will), but homosexuals and women lack „de[n] Wille[n] oder wenigstens die Fähigkeit fehlt, ihren Willen zu äußern, ihn in der bestimmten Form auszudrücken" (lack the will or at least the ability to express that will in a specific form).[61] True music was thus a matter of form and of *Lebenskraft* in terms of will power, not of nervous stimulation, even in the context of physical categories such as gender and sexuality. As masculine strength of will became increasingly politicized in the early twentieth century, this view of music and will power came to be part of a defense of high culture against modernity.

Will Power, Music and Politics 1900-1945

German music theory of the early twentieth century often continued to take *Lebenskraft* in terms of will power as a central principle. For example, musical will power became an important idea in the defence of Absolute Music in the hands of music theorists August Halm and Ernst Kurth, as Lee Rothfarb has noted. Rejecting the musical hermeneutics of Hermann Kretschmar and the narrative approach of Paul Bekker, they advocated a view of music that focused on the dynamic strong-willed nature of serious music.[62] Although they owed much to Hanslick, their view of the will in music saw it not only as the legitimate locus of serious listening but as actually immanent in the music itself. They saw it as manifest in the sense of energetic teleological motion, of being musically oriented towards a goal.[63] August Halm, in his 1913 book *Von zwei Kulturen der Musik* (The Two Cultures of Music) spoke at length of the will of specific pieces, and ascribed volition to tones and praised Bach's *Lebenskraft*.[64] The metaphysical will of the music itself was implicitly related to the will power of the listener. To strengthen society, it was crucial, Halm argued, to turn the debate on musical aesthetics away

60 See Derek Scott, 'The Sexual Politics of Victorian Musical Aesthetics', *Journal of the Royal Musical Association*, 119 (1994), 91-114, and Ian Biddle, *Music, Masculinity and the Claims of History* (Farnham: Ashgate, 2011).

61 Hanns Fuchs, *Richard Wagner und die Homosexualität* (Berlin: Behr, 1902), p.50.

62 Lee Rothfarb, "Hermeneutics and Energetics: Analytical Alternatives in the early 1900s", *Journal of Music Theory* 36.1 (1992), 43-68.

63 Despite his choice of the Schopenhauerian title *Tonwille* for his journal, Heinrich Schenker's feelings about the philosopher and will in music were complicated. Heinrich Schenker, *Counterpoint*, 2 vols (New York: Schirmer Books, 1987), I, p.16.

64 August Halm, *Zwei Kulturen der Musik* (Stuttgart: Klett, 1947), p.206, p.218.

from emotions and the nervous system and towards musical laws and to recognize the will of the music. The implicit political agenda of order and self-control behind the language of musical will is not hard to make out.

Similarly, in his influential *Romantische Harmonik und ihre Krise in Wagners Tristan* (Romantic Harmony and its Crisis in Wagner's Tristan, 1920), the Viennese music theorist Ernst Kurth explicitly defined music as "eine Dynamik von Willensregungen" (a dynamic of volitional impulses).[65] For him, the sense of willed direction in tones was central:

> Hiebei zeigt sich aber von neuer Seite die Bedeutung vom Willen an sich für die Psychologie des gesamten Musikhörens; den auch was diese tief in die Klangfarbe eindringenden Licht- oder Schattenwirkungen selbst ausmacht, ist gar nicht eine wirklich in größerem melodischen Zuge geschechende Bewegung, sondern das bloSSe Andrängen zur Bewegung, wie es in der chromatischen Abänderung liegt.[66]

> (Here the significance of *will* as such for the psychology of all musical hearing becomes clear from a new perspective. For what these effects of luminescence and shading amount to, penetrating deep into the harmonic colour, is not at all actually *kinetic* motion in the larger melodic current but rather the mere *urge* toward motion, as evident in chromatic alteration.)

That is to say that serious music is not about the mere physical impact of sound ("Klang") but rather about a sense of forward motion ("Drang") or will in the music itself. Elsewhere he described the role of the will in explicitly Nietzschean terms: "Der Klang ist tot; was in ihm lebt, ist der Wille zum Klang" (Tone is dead, what lives in it is the will to tone).[67] One can see the work of Halm and Kurth as an attempt to save the German musical tradition from modernity and the threat of materialist interpretations by turning back to the metaphysics of will. This agenda tended to mean that their aesthetic remained strangely disembodied, as is reflected in Kurth's seemingly paradoxical view "daß wir […] mit dem Willen, in letzter Linie erst mit dem Ohr hören" (that we hear with the will, and only lastly with the ears).[68]

The atmosphere of impotence and crisis after the First World War made will power, *Lebenskraft*, and their pathologies and musical analogues a central issue. Beginning in 1919 there was even a German periodical dedicated to the idea of strength of will in music called *Der Energetiker. Monatsschrift für musikalische Hand- und Willenskultur* (Energetics: Monthly Journal for the Cul-

65 Ernst Kurth, *Romantische Harmonik und ihre Krise in Wagners Tristan* (Hildesheim: Georg Olms, 1975), p.3.

66 Kurth, *Romantische Harmonik*, p.147.

67 Kurth, *Romantische Harmonik*, p.4.

68 Kurth, *Romantische Harmonik und ihre Krise in Wagners Tristan*, p.15.

ture of Musical Hand and Will Power). Theodor Ritte, who edited it, expressed his views on the connection between music, will power, pathology and the political revival of Germany in an article about the disease of "Chopinitis." Drawing on decades of observers contrasting the stimulation of the nerves with music for the will, he wrote that "nerve-music" such as that of Chopin was entirely wrong for the ruined nerves of post-First-World-War Germany. No "nerve-music for those whose nerves are shattered," he demanded, and recommended Bach, who was so German and so strong in musical form, as to be a cure for musical sickness.[69] The times demanded the musical expression of manly will power: "Also keine Nervenmusik dem Nervenzerrüttenden, sondern Kraftkunst, die uns Willenskrupel über uns selbst wieder hinauszureissen vermag! Das kann nur ein Tondichter voll männlicher Würde" (Only power-art that can pull us cripples of the will out of ourselves again! Only a composer full of masculine dignity… can do that).[70] The right kind of music, focused on clear forms, would restore the strength of will vital for individual health and the reassertion of national power.

This context is clear in discussions of music as a matter of strength of will in the music criticism and theory of the interwar period. In particular, many critics emphasized the connection between a lack of will power and pathological music. To give just a couple of examples, in 1926 the moderate critic Adolf Weissmann argued that form and proper rhythm, as a model of healthy will power, was being undermined by pathological "emotional excitement."[71] Such focus on the physical impact of music would weaken listeners' minds and nervous systems, leaving them neurotic. Adding a more nationalist tone, the prominent conservative critic Walter Niemann suggested that it was French Impressionism that was the danger to will power. He portrayed as a spiritual and political catastrophe what he took to be a shift from music directed to the "langsame Umwandlung der vergeistigten und verinnerlichten Seelen- und Herzenskunst" to one of "veräußerlichten und artistischen Nerven- Klang- und Stimmungskunst, die Tonmalerei, der Naturalismus und Realismus" (redirecting the spirit and the inner soul to a

---

69 Theodor Ritte, "Chopinitis," *Der Energetiker*, 1 (1919), 5-7. See also Walter Frisch, *German Modernism: Music and the Arts* (Berkeley and Los Angeles: U of California Press, 2005), pp.141-2.

70 Ritte, "Chopinitis," *Der Energetiker*, p.7. See also Hans Walter Schmidt, 'Nerven und Musik: eine psychologisch-physiologische Betrachtung von heute,' *Internationale Reform-Zeitschrift für Gesang, Klavier und Geige: Der Energetiker*, 3.8 (1922), 10-11.

71 Adolf Weissmann, *Problems of Modern Music* (London: J. Dent, 1925), pp.14-15.

foreign externalised art of nerves, sound and mood, of sound painting, naturalism and realism).[72]

Commentators had expressed similar anxieties about the effect of Wagner's music on will power in the nineteenth century, but after the emasculating humiliations of Germany's defeat in the First World War, the Treaty of Versailles, and long years of economic chaos, these fears became explicitly politicized. During the interwar period, the language of will power and *Lebenskraft* in music clearly takes a step away from the abstract disembodied listener towards a more sinister, amoral and racialized conception of will power such can be seen in innumerable books such as Erich Grassl's *Die Willensschwäche* (Weakness of Will, 1937) and Max Wilfert's *Psychologie und Pädagogik der Selbstbeherrschung* (Psychology and Pedagogy of Self Control, 1936) not to mention Leni Riefenstahl's *Triumph des Willens* (Triumph of the Will, 1935)[73] Aesthetic disputes about the direction of German music were often put into moral and medical terms, portraying weak will as a sickness that threatened both the spiritual health of music and of the nation. This mixture of the language of morality, health, *Lebenskraft*, and music was nothing new. It had been going on for over a century already, and, as Susan Sontag argued in *Illness as Metaphor*, "Nothing is more punitive than to give a disease a meaning – that meaning is invariably a moralistic one."[74] Indeed, by the 1930s and 1940s talk of music, *Lebenskraft* and strength of will had taken on an altogether more sinister aspect. Reflections on the relationship between music and *Lebenskraft* thus provide clear evidence of the wide reach of the *Lebenskraftdebatte*.

72 Walter Niemann, *Musik der Gegenwart* (Berlin: Schuster & Loeffler, 1921), p.209.

73 Erich Grassl, *Die Willensschwäche* (Leipzig: Johann Ambrosius Barth, 1937), Max Wilfert, *Psychologie und Pädagogik der Selbstbeherrschung* (Langensalza: Hermann Beyer & Söhne, 1936). See also Daniel Wildmann, *Begehrte Körper: Konstruktion und Inszenierung des Männerkörpers im "Dritten Reich"* (Würzburg: Königshausen & Neumann, 1998).

74 Susan Sontag, *Illness as Metaphor* (New York: Farrar, Strauss and Giroux, 1978), p.57.

Ritter's Musical Blood Flow Through Hoffmann's Kreisler

Alexis B. Smith

Abstract: This chapter investigates E.T.A. Hoffmann's descriptions of the music that flows through the blood in the *Kreisleriana* texts (1810-1814) and the fragmentary Kreisler biography in *Lebens-Ansichten des Katers Murr* (1819/1821). Johannes Kreisler, Hoffmann's pseudonym and alter-ego, emphasizes the repeated Romantic notion that music, in its eternal quality, exists outside of anything tangible. But he breaks with this tradition by also bringing music down to earth, claiming that it exists *in all things*, including the human body. Kreisler's physiological experiences of music are not only poetic metaphors, but also strongly echo the scientific and philosophical investigations of Romantic physicist Johann Wilhelm Ritter (1776-1810), whose *Fragmente aus dem Nachlasse eines jungen Physikers* (1810) has largely been ignored in the Humanities until around the last decade. In this collection of fragments, Ritter provides scientific reasoning for how and why music comes from the spirit realm into the blood as directed by animal magnetism. He claims that music has its own consciousness and is a creative life force that runs through human blood. Hoffmann adopts and expands Ritter's contemporary, intricate, and yet speculative theory as an explanation for the power of music over musicians, composers, and those initiated to sense its flow.

By the time E.T.A. Hoffmann adopted the terminology in his musical critiques for the *Allgemeine musikalische Zeitung* in 1810, the portrayal of music as the reflection of a "higher spiritual realm" was already firmly established in the idealist philosophies of Wilhelm Heinrich Wackenroder, Ludwig Tieck, Friedrich Schelling, A.W. Schlegel, Johann Gottfried von Herder, and others.[1] Representative is Schelling's comment in *Philosophie der Kunst* (1802-03):

> Die Formen der Musik sind Formen der ewigen Dinge, inwiefern sie von der realen Seite betrachtet werden [...] so bringt die Musik die Form der Bewegungen der Weltkörper, die reine, von dem Gegenstand oder Stoff befreite Form in dem Rhythmus und der Harmonie als solche zur Anschauung. Die Musik ist insofern diejenige Kunst, die am meisten das Körperliche abstreift, indem sie die reine Bewegung selbst als solche, von dem Gegenstand abgezogen, vorstellt und von unsichtbaren, fast geistigen Flügeln getragen wird.[2]

1 Mark Evan Bonds provides a contextual history regarding the repeated Romantic vocabulary and illustrates where Hoffmann adopted and in many cases transformed some of the descriptions by these philosophers. In: "Idealism and the Aesthetic of Instrumental Music at the Turn of the Nineteenth Century," *Journal of the American Musicological Society*, 50 (1997): 387-420 (pp.411-12).

2 Friedrich Wilhelm Joseph von Schelling, *Sämtliche Werke*, ed. K.F.A. Schelling, 14 vols. (Stuttgart and Augsburg: J.G. Cotta, 1856-61), vol. V: *Philosophie der Kunst*, pp.501-03.

> (The forms of music are the forms of eternal things in so far as they can be contemplated from the perspective of the real [...]. Thus music manifests in rhythm and harmony, the pure form of the movements of the heavenly bodies, freed from any object or material. In this respect, music is the art which casts off the corporeal, in that it presents movement in itself, divorced from any object, borne on invisible, almost spiritual wings.)[3]

Music, in its eternal quality, exists outside of anything tangible.[4] In fact, one aspect that links together all of the commentators before Hoffmann is their focus on the inherent incorporeality of music – that it does not exist in the physical, material world, but rather purely as "die reine, von dem Gegenstand oder Stoff befreite Form" ("the pure form freed from any object or material"). Johannes Kreisler, Hoffmann's pseudonym and alter-ego beginning with his articles in the *Allgemeine Musikalische Zeitung*, several of which later became the *Kreisleriana* (1810-1814), emphasizes this too. But he breaks with this tradition by also bringing music down to earth, claiming that it exists *in all things*.[5] Kreisler's physiological experiences of music are not only poetic metaphors, but also strongly echo the scientific and philosophical investigations of Romantic physicist Johann Wilhelm Ritter (1776-1810). By reading Ritter and Hoffmann together, music enters the *Lebenkraft* debate in an unsuspected way: it is not only the force giving life to the creative genius of composers and musicians, it is also conscious – takes on a life of its own. Music, which comes from the spiritual realm, is also internalized in all things, including the human body; it is connected not only to the electrical current in the nerves, but is also inside the blood. Blood becomes the carrier of music that transfers from one human being to another, and to and from objects of nature, signaling a universal, musical life force prevalent in everything.

---

3 Translation available in: Bonds, pp.403-404.

4 This passage also strongly echoes the words of Wackenroder and A.W. Schlegel. In "Die Wunder der Tonkunst," Wackenroder writes: "Die Musik aber halte ich für die wunderbarste dieser Erfindungen [the fine arts], weil sie menschliche Gefühle auf eine übermenschliche Art schildert, weil sie uns alle Bewegungen unsers Gemüths unkörperlich, in goldenen Wolken luftiger Harmonieen eingekleidet, über unserm Haupte zeigt." In: *Sämtliche Werke und Briefe*, ed. Silvio Vietta and Richard Littlejohns, 2 vols. (Heidelberg: Carl Winter Universitätsverlag, 1991), vol. I: *Phantasien*, p.407. See also A. W. Schlegel, *Kritische Ausgabe der Vorlesungen*, ed. Ernst Behler, 2 vols. (Paderborn: Ferdinand Schöningh, 1989-), vol. I: *Vorlesungen über Schöne Literatur und Kunst. Erster Teil: Die Kunstlehre* (1801-02), pp.375-81.

5 Kenneth Negus writes: "Hoffmann bestowed highly finite forms on this infinite art, although they are not those of common experience. This is illustrated in his discussions of specific works and composers, where he uses poetic images to describe them." In: *E.T.A. Hoffmann's Other World: The Romantic Author and his "New Mythology"* (Philadelphia: U of Pennsylvania Press, 1965), p.43.

Kreisler establishes the theory that music exists in all things by first recycling the Romantic ideas, including music's role as a universal, yet indecipherable language. In "Gedanken über den hohen Wert der Musik" he begins by criticizing and questioning artists, especially musicians when he remarks:

> [S]ie nennen sie [die Musik] die romantischste aller Künste, da ihr Vorwurf nur das Unendliche sei; die geheimnisvolle, in Tönen ausgesprochene Sanskritta der Natur, die die Brust des Menschen mit unendlicher Sehnsucht erfülle, und nur in ihr verstehe er das hohe Lied der – Bäume, der Blumen, der Tiere, der Steine, der Gewässer![6]
>
> (They call it [music] the most romantic of all the arts since its only subject-matter is infinity; the mysterious Sanskrit of nature, translated into sound that fills the human breast with infinite yearning; and only through it can they perceive the sublime song of – trees, flowers, animals, stones, water!)

While its subject matter is infinity, music's role as the language of nature can also bring one closer to the songs of earthly things. Kreisler takes this critical statement and claims it as his own in the following text, "Beethovens Instrumental-Musik," contending with confidence that music is, indeed, the most romantic of all arts, but that it is also more than that: "Die Musik schließt dem Menschen ein unbekanntes Reich auf, eine Welt, die nichts gemein hat mit der äußern Sinnenwelt, die ihn umgibt, und in der er alle *bestimmten* Gefühle zurückläßt, um sich einer unaussprechlichen Sehnsucht hinzugeben" (FN, p.41). (Music reveals to man an unknown realm, a world quite separate from the outer sensual world surrounding him, a world in which he leaves behind all precise feelings in order to embrace an inexpressible longing. MW, p.96).

While Kreisler agrees that music's subject matter is infinity, it is also contained in earthly things. However, he does not stop there for it would, in a matter of words, place a limit on infinity. In order to maintain the infinite quality of music, he describes it as also existing as something much closer, and yet separate from the outer sensual world. This "embrace" of inexpressible longing is experienced *inside* the human body. As Kreisler conjectures in "Johannes Kreislers, des Kapellmeisters, musikalische Leiden:"

---

6 E. T. A. Hoffmann, *Fantasie- und Nachtstücke, Sämtliche Werke*, ed. Hartmut Steinecke and Wulf Segebrecht (München: Winkler Verlag, 1960), III. *Kreisleriana*, 49. Further citations will be referred to with the initials of the volume, FN. E. T. A. Hoffmann, *E. T. A. Hoffmann's Musical Writings: Kreisleriana, The Poet and the Composer, Music Criticism*, ed. David Charlton (Cambridge: Cambridge UP, 1989), p.94. Subsequent translations will be cited as MW.

> Wie ist doch die Musik so etwas höchst Wunderbares, wie wenig vermag doch der Mensch ihre tiefen Geheimnisse zu ergründen! – Aber wohnt sie nicht in der Brust des Menschen selbst und erfüllt sein Inneres so mit ihren holdseligen Erscheinungen, daß sein ganzer Sinn sich ihnen zuwendet und ein neues verklärtes Leben ihn schon hienieden dem Drange, der niederdrückenden Qual des Irdischen entreißt? (FN, p.25)
>
> (What an utterly miraculous thing is music, and how little can men penetrate its deeper mysteries. But does it not reside in the breast of man himself and fill his heart with its enchanting images, so that all his senses respond to them, and a radiant new life transport him from his enslavement here below, from the oppressive torment of his earthly existence?) (MW, p.88)

That music exists in the human breast, or the heart (*Inneres*), as suggested by the translation, is metaphorically not out of the ordinary. However, Kreisler appears to be able to experience and in some ways *embody* music in a unique way. As the reader is drawn more deeply into Kreisler's world, the line in the relationship between madness and creativity is not clear. Precisely this blurring opens up the possibility for music to reach new areas in Kreisler's way of thinking.[7] The narrator, whose voice is distinct from Kreisler's, explains at the end of the last *Kreisleriana* text, titled "Johannes Kreislers Lehrbrief":

> Der Ton wohnt überall, die Töne, das heißt die Melodien, welche die höhere Sprache des Geisterreichs reden, ruhen nur in der Brust des Menschen. – Aber geht denn nicht, so wie der Geist des Tons, auch der Geist der Musik durch die ganze Natur? Der mechanisch affizierte tönende Körper spricht ins Leben geweckt sein Dasein aus, oder vielmehr sein innerer Organismus tritt im Bewußtsein hervor. Wie, wenn ebenso der Geist der Musik, angeregt von dem Geweihten, in geheimen, nur diesem vernehmbaren Anklängen sich melodisch und harmonisch ausspräche? Der Musiker, das heißt, der, in dessen Innerem die Musik sich zum deutlichen klaren Bewußtsein entwickelt, ist überall von Melodie und Harmonie umflossen. Es ist kein leeres Bild, keine Allegorie, wenn der Musiker sagt, daß ihm Farben, Düfte, Strahlen, als Töne erscheinen, und er in ihrer Verschlingung ein wundervolles Konzert erblickt. (FN, pp.325-26)
>
> (Sound resides in all things; but notes, that is to say melodies, which speak the higher language of the spirit-realm, repose only in the breast of men. But like the spirit of sound, does not the spirit of music also permeate the whole of nature? A sounding body, when it is mechanically animated and thus brought to life, gives expression to its existence, or rather its inner organism consciously emerges. Cannot the spirit of music, when

---

7 John T. Hamilton suggests, e.g.: "In referring to aesthetic and psychic experiences that are conceived as fundamentally nonrepresentable, music and madness, as metaphors, open a text up to what no text can properly contain." In: *Music, Madness, and the Unworking of Language* (New York: Columbia UP, 2008), p.195.

> it is awakened by its votaries, similarly express itself melodically and harmonically in secret resonances intelligible only to them? Musicians, that is to say those in whom music becomes a condition of total awareness, are surrounded on all sides by melody and harmony. It is no empty metaphor, no allegory, when the musician says that colour, fragrance, light appear to him as sounds, and that in their intermingling he perceives an extraordinary concert.) (MW, pp.163-64)

The unnamed narrator explains that the tone lives everywhere. In fact, everything in the world contains sound. Melody, moreover, resides not only in the breast of mankind, but also in all of nature. That the secret resonances of these melodies and harmonies can only be intelligible *to them* suggests that music itself develops its own consciousness. As the agent, then, music uses each body and object of the material world as an instrument – "[d]er mechanisch affizierte tönende Körper." Surrounded on all sides by music, both the outer world and the inner body, the musician or composer's senses mingle with the larger environment, and music thereby extends and becomes audible in the world.

In order to enter Kreisler's synesthetic world (that is, whether we view him as being deranged or not), one must consider for a moment that these descriptions are to be read not only as metaphors, but also as his organic experiences. Music *is* the substance of the creative energy in composers, musicians, and those *initiated* to sense its flow. Kreisler senses the movement of music inside his body itself, thereby going beyond the five senses attuned to the outer world. The creative flow that Kreisler experiences can be broken down into two important modes of interaction: electrical current and blood flow.

It is first experienced in connection with the electrical current that travels through the nerves.[8] This is a direct allusion to the then much debated alternative science known as animal magnetism, based on the discovery of animal electricity by Luigi Galvani (1737-98) that referred to muscle contraction stimulated by an electric current. Thus, animal magnetism connotes the electric current that flows through the organic body.[9] Today, one speaks of

---

8 R. Murray Schafer explains: "For Hoffmann, the nervous system of music was stretched over the visible, tactile, and olfactory world, reconstituting the elements in dazzling but irrational patterns. But music itself, otherwise invisible, now gains other sense-dimensions so that it begins to take on the appearance of a recognizable though confused cosmology." In: *E.T.A. Hoffmann and Music* (Toronto and Buffalo: U of Toronto Press, 1975.), pp.153, 156.

9 Luigi Galvani began conducting experiments with electricity in 1780, convinced that it was a property of the animal kingdom, and therefore purely organic. For more history on the developments in organic electricity see: Siegfried Zielinski, "Electrification, Tele-Writing, Seeing Close Up: Johann Wilhelm Ritter, Joseph Chudy, and Jan

galvanic effects in biology as electrophysiology. Franz Anton Mesmer (1734-1815) developed a healing practice based on animal magnetism, which he called *Mesmerismus*, to treat patients through electrical/magnetic cures. Music was used during these sessions to create a soothing atmosphere for the patients, but it was not considered the cause or related to the substance of the electrical current. The credibility of his method and the effectiveness of his "cures" were widely questioned and debated, even though he supported his work based on existing theories on the physics and physiological aspects of mesmerism.[10] Hoffmann's interest in and creative application of animal magnetism is evident throughout his writings. And analyses of the influence of this practice have appeared in many publications.[11] What informs the following analysis is rather the work by physicist Johann Wilhelm Ritter (1776-1810), whose *Fragmente aus dem Nachlasse eines jungen Physikers* (1810), Hoffmann also studied.

Even though the natural sciences were an integral part of early Romanticism, scholars in the Humanities have largely ignored Johann Wilhelm Ritter's role until the last decade or so.[12] In his study of seeing and hearing in Ritter's work, Thomas Strässle explains: "The rather abstruse and scientific style of much of Ritter's work has hitherto prevented scholars from accessing his speculative philosophy, and particularly from recognizing his highly original contributions to questions of musical aesthetics."[13] While a highly reputed scholar for the majority of his life, the speculative nature of Ritter's

---

Evangelista Purkyne," *Deep Time of the Media: Toward an Archaeology of Hearing and Seeing by Technical Means*, trans. Gloria Custance (Cambridge: The MIT Press, 2006), pp.159-203 (p.167).

10 For more information see: Jurgen Barkhoff, *Magnetische Fiktionen: Literarisierung des Mesmerismus in der Romantik* (Stuttgart and Weimar: Metzler, 1995).

11 See for example: Rupert Gaderer, *Poetik der Technik: Elektrizität und Optik bei ETA Hoffmann* (Freiburg: Rombach Verlag KG, 2009). Monika Schmitz-Emans provides an overview of Hoffmann's "Der Magnetiseur" and includes the reception of Mesmerism by contemporaries in different disciplines. In: "Magnetische Phantasien: Zur Bedeutung des Mesmerismus für Autoren der Romantik," *Der Deutschunterricht*, 57.3 (2005), 55-62. Ingrid Kollak offers a detailed description of Hoffmann's writing style as a result of his interest in Mesmerism, as well as the reception of Mesmerism by other authors. In: *Literatur und Hypnose: Der Mesmerismus und sein Einfluß auf die Literatur des 19. Jahrhunderts* (New York: Campus Verlag, 1997).

12 With the notable exception of Walter Benjamin who endorsed Ritter's work in his *Ursprung des deutschen Trauerspiels* (1928), (Frankfurt am Main: Suhrkamp Verlag, 1963), pp.240-43.

13 Thomas Strässle, "'Das Hören ist ein Sehen von und durch innen': Johann Wilhelm Ritter and the Aesthetics of Music," *Music and Literature in German Romanticism*, eds. Siobhán Donovan and Robin Elliott (Rochester: Boydell & Brewer, 2004), pp.27-42 (p.28).

writings caused him to lose credibility in his contemporary circles, as he often neglected to supply scientific proof for his theories. His scattered aesthetic reflections in *Fragmente aus dem Nachlasse eines jungen Physikers* also lack a coherent philosophical system. However, he has since been credited with crucial advances in the understanding of galvanism and in the discoveries of electrochemistry and the ultraviolet spectrum.[14] Subsequent scientific experiments have proved many of his speculative theories to be true.

Perhaps because of Ritter's minimal reception in the Humanities, the influence of his scientific and aesthetic theories on Hoffmann has largely gone unnoticed. The commentary in Hartmut Steinecke and Wulf Segebrecht's critical edition of *E.T.A. Hoffmann's Sämtliche Werke* offers brief notes on just three passages in the last *Kreisleriana* text, "Johannes Kreislers Lehrbrief."[15] In the footnotes to Martyn Clarke's translation of *Kreisleriana*, David Charlton comments on the same three passages, beginning with Hoffmann's appellation, "the brilliant physicist." The passage reads:

> So wie nach dem Ausspruch eines geistreichen Physikers Hören ein Sehen von innen ist, so wird dem Musiker das Sehen ein Hören von innen, nehmlich zum innersten Bewußtsein der Musik, die mit seinem Geiste gleichmäßig vibrierend aus Allem ertönt was sein Auge erfaßt.[16] (FN, p.453)
>
> (Just as hearing, in the words of a brilliant physicist, is seeing from within, so to the musician seeing is hearing from within, attainable only through the profoundest awareness of music, which radiates from everything his eye falls upon, and vibrates in sympathy with his spirit.) (MW, p.164)

In his *Fragmente aus dem Nachlasse eines jungen Physikers* Ritter remarked: "Das Hören ist ein Sehen von innen, das innersTinnerste Bewusstseyn" (Hearing is an internal seeing, the most innerly internal consciousness).[17] Charlton follows up with comments on Hoffmann's allusions in the text that the musi-

14 He performed dangerous experiments on himself. See Zielinski, pp.175-77.

15 *E.T.A. Hoffmann Sämtliche Werke in sechs Bänden*, ed. Hartmut Steinecke and Wulf Segebrecht (Frankfurt am Main: Deutscher Klassiker Verlag, 1993), vol. II.1, pp.857-58. The Hoffmann text is *Fantasiestücke in Callot's Manier.*

16 For the footnote commentary, see Charlton, p.164.

17 Both the original text and the translation are available in: Johann Wilhelm Ritter, *Fragmente aus dem Nachlasse eines jungen Physikers, Key Texts of Johann Wilhelm Ritter (1776-1810) on the Science and Art of Nature*, trans. and ed. Jocelyn Holland (Leiden and Boston: Brill NV, 2010), pp.122-507 (pp.300-01). See also Fragment Nr. 429, "Wie Gesicht zu Gehör, so Geschmack zu Geruch. Der Geruch ist eine Art Gehör." ("As sight to hearing, so taste to smell. Smell is a kind of hearing.") (pp.334-35).

cian is like a Mesmerist and that music is a universal language.[18] Even though Hoffmann's reference to Mesmerism could be attributed to Franz Anton Mesmer himself, Mesmerism does not have a direct connection to music. However, Ritter integrates concepts of music into his Romantic philosophy of physics, and in particular, into his study of electrophysiology. Ritter's influence on Hoffmann's texts (whether "by Kreisler" or about him) is much more involved than is apparent in these footnote commentaries. For example, in "Johannes Kreisler im Dialog mit einem 'geistreichen Physiker': Zu E.T.A. Hoffmanns Auseinandersetzung mit Johann Wilhelm Ritter," Thomas Strässle analyzes "Johannes Kreislers Lehrbrief" as Hoffmann's *detailed dialog* with Ritter. He illustrates how extensive imagery in the story of Chrysostomus directly corresponds to Ritter's philosophy of physics, specifically in the translations between the senses, the morphing of seeing into hearing and of hearing into seeing.[19] There are additional similarities to consider in regard to Ritter's influence on Hoffmann's writing. Firstly, they are present not only in "Johannes Kreislers Lehrbrief," but also in other *Kreisleriana* texts, which were written as early as 1810, the same year in which Ritter's *Fragmente* was published. Then, Hoffmann recapitulates these similarities and expands them in the fragmentary Kreisler biography in *Lebens-Ansichten des Katers Murr* (1819/1821). Ritter's work offers scientific reasoning as to why Kreisler experiences music in the blood.

As is clear in *Kreisleriana*, the electrical current Kreisler experiences comes from the spirit realm, which directs the creative energy of composers and musicians. For example, Kreisler writes about the "electric flow" through his fingers while playing the piano: "Die Noten wurden lebendig und flimmerten und hüpften um mich her – elektrisches Feuer fuhr durch die Fingerspitzen in die Tasten – der Geist, von dem es ausströmte, überflügelte die Gedanken" (FN, p.30). (The notes come to life and flutter and dance around me; electric sparks flow through my finger-tips into the keys; the spirit generating them overtakes my thoughts. MW, p.85). The music itself, however, appears in a second, altogether different channel: *in the blood.* In "Ombra adorata!" Kreisler asserts:

> [A]lle Töne, die in der wunden Brust im Blute des Schmerzes erstarrt, leben auf, und bewegen und regen sich und sprühen wie funkelnde Salamander blitzend empor; und ich vermag sie zu fassen, zu binden, daß sie wie in einer Feuergarbe

18 See Charlton's additional footnotes on pp.164-65. It is important to note that the composer as a Mesmerist also occurs in the first volume of *Kreisleriana.* (FN, pp.47-48)

19 Thomas Strässle, "Johannes Kreisler im Dialog mit einem 'geistreichen Physiker': Zu E.T.A. Hoffmanns Auseinandersetzung mit Johann Wilhelm Ritter," *E.T.A. Hoffmann-Jahrbuch* 10 (2002), 96-119.

zusammenhaltend zum flammenden Bilde werden, das deinen Gesang – dich – verklärt und verherrlicht. (FN, pp.35-36)

(Every sound congealed in the blood from my wounded breast will be revivified, and stir and spring up and throw out glittering sparks like fire-breathing salamanders. And I shall contrive to harness them, and combine them, so that as if merging into a single bursts of fire they form a single flaming image, that transfigures and glorifies your singing and yourself.) (MW, pp.90-91)

Just as veins and the heart experience blood clots restricting the movement of blood, so does the music; the creative force of the composers becomes stuck, as though dead, inside of the veins. Kreisler describes how the music must be revivified ("leben auf"), brought back to life and movement.[20] While melody then resides in the human heart and is transported and transmitted through the veins, the composer is ultimately the one who has the ability to bring this music to other human hearts. Kreisler wishes to "harness" ("fassen") this power, which must be learned. He is aware that the invisible, electric current ignites this exchange (but not everyone is).[21] In this way, some composers and musicians, by way of performing music, can, indeed, be like Mesmerists. However, Mesmerists do not recognize music as *inside* blood. That is the move Johann Wilhelm Ritter makes.

There is scientific evidence for why Ritter situates the flow of music in the blood, as directed by animal magnetism. In Fragment 51, Ritter makes the claim that "Alle Stoffe auf Erden scheinen zerlegtes Eisen zu seyn. Eisen ist der Kern der Erde, 'der sichtbare Quellgeist der Erde' (*Jac. Böhme*)." (All materials on earth seem to be decomposed iron. Iron is the kernel of the earth, 'the visible spirit of the earth' (Jac.[ob] Böhme). pp.146-47). Iron, which is prevalent in all earthly things, is an important element in animal

---

20 The reference to the "funkelnde Salamander" ("fire-breathing salamanders") coming from the sparks resonates with another concept Ritter used. In the Appendix to his *Fragmente* he writes: "Schön wäre es, wie, was hier äußerlich klar würde, genau auch wäre, was uns die Klangfigur *innerlich* ist: – *Licht*figur. *Feuerschrift.*" In Hoffmann's literary texts, the fire metaphor also plays a role. Ritter explains that when bodies vibrate extremely quickly, they glow, which he calls "Leuchten durch elektrische Schläge." In his speech for the academy entitled, "Physik als Kunst," delivered March 28, 1806, Ritter explains that fire is a universal life phenomenon, and reveals itself in all of the senses. There are two kinds of fire, chemical and electrical. See especially pp.558-99 in: *Key Texts of Johann Wilhelm Ritter (1776-1810) on the Science and Art of Nature*, pp.524-83.

21 Strässle points out that Hoffmann adds this limitation to Ritter's theory: "*Dem Musiker* wird das Sehen ein Hören von innen, 'zum innersten Bewußtsein der Musik'. Hoffmanns Adaptation von Ritters Axiom vollzieht dadurch, dass es in einer seiner möglichen Formulierungen am Typus des Musikers zumindest spezifiert, wenn nicht gar auf ihn reduziert wird." ("Johannes Kreisler im Dialog," p.101)

magnetism, as it is the strongest magnetic metal, and it is the *only* metal that is tolerated *inside* the human body, according to Ritter. He writes in Fragment 367: "Ist es nicht sonderbar, wie das Eisen das einzige Metall ist, welches organisch-thierische Körper in sich dulden?" (Is it not strange how iron is the only metal which organic-animal bodies tolerate within themselves? pp.308-309). Therefore, the reason music enters the musician and composer from the higher, spiritual realm through the veins is due to the magnetic nature of the iron in the blood. The music is then directed to the human heart, because the heart itself is also a magnet: "Der wahre Magnet aller Weltkörper muß ein Herz seyn, und im Menschenkörper, u. a., ist das Herz *der Magnet*." (The true magnet of all planetary bodies has to be a *heart*, and in the human body, for example, the heart is the *magnet*"; pp.314-15).[22] Bringing this all together, Ritter claims in Fragment 357: "(1801.) Der Begriff, Discant: Bass = –: + zeigt: daß alle rechte Musik nichts seyn könne, als eine Potenzirung beyder Gegensätze. Die vollständigste Musik ist die, welche vom Eisen bis zum Menschen geht." (1801.) The concept, descant: bass = –: + shows: that all correct music can be nothing other than a potentializing of both oppositions. The most complete music is the kind which goes from iron to the human" (pp.300-01). Ritter explicitly claims that music reaches the human via the iron in the blood. Therefore, according to Ritter's theory and Hoffmann's application of it, music is first brought from the spirit realm into the blood via iron, which then travels to the human heart – the "true" ("wahre") magnet directing the electric current. In essence, the human body becomes the instrument for the music from the spirit realm, and the blood is the channel through which it travels. As iron is also in all earthly things (as Ritter claims in Fragment 51), this implies that music resides in and travels through nature, too.

The importance of blood as the carrier of music is first clear in Hoffmann's story from Chrysostomus' childhood in the final *Kreisleriana* text, "Johannes Kreislers Lehrbrief." The narrator retells the story that Chrysostomus heard from his father about a rock covered with strange mosses and "reddish-colored veins." A brief summary of this tale is necessary to grasp the importance of these images: A stranger comes to the nobleman's castle. Similar to the uncomfortable feeling one would get in the presence of a Mesmerist (as Hoffmann expresses in other stories), everyone cannot look at him *without an inner shudder*.[23] Yet, this man is intriguing, and people are drawn to him. The beautiful young daughter of the castle is at-

---

22 Ritter goes on to relate the magnetism of the human heart to the magnetism within the earth.

23 Including *Fantasiestücke in Callots Manier* (1813-15) (see especially "Der Magnetiseur"), *Nachtstücke* (1816-17), and *Serapions-Brüder* (1819-21).

tracted to his irresistible music, as though under a spell, and he gives her singing and lute lessons. They always meet at the large rock to play music.

One morning, however, the stranger is gone, and the daughter is missing. The father goes to the rock in search for her:

> Der Junker glaubte, das Pferd scheue sich vor der wunderlichen Form des Steines, er stieg daher ab, um es vorüber zu führen, aber im Starrkrampf des Entsetzens stockten seine Pulse, und er stand regungslos, als er die hellen Bluttropfen erblickte, die dem Stein häufig entquollen. [...] aus dem Blute entstanden aber die wunderlichen Moose und Kräuter, die jetzt auf dem Steine in seltsamlichen Farben prangen. (FN, p.323)

> (The nobleman thought his horse had been frightened by the weird shape of the rock, and he therefore dismounted in order to lead it past. With a sudden spasm of horror, however, his heart stopped and he stood paralysed, as he noticed bright gouts of blood copiously dripping from the rock. [...] And from the blood sprang strange mosses and wild flowers that bedecked the rock with incredible colours.) (MW, p.161)

It is suspected that the stranger killed her with his lute, which they found splintered, covered with blood on the ground. As a young boy, Chrysostomus is drawn to this rock in the forest. As his love for music grows, he begins to take instruction in composing, and only reaches utter frustration that he cannot reproduce the music he hears in his mind. This music he hears from the rock is as though cast from the mosses and flowers that grew from the girl's blood. The red-veined covered rock becomes his source of musical inspiration. With this, the story ends, and the narrator explains to Kreisler that there are many lessons to be learned from it. He proffers:

> [U]nd erläutere ebenfalls denen, die mit der Geschichte vom bösen Fremden und dem Burgfräulein nichts Rechtes anzufangen wissen, die Sache dahin, daß das wunderlich Abenteuer, das so in das Leben des Chrysostomus einwirkte, ein treffendes Bild sei des irdischen Unterganges durch böses Wollen einer feindlichen Macht, dämonischer Mißbrauch der Musik aber dann Aufschwung zum Höheren, Verklärung in Ton und Gesang! (FN, p.327)

> (And to those who do not know what to make of the story of the evil stranger and the noble young lady explain its point: that the extraordinary adventure that exerted such an influence on the life of Chrysostomus is a telling allegory of earthly extinction by the evil workings of hostile powers, of the demonic misuse of music, but finally of the ascent to higher things and of transfiguration through music and song!) (MW, p.165)

The narrator implies here that the stranger had a demonic power over the young girl through music, which could be equated to the powerful spell of a Mesmerist. In the Appendix to his *Fragmente*, Ritter explains the manipulative power of composers: "Componisten [...] verwalten ein ganzes dem Menschen verwandtes Geschlecht; seine Diener und seine Engel lassen sie erscheinen, und auch seine Teufel können sie aufrufen." (Composers [...] manage an entire race related to mankind; they allow its servants and angels to appear and they can also summon its devils. *Fragmente*, pp.478-79). As is the case with Mesmerism, the powerful influence of music can be used for good or evil intentions.

Strikingly, Kreisler experiences a similar musical journey as in Chrysostomus' story, when Hoffmann returns Kreisler from his literary absence in the biographical fragments "mistakenly printed" in *Lebens-Ansichten des Kater Murr* (published in two volumes in 1819 and 1821).[24] As a result Kreisler, too, has the opportunity to find peace in his musical explorations.

Kreisler's first encounter with Princess Hedwiga and Julia as "the stranger" marks the beginning of this parallel. They are both drawn to the stranger they encounter in the woods, however, rather than producing beautiful music, as in Chrysostomus' account of "the stranger," Kreisler struggles to even play the guitar, and later casts it into the bushes. Instead, it is Julia who discovers her own musical spark when she picks up the guitar from the bushes, and involuntarily begins to sing. This instrument gives her "free rein to the full richness of the notes dwelling in her breast:" "Sie begann eine bekannte italienische Canzonetta und verlor sich in allerlei zierliche Melismen, gewagte Läufe und Capriccios, Raum gebend dem vollen Reichtum der Töne, der in ihrer Brust ruhte" (LAKM, p.64). (She began a well-known Italian canzonetta, and lost herself in all manner of delicate melismas, daring runs and capriccios, giving free rein to the full richness of the notes dwelling in her breast. LOTM, p.41). The music was already inside of her, just waiting to be released. The magical guitar was the catalyst. Of course, Princess Hedwiga and Julia interpret this later as a result of Kreisler's bewitching powers. Toward the end, Princess Hedwiga concludes: "Und die Musik war das Zaubermittel mit dem er uns umstrickte" (LAKM, p.329). (And music was his means of ensnaring us. LOTM, p.231). Hoffmann presents Kreisler's power over the women as problematic in the text, because Kreisler is equally attracted to the music evoked by women. Ritter offers one helpful explanation for this phenomenon: "Bey der höchsten, also bey den schnellsten Schwingungen, oder die den höchsten Ton geben, wird gewiß *am stärks-*

24 Hoffmann ceased using Kreisler as a pseudonym when the *Kreisleriana* was completed in 1815, and the character Kreisler seemingly disappeared from the literary and music critic world.

ten magnetisirt." ("The ones with the greatest [*better translation*: highest], and thus with the fastest vibrations, or the ones which yield the highest tone, certainly become *magnetized* the *most strongly*"; pp.502-03). Higher tones have a stronger magnetic pull; this could explain Kreisler's attraction to the female voices, but it does not explain the women's attraction to Kreisler, which in the end remains a mystery, like the magical spell of a Mesmerist. But Kreisler is not controlling them with his music – the *music* has control over them, even Kreisler.

This version of the "Chrysostomus story" includes an additional recapitulation of Kreisler's electrical observations and experiences in music from the *Kreisleriana*, which again, he claims come from the spiritual realm. Because he is not the narrator of this story, as he often is (or appears to be) in the *Kreisleriana*,[25] the reader has the opportunity to observe the experiences of the other characters – it is not just crazy Kreisler's internal synesthetic experiences. For example, the princess receives what seems to be an electric shock when she addresses him, which causes her face to become "blood red."[26] The princess is in general a primary receiver and transmitter of this electrical shock in this story, and for this reason Kreisler thinks to himself after experiencing the "pulse-beat" ("Pulsschlag") throbbing while holding her hand later on that she is a kind of organic Leyden jar: "Am Ende ist die Gnädigste eine Art von Leydner Flasche, und walkt honette Leute durch mit elektrischen Schlägen nach fürstlichem Belieben!" (I suppose her Highness is a kind of Leyden jar and sends electric shocks through honest folk at her gracious pleasure! LOTM, p.106).[27] Ritter himself hypothesized about the possibility of an organic Leyden jar in Fragment 265, speculating whether or not the eye could be considered a Leyden jar for light.[28] In combination with

---

25 With the exception of most of the *Kreisleriana* texts from volume two: "Brief des Baron Wallborn an den Kapellmeister Kreisler," "Kreislers musikalisch-poetischer Clubb," and "Nachricht von einem gebildeten jungen Mann" (or specifically the subtext "Schreiben Milo's, eines gebildeten Affen, an seine Freundin, Pipi, in Nord-Amerika"). In "Der Musikfeind," one thinks that the narrator is perhaps Kreisler until he is referred to in the third person (p.437), and, finally, "Johannes Kreislers Lehrbrief" is specifically addressed to Kreisler.

26 "Die Prinzessin blieb, als träfe sie ein elektrischer Schlag, mitten in der Rede stecken, und schlug, blutrot im ganzen Gesicht, die Augen nieder" (LAKM, p.66). (The Princess stopped in mid-speech as if she had felt an electric shock and cast her eyes down, her whole face red as blood." LOTM, p.42).

27 The Leyden jar was invented in 1745 by Ruggiero Giuseppe Boscovich: "This electrical capacitor realized the principle of storing and boosting electricity, discovered by Pieter van Muschenbroek and Ewald Jürgen von Kleist (an ancestor of the German poet Heinrich von Kleist)" (Zielinski, p.160).

28 "Ist das Auge eine Art von Leidner Flasche für das Licht?" (Is the eye a kind of Leyden jar for light?); pp.260-61.

the princess, herself, as an organic Leyden jar in Hoffmann's story, this "pulse-beat" represents an important connection to the pulse of the heart, pumping blood through the veins to her hand, a creative association between the electrical current that directs the blood flow through the veins. Kreisler experiences this sensation again later in the story, this time in the form of a dream.[29] Her strange look ("seltsamen Blick") transmits the pulsation ("Pulsschlag") to course through Kreisler's nerves. It is through the eyes, then, analogous to Ritter's speculative question, that the electric shock is transmitted to Kreisler. In addition to the motif of electrical current returning from the *Kreisleriana*, we find here a deeper emphasis on blood as the medium of music in the body, which only became apparent in the combination of blood and electricity in the "Pulsschlag" exchanged from Princess Hedwiga to Kreisler.

When Kreisler (as the stranger) returns after hearing Julia's music, Julia sheds tears, and Kreisler asks whether her music stops at the sight of him and turns into tears.[30] As music flows into, through, and out of the body, it changes form: the music-filled blood now turns into tears. This imagery returns later in the novel, when the narrator describes Julia and the Princess' reaction after receiving a blessing from the priest; the tears that flow from their eyes are drops of blood from their wounded hearts.[31] In Fragment 603, Ritter suggests the other way around, that as water disappears from the

---

29 "Prinzessin Hedwiga stand plötzlich vor ihm, starrte ihn an mit jenem seltsamen Blick der ihr eigen, aber in dem Augenblick dröhnte ein Pulsschlag ihm durch alle Nerven, wie damals, als er zum erstenmal der Prinzessin Hand berührte. Doch war ihm auch nun jene unheimliche Angst entnommen, er fühlte eine elektrische Wärme wohltätig sein Inneres durchgleiten" (LAKM, p.312). (Princess Hedwiga suddenly stood before his mind's eye, gazing at him with the strange glance characteristic of her, and at that moment a pulse-beat throbbed through all his nerves, as it had when he first touched the Princess's hand, yet now that uncanny fear was lifted from him. He felt electrical warmth spread benignly through his inner being.) (LOTM, pp.219-20).

30 "Und diese Himmelstöne schweigen vor meinem Anblick und zerfließen in Tränen?" (LAKM, p.65). ((…) do those heavenly notes cease at the sight of me and dissolve into tears?) (LOTM, p.42)

31 "Ein namenloses Weh, aus Entzücken und Schmerz gewoben, schien gewaltsam sich loswinden zu wollen aus ihrer Brust, und Blutstropfen, dem wunden Herzen entquollen waren die heißen Tränen, die aus ihren Augen stürzten" (LAKM, p.216). (A nameless melancholy, woven of pleasure and pain, seemed to be trying to wrench itself violently from their breasts; the hot tears that flowed from their eyes were drops of blood gushing from their wounded hearts. LOTM, p.149). Directly following these tears, they blame Kreisler: " >>Das war *er*<<, lispelte die Prinzessin leise. >>*Er* war's<<, erwiderte Julie. – Sie verstanden sich." ('It was *he*,' whispered the Princess softly. '*He* it was indeed,' replied Julia. They understood one another.) (LAKM, p.216; LOTM, p.149)

earth, the number of people increase, suggesting that water itself becomes blood.[32] Hoffmann therefore seems to reverse the process that Ritter suggests, offering a complete cycle – not only does water turn into blood, but also blood into water, and specifically *musical blood* into *musical tears*.

Another interesting connection to musical blood in *Lebens-Ansichten des Katers Murr* is the description of Kreisler's *biological* musical blood heritage – implying that musical blood can be passed down generation to generation – which can be found in a conversation with the Privy Councilor who asked about his youth. Kreisler explains that in a biopsy, or a "biographical dissection," one would find his musical blood running through his veins and arteries.[33] The narrator offers a related report that Aunt Tootsie's musical influence can be found in Kreisler's breast, branching out into his veins in gushing hot blood:

> Wenigstens mag man nach dem, was Kreisler von Tante Füßchen und ihrer Laute erzählt, nicht daran zweifeln, daß die Musik mit all' ihrer wunderbaren Wehmut, mit all' ihrem Himmelsentzücken, recht in die Brust des Knaben mit tausend Adern verwuchs, und nicht zum verwundern mag's darum auch sein, daß eben dieser Brust, wird sie nur leise verwundet, gleich heißes Herzblut entquillt. (LAKM, p.125)
>
> (At least there can be no doubt, from Kreisler's account of his Aunt Tootsie and her lute, that music with all its wonderful melancholy, all its heavenly delights, took root in the boy's breast, putting out a thousand branching veins, and so it is not surprising to find that this same breast, if only slightly wounded, immediately gushes hot blood.) (LOTM, p.83)

32 "'Das Wasser auf Erden nimmt ab, die Menschen dagegen nehmen zu', – heißt mit andern Worten: das Wasser wird *Blut*." ('Water on earth decreases, humans however increase' – means in other words: water becomes *blood*. pp.442-43). The fragment continues: "Hieraus konnte man eins der Worte des Abendmahls deuten. Doch sagt Christus: Dies *ist* mein Leib, dies *ist* mein Blut. Eher also zeigte er damit seine Identität mit dem höhern Organismus an. Die Erde selbst scheint Christus zu seyn, und in Christo die Erde zu erscheinen." (From here one could interpret one of the sayings of the Last Supper. Yet Christ says: this *is* my body, this *is* my blood. Before, however, he showed thereby his identity with the higher organism. The earth itself seems to be Christ, and in Christ the earth seems to appear. pp.442-43).

33 "Der Prosektor könnte sagen: Es ist gar nicht zu verwundern, daß in dem Innern dieses jungen Mannes durch tausend Adern und Äderchen lauter musikalisches Blut läuft, denn das war der Fall bei vielen seiner Blutsverwandten, deren Blutsverwandter er eben deshalb ist" (LAKM, p.108). (The demonstrator might say: 'It is not surprising that pure musical blood runs through a thousand veins and arteries in this young man's body, for such was the case with many of his blood relations, and for that reason he *is* their blood relation.' LOTM, p.72).

Aunt Tootsie's biological transfer of musical blood to Kreisler is reminiscent of the transfer of blood in Chrysostomus' story from the girl to the rock, which appears to have acquired her music in the formation of its own red veins.

The final element connecting Kreisler to the Chrysostomus story, and in turn, Hoffmann's application of the "brilliant physicist's" theory, is found in Kreisler's letter to Master Abraham, which he wrote after having fled. He explains therein that he heard the singing of a powerful male voice walking in front of him, a Benedictine monk, who turned and left the path. Compelled by the music, Kreisler followed him and observed how the monk sat down "auf einen dick bemoosten Stein" (LAKM, p.275) ("a rock thickly overgrown with moss"; LOTM, p.193). No further explanation of the rock's appearance is offered, but immediately afterward, Kreisler recognized the monk as Father Hilarius, who invited him to stay at the Abbey "als lebendige Musik" (LAKM, p.278) ("as a piece of living music"; LOTM, p.195).[34] At the Abbey Kreisler was well received by none other than *Abbot Chrysostomus*. Suddenly, the character from the story in "Johannes Kreislers Lehrbrief" is in Kreisler's real life. Moreover, Abbot Chrysostomus is responsible for the creative environment established at the Abbey. As living music, Kreisler is finally able to calm himself to the point that he is able to compose. Having *embodied* the music he can now produce, Kreisler is in flow with the positive creative life force of music.

Hence, Kreisler's encounter with Father Hilarius at the moss-covered rock was a turning point in the return of his creative (st)ability, for the Abbey environment opened his musical channels, allowing the melodies to issue from him. In a sense, Kreisler is both the stranger in the Chrysostomus story and Chrysostomus himself; in this dual capacity, Kreisler appears as his own *Doppelgänger*. Kreisler is perceived as the mysterious stranger in the beginning, and ultimately the one to find musical peace after visiting the moss-covered rock with the singing holy man. Viewed in this manner, he represents both sides of the creative energy in music, about which the story of Chrysostomus precisely warns. Firstly, music can be misused to exert negative influences on the minds of other people, as in Princess Hedwiga's inter-

34 In the prologue to his fragments, while explaining Herder's influence on Novalis and specifically his role in verbalizing his *Die Älteste Urkunde des Menschengeschlechts* with Novalis, Ritter explains that Herder became "the living hieroglyph of the word:" "Aber Er *Selbst* in diesem Augenblicke, Sein ganzes Wesen, Sein Auge, Angesicht und Seyn, wurde zur lebendigen Hieroglyphe des Worts, für welches die Zunge das zureichende Organ nicht mehr war." ("But in this moment he himself – his entire being, his eye, his countenance and existence – became the living hieroglyph of the word for which the tongue was no longer the sufficient organ.") (pp.50-51).

pretation of Kreisler's bewitching influence over Julia and herself. Secondly, music also offers access to "higher things and [corporeal] transfiguration." The latter Kreisler achieves due to Father Hilarius and Abbot Chrysostomus. Kreisler's fragmentary biography ends with some ambivalence, and it is not clear what becomes of him or of the other characters. The important point is that his little adventure in this story led him to inner musical peace: "Musik – fromme Musik war aufgegangen in ihm, Julia hatte gesungen und nicht mehr brauste der Sturm in seinen Innern" (LAKM, p.447). (Music – the spirit of sacred music had risen within him, Julia had sung, the storm no longer raged in his heart. LOTM, p.315).

In the foregoing I have endeavored to demonstrate how Johann Wilhelm Ritter's fragmented meditations offer a little regarded model for E.T.A. Hoffmann's exploration of the world of the senses and music. Ritter's influence is apparent not only in the language Hoffmann adopts to translate between the senses, but also in his reconceptualization of the *system* through which the creative musical force flows. Although Ritter's *Fragmente aus dem Nachlasse eines jungen Physikers* lacks a complete philosophical system, the relationship between electric current, blood, and musical flow is clearly described in multi-layered form. He offers a detailed explanation of the path music takes from the spiritual realm into the human via the magnetic pull of iron in the blood. In this way, music reaches the heart and ultimately also the brain – the two major organs of *Lebenskraft* – and ceases to be just a metaphor for an out-of-body experience. As Ritter explains in his Appendix, which clearly echoes Kreisler's claims in the early *Kreisleriana* texts regarding the communicative interconnection between tone and consciousness: "*Ton* [ist] *Bewußtsein.* [. . .] Töne sind Wesen, die einander verstehen, so wie wir den Ton. [...] Alles Leben ist Musik, und alle Musik als Leben selbst – zum wenigsten sein *Bild.*" ("*tone* [is] *consciousness.* [. . .] Tones are beings which understand each other, as we do tone. [. . .] All life is music, and all music as life itself – at least its *image*" [pp.476-77, 478-79].)[35]

Hoffmann adapted this theory to the depiction of Kreisler, beginning with the *Kreisleriana* texts and continuing with *Lebens-Ansichten des Katers Murr*, where he also applied the concepts to other characters and their environment. Together, the images of electricity and blood describe, then, the musical flow and connection between the characters. They also highlight the similarities between Kreisler's biographical adventure and Chrysostomus' childhood story.[36] By describing the medium by which music travels, Ritter's

35 Ritter credits Herder's claim that all consciousness was sound or tone in his *Älteste Urkunde* and *Abhandlung über den Ursprung der Sprache.*

36 Although Kreisler kills neither Julia nor Princess Hedwiga, as would be expected in a direct parallel to the Chrysostomus story, he is almost murdered himself when he in-

scientific speculations and investigations also support the lofty claims of Romantic writers for whom music was the expression of a higher spiritual realm. But he did not stop there. According to Ritter (and Kreisler following him), music is a fundamental substance in nature. For that very reason it has the power to overcome the fragmentation of the universe. What many early Romantics considered an intangible good outside of the earthly realm now appears to exist in all things, even within the human body itself. In Ritter, Hoffmann found evidence that music permeates all matter as a special kind of energy, capable of connecting and giving life to phenomena that otherwise appear unrelated. Music, which comes from the spirit realm and travels through the blood, is the creative life force of musicians, composers, and those initiated to sense its flow.[37]

---

tervenes with Hector courting Julia. Instead he stabs his male attacker to death and subsequently leaves town, never to appear again, as the narrator explains. The only remainder of Kreisler is the bloodstained hat the Gardener finds. (LOTM, p.162; LAKM, p.234)

37 I would like to extend my gratitude to Prof. Stefan Höppner for sharing his expertise and offering me feedback on this chapter, and my friend and colleague, Eva Hoffmann, for her commentary and our many discussions that helped bring this chapter to life. No less valuable was the feedback the editors provided.

Romantic Vitalism and Homeopathy's Law of Minimum

Alice A. Kuzniar

Throughout this varied and eternal world
Soul is the only element: the block
That for unaccounted ages has remained
The moveless pillar of a mountain's weight
Is active, living spirit. Every grain
Is sentient and both in unity and part,
And the minutest atom comprehends
A world of loves and hatred.[1]

Abstract: This contribution looks at synchronistic systems of thought that contributed to the rise of homeopathy around 1800, in particular, its Law of Minimum, which stipulated that the more diluted a remedy the more potent its effectiveness. Homeopathy claims that this exponentiation (*Potenzierung*) of a dose endows it with spirit-like, dynamic qualities. Similarities can be found between the founder of homeopathy, Samuel Hahnemann, and other contemporaries who deployed vitalistic principles, such as Christoph Wilhelm Hufeland, Johann Wilhelm Ritter and Friedrich Wilhelm von Schelling, who imparted properties of life to nonliving matter. Hahnemann increasingly turned to the notion of *Lebenskraft*, positing that the human body was not only governed by such a force, but that it was also responsive to the imperceptible powers in homeopathic medicine, believing that he had found visible proof of the invisible workings of nature. He introduced mesmerism and innoculation as forms of clinical treatment. At first glance homeopathy may seem to resemble innoculation for both acknowledge a principle of "like curing like." But a differentiated understanding of *Lebenskraft* renders the way homeopathy works vastly different from innoculation; homeopathy relies on opening oneself up to the slightest modification offered by the dynamized medicine, while innoculation permits one to seal off and protect oneself from susceptibility to toxic influence. This article contributes to understanding homeopathy within the historical, cultural context in which it arose, rather than as a unique, original praxis.

The word homeopathy stems from the Greek *homoios* (= similar) and *pathos* (= sickness or feeling). The term was coined in 1810 by the founder of homeopathy Samuel Hahnemann (1755-1843) in his major work, *Organon der Heilkunst* (Organon of the Art of Healing), a treatise which underwent six editions from 1810 to 1842. But as early as 1796 in his 'Versuch über ein neues Prinzip zur Auffindung der Heilkräfte der Arzneisubstanzen' (Essay on a New Principle for Ascertaining the Curative Powers of Drugs), Hah-

---

[1] Percy Bysshe Shelley, *Queen Mab,* in *Poetical Works*, ed. by Thomas Hutchinson and G.M. Matthews (London: Oxford UP, 1970), pp.762-800, p.776.

nemann laid down his Law of Similars which was to guide all his subsequent findings:

> *We should imitate nature*, which sometimes cures a chronic disease by superadding another, *and employs in the* (especially chronic) *disease we wish to cure that medicine that is able to produce another very similar artificial disease*, and the former will be cured; *similia similibus*.[2]

This Law of Similars is known as the first law of homeopathy. Refining this principle in his 1805 essay, 'Heilkunst der Erfahrung' (The Medicine of Experience) Hahnemann wrote that there were two incompatible responses residing simultaneously in one body, the primary effect and the after-effect. Unlike with the effects of drugs working according to the principle of *contraria contrariis*, which can aggravate the original disease, the cure according to *similia similibus* would produce a slight aggravation only *resembling* the original disease. This slight aggravation would cause the body's own vital force to overcome the original illness, resulting in a permanent cure. Hahnemann writes in "Heilkunst der Erfahrung":

> *In order therefore to be able to cure, we shall only require to oppose to the existing abnormal irritation of the disease an appropriate medicine*, that is to say, *another morbific power whose effect is very similar to that the disease displays* [and i]*t is only by this property of producing in the healthy body a series of specific morbid symptoms, that medicines can cure diseases, that is to say, remove and extinguish the morbid irritation by a suitable counter-irritation.* (LW 451)

In the *Organon* Hahnemann rephrased this curative action by noting that

> a remedy [...] closely resembling the natural one against which it is employed [...] excites [...] the artificial disease [... and], by reason of its similitude and greater intensity, now substitutes itself for the natural disease.[3]

Already in this essay of 1796 Hahnemann was reacting against the heroic medicine of his day – the prevalent use of leeches, bloodletting, opium, and drastic emetics and purgatives. By contrast, he noted the effectiveness not

---

2 Samuel Hahnemann, "Essay on a New Principle for Ascertaining the Curative Powers of Drugs: with a few glances at those hitherto employed," in *The Lesser Writings of Samuel Hahnemann*, ed. by Robert Ellis Dudgeon and Erastus Edgertin Marcy (New York: William Radde, 1852), pp.249-303, p.265. Further references to the *Lesser Writings* will be annotated as LW.

3 Hahnemann, *Organon of Homeopathic Medicine* (1849), ed. by Constantine Hering (New York: William Radde, 1849), p.171. The English edition from which I am citing is based on the fifth German edition from 1833. Further references to the *Organon* will be annotated as *Organon*.

just of moderation, but specifically of the small dose. Then in 1799 he announced his principle of the infinitesimal dose, and after 1800, respecting what was to be termed homeopathy's Law of Minimum, he gradually reduced dose sizes. The impact of the catalyst was present even though the toxicity of the substance had disappeared. This theory is known as homeopathy's second law, and it will be the focus of this article. According to this principle, the homeopathic pharmacological remedy is dynamized by a series of dilutions on the scale of one part of original substance to nine parts of alcohol (or water) or one part to ninety-nine. This computation creates either the decimal scale ("X"), or the centesimal scale ("C"). Potencies are named according to the number of times they have been diluted in the X or C scale (for example, 6X, 30C, 100C). Less of the original substance means a more profound effect as a remedy becomes increasingly more energetic: the higher the number of dilutions the stronger and deeper the remedy acts. The living spirit within it becomes ever more active.

Although with these laws Hahnemann established a new medical paradigm, his innovation cannot be contemplated outside related systems of thought current at the time, as one can see in the above cited passage from Shelley's *Queen Mab*. Shelley demonstrates a belief in an "active, living spirit" or "soul" that permeates even the inorganic world, turning either a mountain or the minutest atom into a sentient being: the part is just as enlivened as the whole of nature.[4] Yet despite these parallels between Romanticism and homeopathy, proponents of this alternative medical practice tend to present its founder as unique and revolutionary, and they are not alone. Historians of medicine invariably allot him a chapter unto himself,[5] constructing

---

4 For recent discussions on vitalism in the British tradition, see Denise Gigante, *Life: Organic Form and Romanticism* (New Haven: Yale UP, 2009); Catherine Packham, *Eighteenth-Century Vitalism: Bodies, Culture, Politics* (New York: Palgrave Macmillan, 2012); and Sharon Ruston,*Shelley and Vitality* (New York: Palgrave Macmillan, 2005).

5 See, e.g., Harris Coulter, *Divided Legacy: A History of the Schism in Medical Thought.* Vol. II, *Progress and Regress: J.B. Van Helmont to Claude Bernard* (Washington, DC: Wehawken, 1977); Roy Porter, *The Greatest Benefit to Mankind: A Medical History of Humanity* (New York: W.W. Norton, 1997); and Karl Rothschuh, *Konzepte der Medizin in Vergangenheit und Gegenwart* (Stuttgart: Hippokrates, 1978). For general histories of homeopathy, see Gottfried Wilhelm Fink, *Kurze Geschichte der Homöopathie: Entdeckung und Entfaltung der sanften Heilkunde und Medizin* (Leipzig: Bohmeier, 2009); Sigrid Heinze,ed. *Homöopathie 1796-1996: Eine Heilkunde und ihre Geschichte* (Berlin: Edition Lit.europe, 1996); Robert Jütte, *Samuel Hahnemann: Begründer der Homöopathie* (Munich: dtv, 2005). Source of English translation for Hahnemann: http://www.igm-bosch.de/content/language1/downloads/SamuelHahnemannTheFounderofHomeopathy.pdf; Hanspeter Seiler, *Die Entwicklung von Samuel Hahnemanns ärztlicher Praxis* (Heidelberg: Haug, 1988); Rudolf Tischner, *Geschichte der Homöopathie* (Leipzig: Schwabe, 1939); and Matthias Wischner, *Kleine Geschichte der Homöopathie* (Essen: KVC Verlag, 2004).

the two-centuries-long narrative of homeopathy with Hahnemann at its start. Unfortunately, the unintended result of this narrative drive to establish clear beginnings is that similarities between Hahnemann and Romanticism are generally overlooked. Hahnemann, though, clearly did not formulate his ideas in a vacuum, was working under the same epistemic conditions as other writers around 1800, and shared their ways of thinking. Furthermore, the emphasis placed on Hahnemann's experience as a physician has led to neglect of the philosophical and poetic aspects of his writings. The purpose of this article is to investigate how, in particular in his Law of Minimum, Hahnemann takes recourse to the notion pervasive at this time of a vital life force (*Lebenskraft*) – or in Shelley's words, an "active, living spirit" – that guarantees health in the individual. This energetic force is present not only in the human body but also throughout nature, indeed even and especially at the infinitesimal molecular level of the remedy.

The concept of *Lebenskraft* is a crucial one for Hahnemann because, although he does not begin to use it with frequency until 1824 (he had previously used the terms *Natur* or *Leben),* it grounds the presuppositions and principles of homeopathy. Howard Tuttle has written, "[r]adical reality is the question about what absolute presuppositions are at work, or in what sort of inquiry these presuppositions are at work."[6] *Lebenskraft* is the underpinning premise that guarantees empirical reality, in fact allows Hahnemann to claim scientific truth and accuracy in his testing, although, paradoxically, *Lebenskraft* is an unverifiable intensity and imponderable power. Tuttle specifies, "[r]adical reality means here that there has been a distinction made not about something which exists, but about the form of existence in which all else appears and from which they [sic] must be explained."[7] The very word *Lebenskraft* was attractive precisely because it gave voice to both a secularized notion of independent reality or *life* and a benevolent yet powerful autonomous *force* operative within nature. It was general enough to be claimed to subtend everything in the organic and inorganic world, yet specific enough to be said to self-regulate all the internal corporeal processes. It was both internal and external, everywhere and nowhere. Most importantly, as opposed to Enlightenment sterile mechanisms, Romantic *Lebenskraft* presented nature as a unified whole that could not be broken down into constituent parts.

The term *Lebenskraft* was coined by the physician and botanist Friedrich Casimir Medicus (1736-1808) and gained widespread use in the eighteenth and early-nineteenth centuries. German scientists – from Johann Friedrich

---

6 Howard N. Tuttle, *Human Life is Radical Reality: An Idea Developed from the Conceptions of Dilthey, Heidegger, and Ortega y Gasset* (New York: Peter Lang, 2005), p.4.

7 Tuttle, p.4.

Blumenbach (1752-1840) and Gottfried Reinhold Treviranus (1776-1836) to Carl Friedrich von Kielmeyer (1765-1844) and Johann Wilhelm Ritter (1776-1810) – demonstrated a firm belief in a vital, organicist, unifying life force. But it raised more questions than it answered: Was vitalism to be explained as irritability (demonstrated in the muscles) or as sensibility (demonstrated in the nerves), as medical specialists after Albrecht von Haller (1708-1777) were to investigate? Blumenbach and Treviranus investigated infusoria and zoophytes under the microscope in order to trace what was invisible to the naked eye and yet underlay processes of generation. Not a term restricted to generative powers, vital forces were studied by Alexander von Humboldt (1769-1859) and Ritter in order to determine whether they were chemical or magnetic in nature. Did the electrical experiments conducted by Luigi Galvani (1737-1798), Alessandro Volta (1745-1827), Humboldt, and Ritter produce animal energy, contact energy, or a combination of both? *Lebenskraft* was debated among physiologists, naturalists, philosophers, and physicists alike. In all these writers *Lebenskraft* was only a postulate, which needed to be confirmed by observational inquiry. Because it could not be pinpointed, it invited scientists to set up rules that governed it, as if these would then provide the key to nature. Although the belief in vitalism was premised on the hiddenness in nature, it invited experimental intervention so as to make its unseen process evident. Thus, on the one hand, it represented the embodiment of nature and hence embeddedness within the environment, yet, on the other hand, it was an abstraction from nature. It was an undefinable term that explored, as Joan Steigerwald has noted, the "border zones of life,"[8] between plants and animals, and between the lifeless and living. De-

[8] Joan Steigerwald, 'Treviranus' Biology: Generation, Degeneration and the Boundaries of Life', forthcoming in *Gender, Race and Reproduction: Philosophy and the Early Life Sciences in Context*, ed. by Susanne Lettow. Walter Botsch (*Die Bedeutung des Begriffs Lebenskraft für die Chemie zwischen 1750 und 1850* [Dissertation, Stuttgart, 1997]) phrased the problem similarly: "[Q]uestions also remained open about the theory of a physical *Lebenskraft:* did *Lebenskraft* reside in certain elementary elements, was it dependent on the form of matter the organization of its atoms, or was it an additional principle to enhance matter?" p.236. For scholarship on medical *Lebenskraft*, see also Maike Arz, *Literatur und Lebenskraft: Vitalistische Naturforschung und bürgerliche Literatur um 1800* (Stuttgart: Verlag für Wissenschaft und Forschung, 1996); Stefan Goldmann, "Von der Lebenskraft zum Unbewussten: Konzeptwandel in der Anthropologie um 1800," in *Homöopathie und Philosophie & Philosophie und Homöopathie*, ed. Rainer G. Appell (Eisenach: Bluethenstaub, 1998), pp.149-74; Max Neuburger, *Die Lehre von der Heilkraft der Natur im Wandel der Zeiten* (Stuttgart: Ferdinand Enke, 1926); and Chapter 11 of Karl Rothschuh, *Physiologie: Der Wandel ihrer Konzepte, Probleme und Methoden vom 16. Bis 19. Jahrhundert* (Freiburg: Karl Alber, 1968). On the French medical tradition of vitalism, see Elizabeth A. Williams, *The Physical and the Moral: Anthropology, Physiology, and Philosophical Medicine in France, 1750-1850* (Cambridge: Cambridge UP, 1994).

spite this uncertainty, however, through *Lebenskraft* the entire organic and inorganic world was permitted to share the same substantiality. It expressed the dream of universality and the concatenation of being. Nature was seen as a whole, and one life force permeated all of her.

In the field of medicine around 1800, the physicians Christoph Wilhelm Hufeland (1762-1836), Andreas Röschlaub (1768-1835), Carl Arnold Wilmans (1772-1848), Conrad Joseph Kilian (1771-1812), Philipp Franz von Walther (1782-1849), and Johann Christian Reil (1759-1813) also subscribed to the notion of a life power that subtends optimal health. Even the later nineteenth-century German zoologist Otto Bütschli (1848-1920) used the terms *Lebensenergie* and *vis vitalis*. Samuel Hahnemann started using *Lebenskraft* with increased frequency from the third edition of the *Organon der Heilkunst* of 1824 on. As deployed by these physicians, the notion of *Lebenskraft* was indebted to the older sense that illness arose from an imbalance in the body. It goes back to the Neoplatonic belief that nature, although functioning according to laws inscrutable to mankind, was a vital whole that reflected the will of God. The human body was an organism that mirrored this unity and that had the natural capacity spontaneously and purposively to regain health. According to Hippocratic tenets, the task of the physician was to assist the *physis* in restoring its balance and maintaining well-being through minimal medical intervention and by advocating lifestyle moderation. As Roy Porter has summarized: "classical medicine taught that the right frame of mind, composure, control of the passions, and suitable lifestyle, could surmount sickness – indeed, presented in the first place: healthy minds would promote healthy bodies."[9] Abiding within this tradition, Hufeland argued that:

> Hippocrates, and all the physicians and philosophers of that period, knew no other method of accomplishing this end than by moderation; the use of free air and purer air; bathing; and, above all, by daily friction of the body and exercise.[10]

Prior to these writers, the physicians who also ascribed to a vital, indwelling spirit that regulated health included Paracelsus (Theophrastus Bombastus von Hohenheim, 1493–1541), who used the term *archeus* that was later taken up by Jan Baptist van Helmont (1579-1644), and Georg Ernst Stahl (1659-1734), who preferred the term *anima*. But whereas these earlier doctors saw the corporeal wellspring of vitality as inseparable from the mysterious workings of the soul (hence they were in opposition to a Descartian mechanistic

---

9 Roy Porter, 'What Is Disease?' in *The Cambridge History of Medicine,* ed. by Roy Porter (New York: Cambridge UP, 2006), pp.71-102, p.80.

10 Christoph Wilhelm Hufeland, *Art of Prolonging Life,* ed. by Erasmus Wilson (New York: Arno Press, 1979), p.4.

division of body from soul), by the late eighteenth century the body was conceived to operate according to its own independent source of vitality, although in harmony with nature.

The most famous medical proponent of *Lebenskraft* was Hufeland, physician to Goethe and founder and editor of the *Journal der practischen Arzneykunde und Wundarzneykunst* (Journal of Practical Medicine and Chirurgy), in which Hahnemann published his first, important article on the Law of Similars in 1796. Hufeland's most significant legacy was his book *Makrobiotik, oder die Kunst, das menschliche Leben zu verlängern* (Macrobiotics, or the Art of Prolonging Life), published in its first version in 1794 and still in print today.[11] In it Hufeland laid down eleven characteristics of *Lebenskraft*, including the body's capacity to react to stimulation from the environment, to fight against destructive influences, such as decay and frostbite, the capacity of *Lebenskraft* to be strengthened or weakened by external influences, and the presence of *Lebenskraft* within every part of the body. In order to guarantee longevity, *Lebenskraft* needed to be regulated and, in particular, not subjected to excessive physical or mental stimulus, which weakened it and hence curtailed one's life. Concurrent with its paradoxical use among scientists, for Hufeland too, *Lebenskraft* was an objective reality although it was beyond human sight and comprehension:

> [...] in this manner has been introduced into physics an infinite variety of powers: the power of gravity, the power of attraction, the electric power, that magnetic power, which, at bottom, signify nothing more than the letter that expresses the unknown quantity in algebra. We must, however, have expression for things whose existence is undeniable though their agency be incomprehensible.[12]

Despite this comparison to other physical forces in nature, Hufeland also writes:

> [...] the vital power is the most subtle, the most penetrating, and the most invisible agent of nature, with which we are as yet acquainted. In

11 For studies on Hufeland, see Irmgard Egger, *Diätetik und Askese: Zur Dialektik der Aufklärung in Goethes Romanen* (Munich: Fink, 2001); Stefan Goldmann, *Christoph Wilhelm Hufeland im Goethekreis: Eine psychoanalytische Studie zur Autobiographie und ihrer Topik* (Stuttgart: M&P, 1993); Joseph N. Neumann, "Christoph Wilhelm Hufeland (1762-1836)," in: *Klassiker der Medizin*, ed. by Dietrich von Engelhardt and Fritz Hartmann (Munich: Beck, 1991), I, pp.339-59; Klaus Pfeiffer, *Medizin der Goethezeit: Christoph Wilhelm Hufeland und die Heilkunst des 18. Jahrhunderts* (Cologne, Weimar, and Vienna: Böhlau, 2000); and Cornelia Zumbusch, *Die Immunität der Klassik* (Frankfurt a. M.: Suhrkamp, 2011).

12 Hufeland, *Art of Prolonging Life*, p.25.

> these respects it exceeds light, electricity, and magnetism, to which, however, it seems to have the closest affinity.[13]

Attenuating these earlier claims about the undeniable existence of *Lebenskraft*, in 1798 Hufeland published an article, 'Mein Begriff von der Lebenskraft' (My Concept of the Vital Force) in which he specified that the concept of *Lebenskraft* was merely a cipher or useful philosophical term, in short, a merely heuristic notion that designated the inner, but ultimately unknowable, fount of vitality and regeneration in the body.[14] Hufeland's hesitation seems to be a tacit admission that seeing *Lebenskraft* as a natural phenomenon hides the fact that it is merely a postulate and has actually little to do with "nature."

In part Hufeland champions the notion of *Lebenskraft* in reaction to the medical theories of John Brown that were starting to gain currency by the mid-1790s.[15] In 1795 Hufeland published in the *Allgemeine Literatur-Zeitung* an essay on the 'Erste Beurteilung des Brownschen Systems bei seiner Erscheinung in Teutschland' (The First Assessment of the Brunonian System upon its Appearance in Germany), the first of a number of polemics against Brunonianism, which, he charged, lacked respect for the independent self-healing capacity of nature.[16] Although Hufeland stood for a pragmatic and empirically-based approach to healing, in opposition to Brown and his followers in Germany, a number of the physicians inspired by Schelling's *Naturphilosophie* also believed in the organic healing powers of nature. For these thinkers, medicine was a speculative science according to which the idea of life was synonymous with the Absolute. Röschlaub, for instance, like Hufeland, parting ways with Brown, spoke of a *vis medicatrix naturae*, while Walther wrote an article, "Über die Heilkraft der Natur" (On the Healing Powers of Nature) in 1808. For Walther, nature herself was the true artist and physician. Indeed, for several writers of this period, nature was a

13 Hufeland, p.26.

14 Hufeland, "Mein Begriff von der Lebenskraft," in *Journal der practischen Heilkunde* 6 (1798), pp.785-96.

15 On Brunonian medicine, see Neubauer; Thomas Hoyt Broman, *The Transformation of German Academic Medicine, 1750-1820* (Cambridge, UK: Cambridge UP, 1996); Hans Joachim Schwanitz, *Homöopathie und Brownianismus 1795-1844: Zwei wissenschaftstheoretische Fallstudien aus der praktischen Medizin* (Stuttgart: Gustav Fischer, 1983); and Urban Wiesing, *Kunst oder Wissenschaft? Konzeptionen der Medizin in der deutschen Romantik* (Stuttgart: Frommann-Holzboog, 1995).

16 At the same time, Broman is correct to point out that "one lingering legacy of Brunonianism after its disappearance as an organized movement was the tendency of German physicians to emphasize illness as a dynamic problem affecting the entire organism, rather than a localized and material affliction" (Broman, p.156).

genius in her capacity to heal; the physician could only aid her in her artistic accomplishments.[17]

In order to contextualize historically Hahnemann's writings, it is important to understand the popular currency in medical discourse that the term *Lebenskraft* enjoyed by 1800, especially the widespread notion that the body could naturally heal itself, either spontaneously or with minimal intervention from the physician. Put simply, when Hahnemann was reacting to the heroic dosages prevalent at the time, especially to copious amounts of opium and wine that were being prescribed according to the medical theories of John Brown, his prescriptions were in line with those of other thinkers of his day. For Hahnemann, less meant more, although he was to carry this Hippocratic recommendation to the most literal level possible. Hahnemann developed the belief in the spontaneous, healing powers of nature in new directions that set him uniquely apart from the physicians mentioned above. Like the *Naturphilosophen* at the start of the nineteenth century, Hahnemann poetically believed in the relatedness of everything. Yet, unlike them, he also set out to empirically prove it. As we shall see, in this he resembles the physicists around 1800 who believed in the transcendent unity of nature as a coherent and meaningful whole whose forces could be uncovered.

Hahnemann joined the voices of Hufeland and others in challenging Brown's disregard of nature. He charged Brown of the "calumniation of nature" because the latter did not trust the powers in nature and believed that they needed to be artificially stimulated or weakened (LW 547). Hahnemann sarcastically wrote of Brown that "no medical sectarian, *apparently,* knew less about nature than he" (LW 545).[18] Because of the all-pervasive, essential energy that conjoined human and other forms of life, the founder of homeopathy insisted that his science obeyed the code of nature. Already from the

17 See Wiesing, pp.263 and 297: "Die Natur wird vernünftig, heilkräftig, sie wird zum Genie. [...] zum eigentlich Handelnden, therapeutisch genial und allmächtig."

18 Hahnemann had other objections to Brown's thinking. All diseases, according to Brown, were only differences of quantity, that is to say, variants of a single state. Hahnemann, by contrast, believed that diseases were infinite in number. Brown claimed he could heal illness by altering by degree external, physical forces of excitement, whereas Hahnemann believed in a spirit-like animating force that had the capacity to heal. Hahnemann insisted that spirit, not just physical reactions, determined health. Both thinkers, however, claimed that balance needed to be restored in the body, although the homeopathic Law of Minimum prescribed only gentle prompting of *Lebenskraft.* Hahnemann's observations on Brown are recorded in his essays "Fragmentarische Bemerkungen zu Browns *Elements of medicine*" (Fragmentary Observations on Brown's *Elements of Medicine,* 1801), in: LW, pp.350-61, and "Monita über die drey gangbaren Heilarten" (Observations on Three Current Methods of Treatment, 1809), in LW, pp.522-52.

opening of his 1796 essay he advocated imitating nature. Then in the 1810 edition of the *Organon der Heilkunst* he wrote that nature possesses her own energy ("eigene Energie der Natur").[19] Later, in the 1822 edition he refers to *Natur-Heilgesetz* (nature's healing law; *Organon-Synopse* 178). In "Heilkunst der Erfahrung" he likewise termed homeopathy a "treatment so conformable to nature" (LW 453) as opposed to the "unnecessary artificial disease" arising from medicinal doses that were too strong (LW 455). It was "the highest aim of the reflecting mind," he writes here, to imitate nature, "our great instructress" (LW 469).[20] In addition, as with Walther, Hahnemann claimed that nature possessed artistic agency: hers was, in fact, a divine art that allowed the smallest remedy to perform gently and inconspicuously the greatest effects. In brief, nature herself was the best homeopath.

As expected, Hahnemann's use of the term *Lebenskraft* parallels that of other physicians of the era to indicate that power that serves to maintain equilibrium within the body as well as a regenerative, healing force that serves to overcome illness. It represents the adaptability of the body and, as such, operates in tandem with nature. As a concept, *Lebenskraft* offered the fantasy of equilibrium. That said, there are moments in the *Organon* where Hahnemann seems to use the term *Lebenskraft* to contradictory purposes. On the one hand, he refers to it as instinctual, mindless, involuntary, or "bloß animalisch" (simply animalistic; *Organon-Synopse* 143). Disease is described as the passive suffering of *Lebenskraft*, and it is that which causes the symptoms of illness in the body. Although in itself invisible, *Lebenskraft* manifested its *Verstimmtheit* (discord; *Organon-Synopse* 299) concretely as ailments. In particular, it had to work against the dire outcomes of allopathy. Thus, Hahnemann tends to use the above adjectives to describe how *Lebenskraft* struggles to operate *without* the assistance of homeopathy. Hence, in the earlier sections of *Organon*, it seems as if Hahnemann is attacking either those physicians, like Hufeland, who prescribed relying almost exclusively on *Lebenskraft* for natural self-healing, or those physicians, like Brown, who would overprescribe medication.

On the other hand, however, while the vital life force can be described as being instinctual and mindless, it is also *geistartig* (a spiritual power; *Organon-*

19 Hahnemann, *Organon-Synopse: Die 6 Auflagen von 1810-1842 im Überblick,* ed. By Bernhard Luft and Matthias Wischner (Heidelberg: Karl F. Haug, 2001), p.620. This edition lists, next to each other for comparison's sake, all six editions of the *Organon.* I cite it in order to refer to the German variants of the *Organon.* Future references are designated as *Organon-Synopse.*

20 To be sure, Hahnemann also attacked those physicians who advocated the use of purgatives and emetics, which were seen to imitate the body's own natural evacuative processes (see *Organon*, pp.40-50).

Synopse 265). Prior to the third edition of the *Organon*, Hahnemann does not use the term *Lebenskraft* extensively. But, to emphasize nature's independent capacity for agency and self-healing, by 1829 the terms *Leben* and *Natur* are rewritten as *Lebenskraft, Lebensprincip, Lebens-Energie, Lebens-Erhaltungs-Kraft,* and *Autokratie.* The vital force can be *umgestimmt* (retuned; *Organon-Synopse* 307) with the assistance of homeopathy. Because *Lebenskraft* is incapable of memory or reason, the homeopathic remedy can trick it into dimming and extinguishing the sensation of *Krankheits-Verstimmung* (*Organon-Synopse* 309). *Lebenskraft* is merely passive, suffering, and receptive with regards to the primary effect of a drug or disease, yet more energetic in the *Nachwirkung* or *Gegenwirkung* (secondary effect of a toxin; *Organon-Synopse* 410-11), which, to repeat, is the impression or influence that gains dominance in the homeopathic, minimal dose.

In addition, Hahnemann introduces a further variation on the concept of a vital force that sets him apart from Reil and Hufeland. Hahnemann sees it operative not just within *living* nature, especially, the human body, but within the *inanimate* world as well, ecologically conjoining the two. Hufeland saw *Lebenskraft* as the wellspring of organic health, as apart from the body's chemical and mechanical processes. We recall that he also hedged his definition of it to indicate it was only a regulatory principle. By contrast, Hahnemann's understanding of vitalism resembles less that of the physicians of his time and more that of physicists, such as Ritter, who saw *Lebenskraft* as an objective reality beyond human sight but nonetheless permeating *all* matter and accessible to the human mind through scientific investigation. In defining the second law of homeopathy, Hahnemann stressed that the mineral or botanical substance in a remedy contained spiritual potency. The animating force that resided, say, in the medicinal herb left its footprint or memory in the minimal dose: the latter was the immanent manifestation of this vitality. Just as external forces affect one's *Lebenskraft* in a dynamic fashion, so too must medicines operate energetically. Put differently, under homeopathic treatment *Lebenskraft* dynamically reacts in synergy with a medication or, as Hahnemann calls it, "*Krankheits-Potenz*" (*Organon-Synopse* 333). The invalidated vital force in the individual was to be stimulated via a medicine that concentrated in itself the nonphysical, spiritual power permeating all of nature: "It is only by means of the spiritual influence of a morbific agent, that our spiritual vital power can be diseased; and in like manner, only by the spiritual (dynamic) operation of medicine, that health can be restored" (*Organon* 101). Likewise in "Heilkunst der Erfahrung" he writes: "The *dynamic* action of medicines, like the vitality itself, by means of which it is reflected upon the organism, is almost *spiritual* in its nature" (LW 466). The impact of the catalyst was present even though the toxicity of a substance had disappeared, precisely because of the dynamic intensity that it maintained.

What is so unique about the invention of homeopathy, then, is that in conjoining the concept of a vitalist force in nature with the idea of infinitesimal dosage, homeopathy develops an organicist notion into a minimalist one. The substance expresses itself wholly in the miniscule part, and, in turn, via the interface with the treatment, the human body responds to infinitesimal prompting. Moreover, according to the later editions of the *Organon*, Hahnemann postulated that a substance would be not just still present but in fact activated after exponential dilution, as well as by trituration (grinding an insoluble dry ingredient with milk sugar) and succussion (vigorous shaking of a liquid). The vibrational energy of the remedy would be thereby heightened:

> [...] this development of the spiritual power of medicines to such a height by means of the multiplied and continued *trituration* and *succussion* of a small portion of medicinal substance [...] deserves incontestably to be recognized *among the greatest discoveries* of this age. (LW, pp.729-30)

And just as the active force within the remedies could be exponentiated, so too could it be malformed: Hahnemann militated against transporting homeopathic remedies for fear that their powers would be altered by unwanted shaking. Then, by the sixth edition of 1842 in paragraph 270, Hahnemann prescribes how to dilute a substance one million times, at which point it will resolve totally into its individual, spiritual essence. As Peter Sloterdijk noted in a keynote address he gave to a homeopathic organization on its 200th anniversary, nature here is subjectivized: the remedies possess a will and power of their own.[21] Another way of saying this is that the vitality in nature brings nature in line with human consciousness, were it not for the fact that the latter, above all in terms of Fichte's idealist ego philosophy, is much more self-centered than this universal dynamism.

Hahnemann's conviction in the power of the so-called Q-potency must paradoxically be seen as adhering to a firm belief in realism. The pure current running through the infinitesimal dose is testimony to its vibratory *life*: it is concrete and materialist in its corporeal effectiveness. The activity in the dose becomes a creative agent that animates, directs, and rearranges bodily health and impels restitution. The force in the remedy flows across material

21 Peter Sloterdijk, 'Die Andersheilenden: Über einige Aspekte der Homöopathie im Lichte des philosophischen Therapiegedankens der Neuzeit: Der Anspruch auf Heilung-Revolution und Romantik in der Medizin um 1800', in *200 Jahre Homöopathie: Festakt in der Paulskirche zu Frankfurt am Main, 14. September 1996* (Heidelberg: Haug, 1996), p.27.

boundaries and thus cannot be conceived as inert matter. Furthermore, the impulse in the microdose is only evident in interaction or communication with the life force in the human being. In fact, its intensity only seems to emerge through interaction with the body, whose own *Lebenskraft* responds and is in turn enhanced. Another way of stating this enigma is that activity and spirituality can be seen as operative par excellence within the mineral or botanical medicinal substance. These minute, varied substances represent the infinite and infinitely active manifold of the natural world. The body, too, is part of this vitalistic flow. Because of the realism inherent within homeopathy, one cannot state that *Lebenskraft* is identical with the soul: whereas the soul can be said to exist without a body and represents opposition to the body, *Lebenskraft* has existence within the body. This dynamic power or intensity is life itself.

What is also curiously paradoxical about homeopathy is that, precisely at the moment when it becomes most speculative in the Law of Minimum, it claims to be based on close observation, empirical testing, and verifiable evidence. Hahnemann writes in the *Organon*:

> By a mere effort of the mind we could never discover this innate and hidden *faculty* of medicines – this spiritual *virtue* by which they can modify the state of the human body and even cure disease. It is by experience only, and observation […], that we can either discover or form to ourselves any clear conception of it. (*Organon*, p.103)

In his own self-testing of medicinal substances and in his observation of the homeopathic effects of his remedies, Hahnemann prods the body to see how *Lebenskraft* manifests itself. This intervention is necessary so as to make the process evident that was otherwise imponderable and imperceptible. Typical for the Romantic move away from the eighteenth-century belief in the reliance on and accuracy of sight, homeopathy thus claims to demonstrate something invisible. Hahnemann's experiments, driven by his hypotheses, purport to show the hidden interiority of how illness and cure works within the body. In short, the efficacy of homeopathic drugs offered, Hahnemann claimed, tangible evidence of the intangible. What he fails to acknowledge was that there could be tension between empirically based scientific experimentation and the philosophy of vitalism.

As such, homeopathy is a movement synchronous with other scientific efforts around 1800 that wanted to bridge the ideal of *Wissenschaft* with the poetic idea of life. It can be compared, in particular, to the experiments that made visible the workings of electricity, beginning with Luigi Galvani who claimed to be the first "to hold in his hands, as it were, this electricity which is concealed in the nerves, and to draw it forth from the nerves and to set it practically before our eyes."[22] Galvani thought he had proven the presence of innate animal electricity in the frogs on which he experimented, although, as Volta later challenged, the electricity was actually produced by contact with the metal. Ritter took these trials one step further by conducting electrical tests on his own body and thus like Hahnemann engaged in self-experimentation. For all these scientists, including Hahnemann, an infinite process undetectable to the naked eye was still able to produce corporeal, physical results, bringing a radical reality to light. As Iwan Morus summarized in *When Physics Became King*, in the nineteenth century there was an "increasing array of experiments and instruments designed to make electricity visible."[23]

The Law of Minimum also seems to be indebted to calculus and the concept of the infinitesimal. Analogous to a function, $f(x)$, that approaches zero as x approaches infinity (i.e., the distance between $f(x)$ and zero can be made infinitesimally small), according to the function operative in homeopathy's Law of Minimum, the medicinal ingredient could still retain certain properties even though the amount of it was quantitatively negligible. Indeed, Newton referred in his *Principia* to vanishing quantities. Homeopathy, too, offered a theory of how substances could exist even though they were so miniscule that there was no way to see or measure them with the naked eye. That is, homeopathy, like calculus, is indebted to a principle of continuity such that reality, like the line of real numbers, is infinitely divisible (or in Leibniz's words, *natura non facit saltus*).[24] In fact, in the 1833 edition of the *Organon*, Hahnemann himself says:

> Mathematicians will inform them, that in whatever number of parts they may divide a substance, each portion still retains a *small share* of the material; that, consequently, the most diminutive part that can be conceived never ceases to be *something*, and can, in no case, be reduced to nothing. (*Organon* 221)

In Spinozistic terms, one would say that there is no nothingness, only substance. Finally, in what seems to be a paradoxical inversion of mathematical principles, the strength or energy of a remedy actually increases exponentially with each dilution. Here one can speak of the incommensurability of the life force in homeopathy: although one cannot say for sure that a life force exists and it cannot be mechanistically explained, it seems nonetheless to be present in the homeopathic dose; the latter prompts and converts energy.

---

22 Luigi Galvani, *Commentary on the Effects of Electricity on Muscular Motion,* ed. by Margaret Glover Foley (Norwalk: Burndy Library, 1954), p.79.

23 Iwan Rhys Morus, *When Physics Became King* (Chicago: U of Chicago Press, 2005), p.98.

24 I wish to thank John H. Smith for our discussions about infinitesimal calculus.

Yet despite these ties to ideas from mathematics, physics, and philosophy circulating at the time, the belief that the elemental, essential action of a substance could be extracted and transmitted seems to place Hahnemann squarely back within the older tradition of alchemy, especially since he claimed that the spectral vibrant force in the remedy was somehow purer in distilled form. The micro-dose displayed a mysterious performative magic. Another way of putting it is to say that, like in alchemy, the process of dilution/distillation leads to the transmutation of the ingredient. The latent energy of a substance unfolds and is refined and clarified. In alchemy, the spirit of metals was supposed to have been released from its base surroundings upon heating: through homeopathic dilution, the spiritual force in the tincture was likewise claimed to be enhanced or, to use Hahnemann's phrase, dynamized. Finally, although homeopathy is not about finding the elixir of life or extracting gold from base metals, the homeopathic operation similarly involves only a few particles of matter.

Moreover, the belief in the infinitesimal presence of the substance seems similar to magical thinking in the Paracelsian belief in the Signature of Things, whereby the appearance of a plant was supposed to be the key to its healing properties. For instance, boneset (*Eupatorium perfoliatum*) was said to mend broken bones because its perfoliate leaves appear to be interlocked or "knitted" at the stem. But as a practical physician whose roots were in the Enlightenment, Hahnemann explicitly rejected the belief in the Signature of Things as well as any faith in astrology.[25] His Law of Similars operates according to the *comparison of symptoms*, and as such is unrelated to the thinking in *analogy between plant appearance and the human body* that characterizes the Signature of Things. In other words, homeopathy's laws are based on a belief in the unchangeable, eternal, and undeviating principles in *nature*, not on the basis of the presence of *universal, divine being* in its smallest manifestations. Hahnemann did believe that nature was indivisible and completely present in each of its parts, given that it was infinite. And he did believe that man and nature worked in accordance with similar laws. But he was too much a product of his secular times to believe that there was a sympathetic, magical correspondence between a universal macrocosm and the microcosm. Furthermore, the efficacy of the individual medicinal element does not so much, as in Paracelsus, rely on its capacity to represent or mirror the godly order of the universe. Instead, the elemental substance is one and the same or, to restate, coextensive with the energy residing within all things. Conceivably the other reason why there are no allusions to Paracelsus in Hahnemann's works is that Paracelsus was decried by such famous physicians as

25 Although at one point in the *Organon* he does allude to "the sway of meteoric and telluric influence" (*Organon*, p.137).

Hufeland, who wrote: "one of the most impudent quacks and greatest boasters among the prolongers of life was Theophrastus Paracelsus."[26] Had Hahnemann located and borrowed Paracelsus's analogous concept of *similia similibus curantur,* he conceivably would never have admitted it openly for fear of exposing himself to even greater derision than he already experienced.[27]

Both laws of homeopathy – the Law of Similars and the Law of Minimum – invite contrast as well to how concepts of inoculation and immunity were conceived around 1800, especially in terms of the workings of *Lebenskraft.* Here Hahnemann differs significantly from Hufeland, a divergence that points to the uniqueness of homeopathy. There are moments in Hufeland's *Die Kunst, das menschliche Leben zu verlängern* that seem to resemble in its call for moderation Hahnemann's notion that a toxin in small doses strengthens the body. For instance, Hufeland believes that exposure of the body to moderate degrees of cold can strengthen or toughen it:

> [...] by such daily enjoyment of air, acquainted and familiar with the free atmosphere [...] people are thus secured against one of the greatest evils that usually afflict mankind, I mean *too much sensibility in regard to all impressions and variations of the weather.* This is one of the most abundant sources of disease; and there is no other method of counteracting it, but to harden oneself by daily exposure to the open air.[28]

Towards the end of *Die Kunst, das menschliche Leben zu verlängern* Hufeland speaks of how to procure a degree of immunity from grave dangers to the body: he recommends "the utmost possible agility and readiness in all bodily exercises" and that one should

> [...] endeavor to render the mind intrepid; to give it strength and philosophical equanimity; and accustom it to sudden and unexpected events.[...] One will thereby guard against the physical injury of sudden and alarming impressions.[29]

To get a little less sleep than normal, to drink a glass more of wine, or to do gymnastic exercises, keeps the body from residing in too passive a state and hence serves as preventative measures against disease.[30] If you will, they of-

---

26 Hufeland, *Art of Prolonging Life,* p 7.

27 Gantenbein focuses on several parallels rather than fundamental differences between Hahnemann and Paracelsus in order to suggest that Paracelsus functions as a dark Jungian shadow to the founder of homeopathy, who resisted acknowledging direct influences. Urs Leo Gantenbein, 'Similia Similibus: Samuel Hahnemann und sein Schatten Paracelsus', in *Nova Acta Paracelsica,* 13 (1990), pp.293-328.

28 Hufeland, *Art of Prolonging Life,* p.259.

29 Hufeland, *Art of Prolonging Life,* pp.315-16.

30 Cf. Hufeland, *Art of Prolonging Life,* p.303.

fer examples of "like *preventing* like." Although Hufeland himself was later to recommend against the smallpox vaccination because it represented to him an invasion of the body's integrity, his notion of steeling *Lebenskraft* parallels the notion of guarding and protecting the self that inoculation represents: a small dose of the pharmakon will boost one's immunity against greater exposure. In sum, for Hufeland, *Lebenskraft* is best sustained and guaranteed equilibrium by the moderation of stimuli. As a final example, for the elderly, who see a decrease in the vital life force, he recommends a moderate degree of wine for stimulation. Yet in a paradoxical example of how a diminished *Lebenskraft* actually sustains and prolongs life, he also says that, because the intensity of the elderly's responsiveness is reduced, they are also less susceptible to infectious diseases than the young.[31]

Although Hahnemann also prescribes moderation and explicitly states that inoculation operates according to homeopathic principles, his notion of *Lebenskraft* actually veers from Hufeland's concept of strengthening immunity. The dichotomy works as follows: moderation, although also a goal in itself for Hahnemann, is a basic prerequisite to homeopathy's proper functioning. In other words, the body will only be sensitive to the homeopathic remedy if it is not exposed to other stimuli such as wine or coffee. What this means, however, is that, rather than toughening one's immunity, homeopathy is about increased receptivity to the gentle stimulus and force field that the microdose has to offer. Stated briefly, inoculation means developing a resistance, a hardening, or habituation, precisely in order to avoid susceptibility that is the precondition for homeopathy's effectiveness.[32] Hahnemann, for instance, claimed that the sick patient is far more impressionable and susceptible to the action of a medicine than a healthy person.[33] In fact, a patient who is sick with an acute fever doesn't [does not] even need to ingest a drop of medicine, for the mere smell of it is enough to produce results.[34] Notwithstanding their differences, however, Hahnemann, Hufeland, and Brown do overlap to the extent that they offer variations on the same eight-

31 Cf. Hufeland, *Art of Prolonging Life*, p.322.

32 Hahnemann includes a footnote in the *Organon* in which he states that testers will not suffer from any detrimental effects to their health: "[o]n the contrary, experience has shown us that they only render the body more apt to repel all natural and artificial morbific causes, and harden it against their influence. The same experience also teaches, that thereby the health becomes more firm, and the body more robust"(*Organon*, p.168).

33 Hahnemann, "The Medicine of Experience," in LW, pp.435-76 (p.464).

34 Hahnemann, "On the Power of Small Doses of Medicine in General and Of Belladonna in Particular." in LW, pp.385-89 (p.388).

teenth-century framework of irritability and sensibility and in their belief that health resided in the body's ability to maintain and restore equilibrium.

Finally, in order to situate Hahnemann's medical usage of the term *Lebenskraft*, it is important to mention that in the third (1824) through the sixth (1842) version of the *Organon* he refers to another popular healing modality, mesmerism, comparing homeopathy to it. Mesmerism, as a form of energy work, underscores and illuminates certain salient characteristics of homeopathy. Like the homeopath, the practitioner of animal magnetism manipulates *Lebenskraft*, that is to say, rights the energetic flows in the body. Hahnemann claims,

> [...] the powerful will of a well-intentioned individual influences the body of the patient by the touch, [and] acts homeopathically by exciting symptoms analogous to those of the malady – and this object is attained by a single transit, [...] gliding the hand slowly over the body from the crown of the head to the soles of the feet [...]. It acts likewise by imparting a uniform degree of vital power [*Lebenskraft*] to the organism when there is an excess of it at one point and a deficiency at another. (*Organon* 297)

He continues by saying that mesmerism "acts by immediately communicating a degree of vital power to a weak part or to the entire organism [...] without interfering with the other medical treatment" (*Organon* 228). As in homeopathy, in mesmerism the disease is subdued, allowing the otherwise suppressed vital life force to kick in.

In addition to the restoration of health occasioned by the revitalized *Lebenskraft*, both mesmerism and the energized homeopathic dose are similar in that they demand a receptive, open patient and respond to the capacity in the individual to be stimulated. Both involve sympathy, flows, and exchange between and across bodies and matter. Both call in various ways upon the invisibility topos. For instance, even with eyes closed, the somnambulist purports to see. In some cases she can even be clairvoyant, seeing parallelisms [parallels] and correspondences that cannot be otherwise explained, in a form of what James Frazer in *The Golden Bough* called "homeopathic magic."[35] Moreover, like the dynamic force in the homeopathic remedy, in mesmerism what heals cannot be seen, yet acts upon the body, specifically a

---

35 Frazer classifies "Homeopathic or Imitative Magic" under the category of "Sympathetic Magic" according to which "things act on each other at a distance through a secret sympathy, the impulse being transmitted from one to the other by means of what we may conceived as a kind of invisible ether." James George Frazer, *The Golden Bough: A Study in Magic and Religion* (New York: Macmillan, 1956), abridged edition, p.14.

force in the body that lies beyond conscious will.[36] The movements of the magnet, a pendulum, or the mesmerizer's hands are governed by natural phenomena that make manifest an eternal but invisible vital principle.

Gotthilf Heinrich Schubert (1780-1860) was one of the most important Romanticists to write on animal magnetism in both *Die Symbolik des Traumes* (Symbolism of Dreams) and *Ansichten von der Nachtseite der Naturwissenschaft* (Perspectives on the Dark Side of the Natural Sciences), and, although he does not refer to homeopathy, there are several moments in his work that evoke this other alternative medical practice.[37] He points out how pure gold, similar to the touch of the mesmerizer, can give a pleasant feeling to the person in a magnetic slumber.[38] In homeopathy, of course, certain substances comparably exert healing properties. Schubert notes the properties of magnetized water,[39] analogous to how, in homeopathic dilution, the menstrum carries and transmits energetic properties. Finally, he compares the state of somnambulism with its deep sleep to that of death:[40] as in homeopathy, healing arises from a form or variant of a toxicological pharmakon. Overall, the reason that mesmerism culminates Schubert's discussion of natural science is that it gives the perfect example of "das ewig harmonische Zusammenwirken des Weltalls in allen seinen Theilen" (the eternally harmonic working together of the World-All in every part).[41] The goal of his study is to point to "die Lebensseele, welche von oben ausgehend, alle Natur bis in das Aeußerste und Kleinste durchdringt" (the vital soul, which descending from above, suffuses all of nature even into the most external and smallest elements).[42]

Conclusion

Samuel Hahnemann vociferously criticized what he saw as the over-speculative *Naturphilosophie* of Friedrich Wilhelm Joseph Schelling (1775-1854) and his followers, writing that their "dualism, their polarization, and representation [...] their potentizing and depotentizing [...] incorporeal and ethereal, [...] soars aloft beyond our solar system, beyond the bounds of the

36 As if trying to visualize or metaphorize the otherwise invisible vital force, Schubert writes that the patient of magnetism experiences light pervading her body. Cf. Gotthilf Heinrich Schubert, *Ansichten von der Nachtseite der Naturwissenschaft: Nachdruck der Ausgabe von 1808* (Karben: Petra Wald, 1997), p.357. The translations are my own.

37 Hufeland, by contrast, warned against the charlatanry of mesmerism as early as 1784 in an essay entitled "Mesmer und sein Magnetismus."

38 Cf. Schubert, Ansichten, p.336.

39 Cf. Schubert, Ansichten, p.336.,

40 Cf. Schubert, Ansichten,, p.359.

41 Schubert, Ansichten,, p.372.

42 Schubert, Ansichten, p.372,

actual."[43] Yet, as we have seen, despite this attack, clear parallels exist between Hahnemann and other important writers of this time, above all in terms of their common belief in nature's vitalism. For Hahnemann, *Lebenskraft* was not just a regulatory force of healing and well-being in the human body; it also completely suffused nature. Clearly, as a physician who conspires with life, Hahnemann, although openly anti-Schelling, notwithstanding closely resembles Schelling, who philosophized that the human subject could not be opposed to nature, because he or she was part of nature. Hahnemann, too, as we have seen, joins a host of scientists, such as Lorenz Oken (1779-1851) and Ritter, as well as physicians who followed Schelling, such as von Walther and Röschlaub, who all supported a unified view of the world that overcame dualisms and saw the infinite in the finite, natural realm.

In closing, I wish to cite, though without going into exegetical detail, a few key passages that Johann Wolfgang von Goethe (1749-1832) wrote in the 1790s that resonate with Hahnemann's vitalist philosophy, most concretely expressed in the latter's Law of Minimum. The idea that a minute substance contains *in nuce* the vital energy present throughout the infinitely manifold natural world can be found patterned in a number of thinkers from this era but, above all, in Goethe. In setting about to study the grammar of botany in *Die Metamorphose der Pflanzen* (The Metamorphosis of Plants, 1790), Goethe believed that an inner law residing within the plant governed its organic growth and manifested itself in various forms. Despite their diversity, the separate vegetal organs are one and the same: from the individual attributes of a plant one can perceive its unifying wholeness. Goethe writes: "Es mag nun die Pflanze sprossen, blühen oder Früchte bringen, so sind es doch nur immer *dieselbigen Organe*, welche, in vielfältigen Bestimmungen und unter oft veränderten Gestalten, die Vorschrift der Natur erfüllen" (Whether the plant sprouts, blooms, or carries fruit, it is always *the same organs*, which, despite manifold purposes and in the guise of often variegated forms, fulfill the prescription of nature).[44] In other words, by paying close attention to the disparate parts of a plant one can intuit the essence behind them. Thus, the distinctive phenomena in nature – think as well homeo-

---

43 Hahnemann, 'On the Value of the Speculative Systems of Medicine, especially as viewed in connection with the usual methods of practice with which they have been associated', in LW, pp.488-505 (p.495).

44 Johann Wolfgang von Goethe, *Werke*, ed. by Erich Trunz (Munich: Beck, 1978) XIII, p.100, §115. The translations of Goethe are my own. Frederick Amrine offers a helpful explanation of this passage: "Goethe reads the process [of anastomosis] as revealing, intuitively, the inner identity of the plant." Frederick Amrine, "Goethean Intuitions," in: *Goethe Yearbook* 18 (2011): 35-50 (p.47).

pathic infinitesimal, potentiated substances – are linked to the entirety of nature and reveal its inner, unseen, teleological forces. And Hahnemann, like Goethe, believed one could come to this essence through close observation of nature; but it was an empirically trained eye that also possessed a gift for perceiving analogies intuitively.

To the same purpose of linking the disparate elements in nature to a unified, dynamic whole, Goethe wrote in his 1792 essay, "Der Versuch als Vermittler von Objekt und Subjekt" (The Experiment as Mediator of Object and Subject):

> [...] in der lebendigen Natur geschieht nichts, was nicht in einer Verbindung mit dem Ganzen stehe. [...] Da alles in der Natur, besonders aber die gemeinern Kräfte und Elemente in einer ewigen Wirkung und Gegenwirkung sind, so kann man von einem jeden Phänomene sagen, daß es mit unzähligen andern in Verbindung stehe, wie wir von einem freischwebenden leuchtenden Punkte sagen, daß er Strahlen auf allen Seiten aussendet. (*Werke*, vol. XVIII, pp.17-18)

> ([...] in living nature nothing occurs that does not relate to the whole. [...] Since everything in nature, but especially the most common forces and elements, are in eternal activity and counter-activity, one can say of every single phenomenon that it stands in relationship with countless others, as one can say of a hovering, illuminating point that it sends its beams in all directions.)

Converting Goethe's words into homeopathic vocabulary, the miniscule dose embodies this freely suspended, irradiating element that sends its rays outwards and re-energizes *Lebenskraft*. Finally, it should be mentioned that the 1790s also saw Goethe develop his theory of the symbol, which is to say, the making visible of the invisible. Like Goethe's concept of the symbol, the infinitesimal homeopathic quantity holds the essence of what it represents. Even in an infinitely small dosage, the homeopathic cure exemplifies the living, active presence in all of nature. Indeed, through its dynamism it bodies it forth rather than merely signifying an unseen presence. In sum, then, the homeopathic dose is radical reality.

Folklore and Physiology: The Vitality of Blood in the Works of the Brothers Grimm

Ann C. Schmiesing

Abstract: This essay explores discursive affinities between the Grimms' assessment of folk beliefs concerning the vitality of blood and the physiological notions of vitality espoused by their contemporaries. The concept of organic assimilation (*Verähnlichung*) not only shapes the Grimms' analyses of the body's ingestion of substances such as blood in folklore and medieval literature, but also yields insights into their desire as editors to gently nurture and rejuvenate folk texts by carefully supplementing appropriate wordings or by intermixing tale variants into an organic whole. These affinities are particularly striking in light of Wilhelm Grimm's medical treatment in Halle by the famed Romantic physiologist Johann Christian Reil. Reil's Romantic notion that healing the body entailed bringing it back into equilibrium with internal and external forces is analogous to the Grimms' vitalistic conception of their editorial project.

In *Von der Lebenskraft* (On the Vital Force, 1795), Johann Christian Reil describes the body's ability to incorporate foreign substances into itself as a manifestation of *Lebenskraft*.[1] The substances the body absorbs through the mouth, lungs, skin, and intestinal tract and then uses to build and repair itself are many, and include "unbekannte Stoffe aus der Luft, das Blut der Mutter, die Muttermilch, die Arzneien und die Nahrungsmittel" (unknown substances from the air, the blood of the mother, the mother's milk, medications, and nourishment).[2] Because of the existing or potential similarity of these substances to the body's components, Reil postulates, the body can use the absorbed substances either to increase its mass or to alter "die Mischung der Materie" (the mixture of its material), explaining:

> Der tierische Körper nimmt Stoffe von außen auf, die entweder seinen Teilen schon ähnlich sind, oder die er ihnen erst verähnlicht. Diese Stoffe werden von seinen Teilen angezogen, und bei der Anziehung wird ihnen zugleich eine zweckmäßige Bildung und Form mitgeteilt. Der tierische Körper macht also fremde Stoffe sich zu eigen.[3]
> (The animal body incorporates external substances which are either already similar to its own components or which the body first makes similar. The body's

---

1 Setting forth his conception of *Lebenskraft*, Reil writes: "Ich habe die Kraft der Materie, die das Pflanzen- und Tierreich charakterisiert, Lebenskraft genannt, und das Wort Leben im weitläufigsten Verstande genommen" (I call the force of the material that characterizes the botanical and animal realms *Lebenskraft* [vital force], and understand the word 'life' in the broadest sense). See Johann Christian Reil, *Von der Lebenskraft*, ed. by Karl Sudhoff (Leipzig: Johann Ambrosius Barth, 1910), p.25. Unless otherwise noted, all translations are my own.

2 Reil, *Von der Lebenskraft*, p.36.

3 Reil, *Von der Lebenskraft*, pp.35-36.

components attract these substances and in this attraction the substances are given an appropriate development and form. Thus the animal body makes foreign substances its own.)

Rejecting Johann Friedrich Blumenbach's categorization of the process by which the body incorporates foreign substances into itself as a formative power or drive (*Bildungskraft* or *Bildungstrieb*), Reil further argues that such terms are too narrow and misleading to describe a process that to him should be understood in the context of *Lebenskraft.* In keeping with the materialistic and mechanistic concept of the vital force that he sets forth in *Von der Lebenskraft*, he thus rejects notions that physiological processes such as digestion and reproduction are functions of anything beyond the basic chemical makeup and structure of the body. Influenced by Romanticism, however, Reil would later embrace the notion of *Bildungskraft* or *Bildungstrieb* and emphasize the organism's striving for the ideal both in its own individual form and as part of the universal organism that is nature.[4]

The body's ability to absorb substances and use these to grow and heal itself figures prominently in many European fairy tales and legends, where fascination with this ability is evident in the magical properties with which it is frequently associated. As part of what we might conceptualize as a folkloric notion of *Lebenskraft* or *Bildungstrieb*, such tales and legends frequently depict characters who are miraculously healed after ingesting another's blood or heart, eating an herb or fruit, drinking special water, or being touched by magical dew. The potentially harmful effects of ingestion are also thematized, as when a character is sickened by the breath of a malevolent character or transformed into a non-human form after ingesting food or water. Taken together, these tales and legends offer insights into folk beliefs concerning which external substances enhance the integrity of the body and which substances imperil it – or to use Reil's terminology, of which substances can and cannot be beneficially *verähnlicht* (made similar) to and by the body.

In this essay, I examine manifestations of this folkloric conception of *Verähnlichung* or assimilation, and the healing and regeneration it enables, in works edited by Jacob and Wilhelm Grimm. I focus on the Grimms not only because Reil himself was Wilhelm Grimm's physician in Halle and Wilhelm frequently writes of Reil's treatment of his heart ailment in his correspondence, but also because the Grimms' fascination with disease and

---

4 For a discussion of Reil's principal criticisms of Blumenbach's concept of the *Bildungstrieb* and his later embrace of the term, see Robert J. Richards, *The Romantic Conception of Life: Science and Philosophy in the Age of Goethe* (Chicago: U of Chicago Press, 2002), pp.255-56 and pp.280-81.

cure, and in particular with folkloric portrayals of the healing powers of blood, is evident in works such as their afterword to the edition they published in 1815 of Hartmann von Aue's medieval narrative poem *Der arme Heinrich* (Poor Henry, ca. 1190). The afterword reveals conceptual affinities between the Grimms' assessment of folk beliefs concerning the vitality of the blood and late eighteenth- and early nineteenth-century physiological notions of vitality. More strikingly, the concept of *Verähnlichung* yields insights not only into depictions of the body's ingestion of substances such as blood within the fairy tales and folk texts the Grimms collect and comment on, but also into their own conception of their editorial process as that of sustaining texts by means of the careful supplementation of appropriate wordings or the intermixing of tale variants into an organic whole. After first considering Reil's treatment of Wilhelm Grimm, I turn to the Grimms' commentary on blood and other substances ingested by characters in the folk texts they edited, and then study the Grimms' notion of the incorporation of textual additions and amendments into the texts as a process akin to organic assimilation.

Baths, Pills, Salves, Magnets, and Electric Shock Therapy: Reil's Treatment of Wilhelm Grimm in Halle

In the spring and summer of 1809, Wilhelm Grimm spent several months in Halle under the care of Johann Christian Reil. He had been in precarious health off and on since the autumn of 1802, when he grew ill with scarlet fever and asthma and as a result was unable to attend school for half a year. His health worsened in 1808, and at this time he suffered in particular from terrifying chest pains. In Halle, Reil diagnosed the cause of these chest pains as atony of the heart muscle, but it has subsequently been assumed that Wilhelm suffered from essential paroxysmal tachycardia, a condition in which the heart suddenly beats at two or three times its normal rate, with episodes lasting from a few minutes to several hours and accompanied by fatigue and lightheadedness.[5] As Wilhelm related to Jacob in a letter from May 1809, Reil prescribed an elaborate regimen of pills, powders, and salves, including medication given to Wilhelm by Johann Wilhelm Heinrich Conradi, his doctor in Marburg:

---

5 See H. C. Fairbrother, "A Remedy for Paroxysmal Tachycardia," *Journal of the American Medical Association* 53:4 (1909), (300); and Steffen Martus, *Die Brüder Grimm: Eine Biographie* (Berlin: Rowohlt, 2009), pp.141-43. For a more detailed study of Wilhelm Grimm's treatment and the effect of his heart ailment on his work, see chapters 1 and 2 of Ann Schmiesing, *Disability, Deformity, and Disease in the Grimms' Fairy Tales* (Detroit, Michigan: Wayne State UP, 2014).

> Arznei muß ich recht viel nehmen, morgens, wenn ich aufstehe, um 1/2 8 reibe ich den Hals ein, jetzt mit einer schwarzen starken Merkurialsalbe, darnach wasch' ich das Herz mit Spiritus. Um 9 Uhr nehm' ich ein Pulver, das mir äußerst zuwider und das ich alle Monate bei abnehmendem Mond erhalte; weil es mir sehr übel darauf wird, eine halbe Stunde später bittere Essenz, um den Magen und Eßlust wiederherzustellen, um 11 Uhr Conradis Pillen, um 12-2 wieder Herzwaschen, Pulver und Essenz, die mir nun einen heftigen, zitternden Hunger macht, der aber durch weniges Essen bald gestillt wird; um 4 Pillen und Waschen und vor Bettgehen noch einmal Pillen und Waschen. Diese Arznei kostet mich sehr viel, schon an 25-30 Taler, da das sprirituöse Wasser so teuer und ein Verschreiben oft ½ Louisdor kommt.[6]
>
> (I have to take quite a bit of medicine. When I get up in the morning, I rub my throat at half past seven with a mercury salve, and after that I wash my chest with spirits. At 9 o'clock I take an extremely abhorrent powder that I receive each month when the moon is waning. Because it makes me feel so awful, I take a bitter essence half an hour later to restore my stomach and appetite, then at 11 o'clock Conradi's pills, and from 12 to 2 o'clock I wash my chest again and take the pills and essence, which this time makes me quiver with such a strong hunger, though my hunger is satisfied after eating just a little. At 4 o'clock pills and chest washing, and then pills and chest washing again before bed. This medical treatment costs me a lot, 25 to 30 talers already, because the spirits are so expensive and a prescription often comes to ½ Louisdor.)

In subsequent letters to Jacob, Wilhelm wrote of the magnet, bath, and electric shock therapies Reil had prescribed. He noted the strange effects that he experienced the first time he wore a magnet, sewn into a black band and worn over his heart: his heartbeat grew irregular, he perceived a metallic taste in his mouth, and he became dizzy, nauseous, and sweaty.[7] He also described holding a very cold sponge over his heart while sitting halfway in hot water, a treatment intended to change the direction of his bloodstream,[8] and he related that at one point Reil had the idea of using the stimulus of the iron and brine baths to activate a rash on Wilhelm's body before he placed the cold sponge on his chest.[9] Invigorated by his bathing regimen, Wilhelm recommended that Jacob, too, try bathing for eight days in a row in warm water into which soap has been dissolved, claiming that if he did this in the morning when the outside temperature was cool it would thin his blood and stimulate his appetite.

---

6 Jacob and Wilhelm Grimm, *Briefwechsel zwischen Jacob und Wilhelm Grimm aus der Jugendzeit*, 2nd edn, ed. by Herman Grimm and Gustav Hinrichs (Weimar: Hermann Böhlaus Nachfolger, 1963), pp.102-3.

7 Jacob and Wilhelm Grimm, *Briefwechsel*, p.113.

8 Jacob and Wilhelm Grimm, *Briefwechsel*, p.118.

9 Jacob and Wilhelm Grimm, *Briefwechsel*, p.123.

As for electric shock treatment, Wilhelm reported that electric shocks were administered to him through a chain worn around his neck while he sat halfway submerged in water;[10] later, he would first bathe and then go to the clinic for electric shock therapy. "Eine prächtige, große Maschine von Mahagoniholz wird da gedreht," he explained, and "auf einen Tisch mit Glasbeinen, worauf ein Armesünderstühlchen, muß ich mich setzen, und mit Ketten werd' ich dann in Verbindung gebracht, und die Electricität strömt durch mich" (a splendid large machine made of mahogany wood is turned, and I must sit on a cutty stool atop a table with glass legs, and I am then attached to the machine with chains and the electricity streams through me).[11] He found this shock treatment increasingly unpleasant and complained that it made his skin blister, but he was in any case able to tolerate it better than some of the other patients, two of whom screamed and jerked terribly as the electric current streamed through their bodies.[12]

Bizarre as they might seem today, the various treatments that Wilhelm underwent in Halle exemplify the Romantic notion that healing the body entailed bringing it back into equilibrium with internal and external forces. As Steffen Martus observes with regard to Reil and Romantic medical theory, "Der Arzt versteht sich weniger als eine Art Mechaniker, der von außen eingreift, sondern arbeitet eher mit diffusen Kräften wie eben Elektrizität oder Magnetismus, die den Körper dazu bringen sollen, seine Funktionen selber zu regulieren" (The doctor sees himself less as a kind of mechanic who intervenes from the outside, but instead works with diverse forces such as electricity or magnetism, which are supposed to bring the body to regulate its functions itself).[13] By the time he was treating Wilhelm Grimm in Halle, Reil's views of the body had changed from the mechanistic and materialistic theories he had expounded in *Von der Lebenskraft* to a much more Romantic conception influenced by Schelling and others. Nevertheless, many of Reil's views in *Von der Lebenskraft* yield insights into what Reil may have hoped to achieve in treating Wilhelm. He writes in *Von der Lebenskraft* of the effects of air, electricity, and nourishment on the body in discussing the human being's ability to adapt to and bring itself into equilibrium (*Gleichgewicht*) with the various forces and substances in its environment, noting that these and other forces and substances both add to and take from the body: "Die Atmosphäre, die Wärme, das Licht, die Elektrizität, die Nahrungsmittel usw. haben einen beständigen Einfluß auf die tierische Maschine, teilen ihm Stoffe mit und entziehen ihm andere"

---

10 Jacob and Wilhelm Grimm, *Briefwechsel*, p.113.

11 Jacob and Wilhelm Grimm, *Briefwechsel*, p.124.

12 Jacob and Wilhelm Grimm, *Briefwechsel*, p.129.

13 Martus, *Die Brüder Grimm*, p.146.

(The atmosphere, heat, light, electricity, nourishment, etc. have a constant influence on the animal machine, imparting substances to it and pulling others from it).[14] The body is thus constantly changing in force and in the composition of its material as it absorbs or interacts with external agents. While this ability is "die Quelle der vorzüglichsten Vollkommenheiten tierischer Körper" (the source of the most exquisite perfections of animal bodies), it is also, in Reil's view, the root cause of illness and old age, insofar as the constant depletion and replenishment of portions of the body's material sooner or later erodes the body's original nature.[15] Reil also observes in *Von der Lebenskraft* that so much is unknown about how the body interacts with external forces, since the laws by which this interaction takes place escape our powers of observation. He notes, for example, that a dram of rhubarb given to a patient might serve as a laxative today but tomorrow cause colic or vomiting.[16]

Reil's admission of the difficulty of determining the laws according to which the body interacts with and absorbs other forces and substances sheds some light on the frequently changing regimen of medications, baths, and magnet and shock therapy he prescribed for Wilhelm Grimm. He believed that glandular issues in Wilhelm's throat were a secondary cause of his heart ailment, and while these glandular issues were in his view treatable, he acknowledged to Wilhelm that the heart condition itself could be only ameliorated, not cured.[17] There was, as Wilhelm wrote to Jacob, no "Radicalkur" (radical cure)[18] for his heart.

Nevertheless, despite the unpleasantness of the medicines, the magnet therapy, and the electric shock treatment, for the first time in two years Wilhelm found that he could fall asleep quickly and without fear at night, whereas at home, he told Jacob, he had gone to bed "wie zur Qual, wo ich in der Stille und in langem Wachen all die seltsamen Gänge des Bluts fühlen und die Angst zum Herzen kommen sah" (as if to be tortured, where in the silence and the long wakefulness I would feel the strange coursings of my blood and fear for my heart).[19] Writing to friend and mentor Friedrich Carl von Savigny, he confided his belief that he would eventually die of his condition, but he also expressed thanks to God for the improvement he had experienced under Reil's treatment and the return to scholarly work that this improvement had enabled. Wilhelm's use of the idiom "aus ganzem

14 Reil, *Von der Lebenskraft*, pp.67-68. See also pp.2-3.

15 Reil, *Von der Lebenskraft*, p.71.

16 Reil, *Von der Lebenskraft*, pp.67-68.

17 Jacob and Wilhelm Grimm, *Briefwechsel*, p.123.

18 Jacob and Wilhelm Grimm, *Briefwechsel*, p.123.

19 Jacob and Wilhelm Grimm, *Briefwechsel*, p.113.

Herzen" (with all my heart) in thanking God for the improvement of his condition underscores his preoccupation with his heart and blood as he copes with his incurable heart ailment.[20] That he thanks God, not Reil, is also characteristic of his view of disease and cure: although he regarded Reil as an extraordinarily skilled human doctor who was nobly attempting to help his patient, in his correspondence he repeatedly suggests that disease and cure are ultimately God-given.[21] The improvement of which Wilhelm wrote to Savigny nevertheless seems to have been short-lived, for in 1811 Wilhelm wrote to Achim von Arnim that he had immersed himself in his philological work in part because his heart ailment had made so many other activities impossible for him. He described himself as someone whose heart beat so rapidly that he could no longer walk far, and he lamented not being able to stroll through the forest or climb trees.[22]

Blood as *Lebensquelle*

Although reality presented Wilhelm with no *Radicalkur* for his heart ailment, he and Jacob found many extraordinary cures in the tales they collected for their *Kinder- und Hausmärchen* (Children's and Household Tales, 1812/1815). Magical water heals characters of their ailments in "Der Herr Gevater" ("The Godfather"), "Der König vom goldenen Berg" ("The King of the Golden Mountain"), and "Das Wasser des Lebens" ("The Water of Life"), while magical dew restores sight to the blinded tailor in "Die beiden Wanderer" ("The Two Travelers").[23] An herb cures the sick in "Der Gevatter Tod" ("Godfather Death"), apples restore health to the ill princess in "Der Vogel Greif" ("The Griffin"), and Saint Peter can heal the sick with a magic salve and even bring the dead back to life in "Bruder Lustig" ("Brother Lustig").[24]

---

20 Jacob and Wilhelm Grimm, *Briefe der Brüder Grimm an Savigny*, ed. by Wilhelm Schoof (Berlin: Schmidt, 1953), p.81.

21 A similar conception of disease and cure appears in the Grimms' fairy tales. Human characters at times deliver cures in *Kinder- und Hausmärchen* tales, but the cures themselves are depicted as bestowed or withheld by supernatural agents such as God, St. Peter, or Godfather Death. The ability of doctors to cure ailments is typically quite limited in these tales. See Schmiesing, *Disability, Deformity, and Disease in the Grimms' Fairy Tales*, pp.50-52.

22 Reinhold Steig and Herman Grimm, eds., *Achim von Arnim und die ihm nahe standen*, 3 vols (Stuttgart: J. G. Cotta, 1894 -1904), vol. 3, p.123.

23 All English titles of the Grimms' fairy tales are from *The Complete Fairy Tales of the Brothers Grimm*, ed. and trans. by Jack Zipes (New York: Bantam, 2002).

24 The harmful effects of these and other substances can also be seen in *Kinder- und Hausmärchen* tales. Magical water transforms the brother into a deer in "Brüderchen und Schwesterchen" ("Little Brother and Little Sister"), for example, and in "Der

The Grimms occasionally comment on these various cures in the appendix to the *Kinder- und Hausmärchen,* but their most extended discussion of cures in fairy tales and folklore appears in their afterword to the edition they published in 1815 of Hartmann von Aue's medieval narrative poem *Der arme Heinrich.* There they consider portrayals in folklore of the purported healing powers of substances and forces including water, dew, mother's milk, and touch. First and foremost, however, they focus on representations of blood as a vital substance in folklore from the medieval period to their own time, and they refer not only to *Der arme Heinrich* but also to *Kinder- und Hausmärchen* tales, tales from other European fairy-tale collections, and German and Scandinavian sagas, myths, and legends. They also write about literal and metaphorical dimensions, discussing, for example, the connection between blood and blood brotherhood in folklore.

Hartmann von Aue's *Der arme Heinrich* portrays the leprosy-stricken Heinrich, who learns that his disease can be cured only by the blood of a virgin who willingly sacrifices her life for his. When a maiden expresses a willingness to sacrifice herself for Heinrich, Heinrich intervenes right at the moment her heart is to be cut out, unwilling to allow her to die for him. Later he is miraculously cured by God and marries the maiden. Because blood is presented as a cure for leprosy in *Der arme Heinrich,* it is not surprising that the Grimms focus on blood cures in their afterword to the work. Striking, nevertheless, is the extent to which they characterize blood there and in other writings as the most powerful of all cures. At times they interpret even cures that have no ostensible relationship to blood as originating in folk beliefs pertaining to blood. Even if there is no direct evidence that Wilhelm's concerns with his heart ailment and the circulation and composition of his blood influenced the Grimms' preoccupation with blood and heart symbolism in folklore and mythology, it seems likely that, in studying blood cures in folklore, he would have reflected on his own physical condition.

The importance that the Grimms ascribe to blood in their edition of *Der arme Heinrich* and elsewhere is nevertheless unremarkable when viewed in the context of late eighteenth and early nineteenth-century notions of the

---

Krautesel" ("The Lettuce Donkey") the ingestion of an herb turns the hunter into a donkey. The breath of witches or alleged witches is viewed as malevolent in "Die Eule" ("The Owl") and "Die Bremer Stadtmusikanten" ("The Bremen Town Musicians"); moreover, in *Deutsche Mythologie* Jacob Grimm refers to superstitions that the breath of dwarfs can cause illness and death. See Jacob Grimm, *Deutsche Mythologie,* photomechanical reproduction of the 4th edn, ed. by Elard Hugo Meyer, 3 vols (Graz: Akademische Druck- und Verlagsanstalt, 1953), vol. 1, p.381.

importance of the blood to the body's functioning. In John Hunter's theory of the "living blood," for example, blood was "the location and source of the vital principle"[25] and possessed "the extraordinary power…to promote healing and the regeneration of organs."[26] At the beginning of a chapter devoted to blood in *Anfangsgründe der Physiologie* (First Principles of Physiology, 1789), Johann Friedrich Blumenbach sets forth his intention to write "von dem Blute als dem wahren Lebenssaft, und dem Urquell aller thierischen Säfte; denn der rohe Saft wird in Blut verwandelt, woraus sodann alle übrigen Feuchtigkeiten entspringen" (of the blood as the true fluid of life and the original source of all animal fluids; for the raw fluid is transformed into blood and from this all other fluids spring).[27] And with an emphasis on matter and mechanics, but still highlighting the universal importance of blood to the body, Reil proclaims in *Von der Lebenskraft*, "Im Blute sind die Grundbestandteile aller Organe des tierischen Körpers enthalten. Das Blut kreist durch Hilfe der Gefäße im Körper, und aus dem Strome des Blutes zieht die tierische Materie eines jeden besonderen Organs solche Bestandteile, die ihm eigentümlich sind, und mit welchen es Verwandtschaft hat, an sich" (In the blood the fundamental components of all the organs of the animal body are contained. The blood circulates through the body with the help of the body's vessels, and from the bloodstream the animal material of each particular organ takes those components that are unique to it and to which it is related). In an essay on illness he further asks, "Wozu das Blut? Woher der gleichzeitige Verlust des Bluts und des Lebens?" (What is blood for? Why the simultaneous loss of blood and of life?).[28]

In some respects, these various descriptions of the vitality of blood merely put in physiological terms religious and cultural associations of blood with life and vitality that had been in existence for millennia. Blood is regarded as "the source of life" in Judaism, and the centrality of blood to Christianity is emphasized not only in the bloody sacrifice of Jesus but also, in the Eucharist, in the association of the ingested wine with the blood of Christ.[29] As Jacob Grimm states in *Deutsche Mythologie* with regard both to

25 Hermione De Almeida, *Romantic Medicine and John Keats* (New York and Oxford: Oxford UP, 1991), p.89.

26 De Almeida, *Romantic Medicine*, p.94.

27 Johann Friedrich Blumenbach, *Anfangsgründe der Physiologie, aus dem Lateinischen übersetzt und mit Zusätzen vermerhrt von Joseph Eyerel* (Vienna: Wappler, 1789), pp.5-6.

28 Johann Christian Reil, "Veränderte Mischung und Form der thierischen Materie, als Krankheit, oder nächste Ursache der Krankheitszufälle betrachtet," *Archiv für die Physiologie*, 3 (1799), 424-461 (p.428).

29 Hugh Aldersey-Williams, *Anatomies: A Cultural History of the Human Body* (New York and London: W. W. Norton, 2013), pp.142-43.

classical and medieval traditions, "im blut hat die seele sitz" (the seat of the soul is in the blood), and he clarifies that the soul is equated with "leben und lebenskraft" (life and *Lebenskraft*).[30]

The primacy of blood in folklore is further emphasized in several of the Grimms' writings.[31] In their afterword to *Der arme Heinrich,* they refer to blood as "Lebensquell" (the source of life),[32] and in *Deutsche Mythologie* (German Mythology, 1835), Jacob speaks of "die verjüngende, heilende kraft des bluts" (the rejuvenating, healing power of blood).[33] Remarkably, they frequently interpret references to the ingestion of materials other than blood as symbolic of drinking blood. In their afterword to *Der arme Heinrich*, they point to the many myths in which flowers and trees spring from drops of blood that have fallen to the ground and conclude:

> Daraus folgt aber auch, wie Pflanzensäfte in Krankheiten gebraucht werden, die heilende Eigenschaft des Blutes. Und die Mythen, daß aus gefallenen Blutstropfen Blumen und Bäume wachsen, wiederum aus den Bäumen Menschen werden, bestätigt sich und dieses also desto stärker.
> (From this the healing quality of blood also follows, like the use of the juices of plants in illnesses. And this is corroborated even more strongly by the myths in which flowers and trees grow from fallen drops of blood and humans in turn form from these trees).[34]

It makes no difference, they claim, whether ancient sacrifices first involved actual blood or plants, because the extracts of plants were themselves

---

30 Jacob Grimm, *Deutsche Mythologie*, photomechanical reproduction of the 4th edn, ed. by Elard Hugo Meyer, 3 vols (Graz: Akademische Druck- und Verlagsanstalt, 1953), vol. 2, pp.689-90. I have retained Jacob's frequent practice of not capitalizing nouns or the first word in a sentence.

31 The fairy tales collected by the Grimms also contain numerous depictions of the vitality of blood. Drops of blood come alive in "Der liebste Roland" ("Darling Roland"), show characters which direction to travel in "Das singende springende Löweneckerchen" ("The Singing, Springing Lark"), flow out of a loaf of bread when it is cut in "Gottes Speise" ("God's Food"), and have healing and protective powers in "Der treue Johannes" ("Faithful Johannes") and "Die Gänsemagd" ("The Goose Girl"). Although these and other examples point to the thematic significance of blood in the *Kinder- und Hausmärchen,* little attention has been paid to blood and blood symbolism in the collection. One of the few studies of blood in fairy tales is Isabel Cardigos's study of the blood of menstruation, defloration, and childbirth in Portuguese fairy tales: Isabel Cardigos, *In and out of Enchantment: Blood Symbolism and Gender in Portuguese Fairytales* (Helsinki: Suomalainen Tiedeakatemia, 1996). According to Cardigos this blood is often central to portrayals of enchantment and disenchantment.

32 Jacob and Wilhelm Grimm, *Hartmann von Aue: Der arme Heinrich*, p.172.

33 Jacob Grimm, *Deutsche Mythologie,* vol. 3, p.96.

34 Jacob and Wilhelm Grimm, *Hartmann von Aue: Der arme Heinrich*, p.203.

viewed as blood. As evidence, they note the similarity of the words "Blut" ("blood") and "Blüte" ("blossom").[35] Similarly, in his essay on Wolfram von Eschenbach's *Parzival* in the first volume of the Grimms' *Altdeutsche Wälder* (Old German Forests, 1813), Jacob explores an etymological connection between the words "Blut," "Blüte" and "Blume" (blood, blossom, and flower),[36] and in *Deutsche Mythologie*, he hypothesizes that beliefs in the soul's ability to immediately assume the form of a flower arise from instances in which flowers are depicted in myths as springing up from drops of blood.[37]

In other writings, the Grimms see blood symbolized not only by plant extracts but also by substances including water and gold. In his essay on *Parzival*, Jacob asserts in a lengthy note that the words "Blut" (blood), "Flut" (flood), "Fluß" (river), and "Frost" (frost) share a common root word. He uses this association to explain ancient descriptions of rain or snow as bloody; for instance, Greenlandic beliefs render snow as the blood of the dead, and other myths associate birth with blood, snow, and ice.[38] For his part, in an essay on the German heroic saga in the same volume of *Altdeutsche Wälder* as Jacob's *Parzival* essay, Wilhelm writes about the description of a flood of blood in a battle scene of the *Nibelungenlied*, observing that the warriors engage in blood-drinking after Hagen tells a thirsty warrior that he should drink the blood of the slain. Wilhelm connects this scene with a passage in which Attila drowns in his own blood and Aldrian exhorts others to drink the gold and silver after which they have thirsted. "Das ist auch wieder Blut" (That is again blood), Wilhelm states, as he interprets the gold and silver as Attila's blood. Moreover, he interprets mythological depictions of melted gold poured into characters' mouths as symbolic of the characters' drinking of blood. Then, too, he points to similes of gold as a drop of blood or of golden ore as bleeding. He also finds a similar relationship between blood and gold in alchemistic descriptions of red gold.[39]

Wilhelm's statement "That is again blood" might in short be viewed as emblematic of the Grimms' interpretation of mythological or folkloric depictions of substances ingested by or absorbed into the body. In their various interpretations, herbs, water, and gold are all at times symbolic of

---

35 Jacob and Wilhelm Grimm, *Hartmann von Aue: Der arme Heinrich*, p.202.

36 Jacob Grimm, "Commentar zu einer Stelle in Eschenbachs Parcifal," in Jacob and Wilhelm Grimm, *Altdeutsche Wälder*, vol. 1 (Kassel: Thurneissen, 1813), p.23.

37 Jacob Grimm, *Deutsche Mythologie*, vol. 2, p.690. See also Jacob Grimm's essay "Bedeutung der Blumen und Blätter," in *Altdeutsche Wälder*, vol.1, pp.131-58 (p.140).

38 Jacob Grimm, "Commentar zu einer Stelle in Eschenbachs Parcifal," in Jacob and Wilhelm Grimm, *Altdeutsche Wälder*, vol. 1, pp.1-30 (p.15).

39 Wilhelm Grimm, "Zeugnisse über die deutsche Heldensage," in Jacob and Wilhelm Grimm, *Altdeutsche Wälder*, vol. 1, pp.195-323 (p.223).

blood. The importance they ascribe to blood in folklore is borne out in particular by their interest in portrayals of blood as capable of curing even the most intractable diseases. As they observe in the afterword to *Der arme Heinrich*, in folk belief certain diseases are regarded as so systemic that they can be healed not with cures that target specific bodily functions or organs, but only with the blood of a child or virgin:

> Die in ihrem Mittelpunkt zerstörte und verdorbene Gesundheit kann nur durch Annäherung und Erquickung mit dem *Reinen* geheilt werden; gewöhnliche Hilfe durch Kräuter, Säfte, Steine, die nur für das besondere wirken, ist umsonst, es wird eine gänzliche Vernichtung des Bösen und ein neues *verjüngtes* Leben erfordert. Als solche insgemein unheilbare, gleichsam durch ein Wunder nur zu hebende, Krankheiten wurden der *Aussatz* und die *Blindheit* angesehen […]. Als der Lebensquell, welcher jene Krankheiten vernichte und ein neu treibendes Leben erzeuge, ward vor allem das *reine Blut* einer Jungfrau oder eines Kindes betrachtet.[40]
>
> (Health that has been destroyed and debased at its very core can be healed only by a convergence and energizing with that which is pure; the usual remedy with herbs, juices, and stones, which have only a particular use, is in vain, and a complete annihilation of the bad and a new *rejuvenated* life is required. *Leprosy* and *blindness* were viewed as such in and of themselves incurable illnesses that could be lifted only with a miracle […].The *pure blood* of a virgin or a child was viewed as the source of life [*Lebensquell*] that could destroy such illnesses and generate and propel a new life.)

The Grimms detect this belief in the rejuvenating power of blood even in the first-edition *Kinder- und Hausmärchen* fairy tale "Blaubart" ("Bluebeard"), where, they admit, no actual reference to disease appears. Maintaining that Bluebeard's odd-colored beard is nevertheless meant to indicate illness, they suggest that his murdering of young maidens is motivated by his desire to use their blood to cure himself. Among the many other examples they give from fairy tales, legends, myths, and religious texts in support of their conclusions, they mention blood as a cure for leprosy in a fairy tale from Basile's *Pentamerone*, Sigurd's acquisition of wisdom after drinking the dragon's blood in the Volsunga Saga, and the fictional Massacre of the Innocents, which they assert could be viewed as a story about Herod's desire to cure himself of illness by bathing in the murdered children's blood.

In their exploration of the regenerative effects of blood in folklore, the Grimms express particular interest in the conditions under which the cure

---

40 Jacob and Wilhelm Grimm, *Hartmann von Aue: Der arme Heinrich*, pp.172-73. In the context of folk cures involving the blood of various animals, Jacob further mentions folk beliefs concerning the healing of lepers with the blood of a human child or virgin in *Deutsche Mythologie* (vol. 2, p.981).

allegedly takes place. In addition to studying the frequent mandate that the blood must be the pure blood of a child or virgin, they draw attention to the stipulation that the sick person must often bathe in or be sprinkled with this blood for the cure to take place.[41] They also focus on the process by which health and youth are purportedly transferred via blood from the sacrificed individual or animal to the ill person. Just as Reil, Blumenbach, and others examined the physiological process by which the body assimilates the various substances it absorbs, the Grimms explore folkloric depictions of the process through which the ill person is made similar to the person whose blood he ingests. The ill person assimilates the youth, health, and spiritual purity of the sacrificed victim through a process of *Übergang* or *Übertragung*, both terms meaning "transfer" or "transmission."[42] A similar transmission occurs in folkloric depictions of malevolent beings who inhale the vitality of innocent youths: thus the Grimms observe that witches coax young maidens into their beds "um deren Gesundheit und Lebenskraft mit dem Athem einzusaugen und auf sich selbst zu übertragen" (in order to absorb their health and *Lebenskraft* with the breath and transfer this to themselves).[43]

In addition to making the recipient similar in age and vitality to the person whose blood he received, sacrificed or exchanged blood can confer similarity in other ways. In a chapter of their afterword devoted to blood brotherhood, the Grimms discuss friends' ingestion of one other's blood as the formalization of their blood brotherhood.[44] The blood brothers would as a result of this ingestion view themselves as united with each other in body and soul, a unity that according to some legends could even enable blood brothers to adopt one another's form at will. From these and other accounts it is clear "daß sich solche Freunde wirklich in allem ähnlich sahen" (that such friends really regarded each other as similar in all respects), and this similarity is further underscored by references to blood brothers as

41 The Grimms frequently refer to the blood of virgins and children as pure or clean, but they do not give much attention to whether this purity is regarded principally in spiritual or physiological terms. As Peggy McCracken observes, medieval sources regarded the blood of virgins as not just indicative of the spiritual purity incarnated in the body of virgins, but also literally different from the blood of nonvirgins. See Peggy McCracken, *The Curse of Eve, the Wound of the Hero: Blood, Gender, and Medieval Literature* (Philadelphia: U of Pennsylvania Press, 2003), pp.2-3.

42 Jacob and Wilhelm Grimm, *Hartmann von Aue: Der arme Heinrich*, p.198.

43 Jacob and Wilhelm Grimm, *Hartmann von Aue: Der arme Heinrich*, p.176.

44 Jacob Grimm further studies blood brotherhood in *Deutsche Rechtsaltertümer* (*German Legal Antiquities*, 1828). See Jacob Grimm, *Deutsche Rechtsaltertümer*, 4th ed., 2 vols., ed. by Andreas Heusler and Rudolf Hübner (Leipzig: Dieterich'sche Verlagsbuchhandlung Theodor Weicher, 1899), vol. 1, pp.264-66.

"milk brothers" (*Milchbrüder*) to indicate that they are so similar that it is as if they had been nursed by the same mother.[45]

Blood brotherhood and blood sacrifice thus result in what we might conceptualize as a magical *Verähnlichung*, whereby the properties of one individual are transferred to and assimilated into another. Although not discussed in the afterword to *Der arme Heinrich*, a related transference and assimilation is also portrayed in a mother's frequent wish in folklore for a child as red as blood. For instance, Jacob refers in *Deutsche Mythologie* to fairy tales in which women wish for a blood-red child, and in his essay on *Parzival* he studies the mother in the *Kinder- und Hausmärchen* tale "Von dem Machandelboom" ("The Juniper Tree"), who similarly wishes for a child as red as blood and white as snow after blood falls on the snow from her cut finger.[46] Such wishes, which in the *Kinder- und Hausmärchen* also appear in "Sneewittchen" ("Snow White"), are suggestive to him of descriptions in myth of individuals being born of blood and milk.[47] The mother's desire for a child who will look like blood and milk thus symbolizes the physical processes by which the child as a fetus is nourished with her blood and as an infant is nourished with her milk. The wished-for child has the colors of her bodily fluids because the child is formed of them. In exploring this connection, Jacob Grimm also associates the frequent red, white, and black color trio in fairy tales with folk notions of the body. These colors appear together in folklore because "in ihnen und in ihrer Mischung der menschliche Leib erscheint, das Weiße in Haut, Nerven, Sehnen, Knochen, das Rothe im Blut, das Schwarze in Haaren und in der Sehe des Auges herrscht, wie den alle drei sich in dem Milchsaft, dem rothen und scharzen

---

45 Jacob and Wilhelm Grimm, *Hartmann von Aue: Der arme Heinrich*, p.186. In *Disability, Deformity, and Disease in the Grimms' Fairy Tales*, I argue that the concept of blood brotherhood could also explain why the discharged soldier in the *Kinder- und Hausmärchen* tale "Bruder Lustig" ("Brother Lustig") begins to refer to the soldier companion who, unbeknownst to him, is really St. Peter as "Bruderherz" ("Brother Heart") only after ingesting a lamb's heart. See pp.71-72.

46 Jacob Grimm, "Commentar zu einer Stelle in Eschenbachs Parcifal," in Jacob and Wilhelm Grimm, *Altdeutsche Wälder*, vol. 1, p.11, and Jacob Grimm, *Deutsche Mythologie*, vol. 2, p.752.

47 "Weil nun gerade die Vergleichung der Schönheit mit *Milch, Schnee* und *Blut* in aller Poesie, die wunderbare Geburt aus Blut oder Milch in den Mythen zusammenfließen, so ergiebt sich leicht ein recht gründliches Verhältniß der Wünsche, welche die Mutter thut, zu der hernach erfüllten Geburt des Kindes" (Because the comparison of beauty with *milk, snow,* and *blood* flows together in all poetry with the wondrous birth from blood or milk in the myths, a fundamental relationship easily follows between the wishes expressed by the mother and the resulting birth of the child). Jacob Grimm, "Commentar zu einer Stelle in Eschenbachs Parcifal," in Jacob and Wilhelm Grimm, *Altdeutsche Wälder*, vol. 1, pp.16-17.

[sic] Blut wieder besonders ausdrücken" (in them and in their intermixing the human body appears: the white in skin, nerves, tendons, and bones; the red in blood; the black in hair and in the sight of the eyes; and all three colors express themselves in particular in the chyle and in the red and black blood).[48]

Anatomy, Philology, and the Grimms' Editorial Practice

Jacob Grimm's associations of blood-red and snow-white children with the mother's blood and milk put the physiological interrelationship of the mother's body with that of her child in folkloric terms. Physiology texts of the late eighteenth and early nineteenth centuries also emphasize this interrelationship, noting, for example, that the body of the fetus easily absorbs and assimilates the mother's blood. As Reil writes in *Von der Lebenskraft*: "Die Mutter teilt der Frucht ihr Blut mit, eine tierische Materie, die nur einen geringen Grad von Verähnlichung nötig hat, um mit der Organisation der Frucht anschließen zu können" (The mother shares with the embryo her blood, an animal substance that requires only a small degree of assimilation, in order to connect with the organization of the embryo).[49] He also postulates in *Von der Lebenskraft* that milk is prepared in the mother's breast in such a way as to make it easily assimilated into the components of the child's body.[50] From a more Romantic viewpoint, Reil also later writes that the vitality of the uterus increases during pregnancy and melds into a single individuality with the life of the egg.[51]

Although there is no evidence that these or other contemporary scientific conceptions of vitality influenced the Grimms in their study of the vitality of blood in folklore, one can nevertheless assume that Wilhelm was familiar with Reil's views on blood and the heart, since the various therapies that he underwent in Reil's care were meant to ameliorate his heart condition. In his correspondence with Jacob he also mentions some of the other physiological topics that he heard Reil discuss or lecture on while in

---

48 Jacob Grimm, "Commentar zu einer Stelle in Eschenbachs Parcifal," in *Altdeutsche Wälder*, vol. 1, p.16. For an analysis of the red, white, and black color symbolism in the Grimms' fairy tales, see Heinz Rölleke, "Weiß – Rot – Schwarz: 'Die drei Farben der Poesie': Zu Farbspielen in Grimms' 'Sneewittchen'-Märchen und anderwärts," *Fabula*, 54 (2013), 214-34.

49 Reil, *Von der Lebenskraft*, p.41.

50 Reil, *Von der Lebenskraft*, p.41.

51 Reil, "Ueber das polarische Auseinanderweichen der ursprünglichen Naturkräfte in der Gebärmutter zur Zeit der Schwangerschaft und deren Umtauschung zur Zeit der Geburt, als Beytrag zur Physiologie der Schwangerschaft und Geburt," *Archiv für die Physiologie*, 7 (1807), 402-501 (p.499).

Halle.[52] Beyond this, it is notable that while working on the edition of *Der arme Heinrich*, Jacob indicated in a letter to Wilhelm of April 1814 that he wanted to find a medical basis for the use of blood in curing leprosy, and that he hoped that an acquaintance in Göttingen could obtain information about this from Blumenbach.[53] But in a letter of May 1814, Jacob wrote Wilhelm that he would search medical books himself for information about the use of blood to cure leprosy.[54] This search appears to have remained largely fruitless or incomplete, for the Grimms state in the afterword to *Der arme Heinrich*:

> Ob unter den alten Aerzten selbst das Blut als ein verjüngendes und überhaupt als Heilmittel gegolten, ist wahrscheinlich, kann aber hier nicht untersucht werden. Noch jetzt suchen die, welche an der fallenden Sucht leiden, das Blut Hingerichteter zu Erlangen, weil dieses allein nach dem Volksglauben sie heilt.
>
> (Whether blood was regarded at all by doctors in olden times as a healing agent, and a rejuvenating one, is probable, but cannot be examined here. Even now people who suffer from epilepsy attempt to obtain the blood of those who have been hanged, because according to folk belief this alone will heal them.)[55]

The Grimms' consideration of possible medicinal uses of blood in the first sentence of this passage and of an example taken from folk belief in the second sentence is noteworthy. The example from folk belief seems to suggest that Jacob did not actually find sufficient medical accounts concerning the use of blood in treating leprosy, but it might also underscore

---

52 See Jacob and Wilhelm Grimm, *Briefwechsel*, p.87.

53 "Über den armen Heinrich finde ich aber noch nichts. Kann denn Bauer in Göttingen (durch Blumenbach etc.) nichts Medicinisches über die alte Heilung der *lepra* durch Blut ausmitteln, ich meine aus Büchern Spuren?" (But I am still finding nothing about *Poor Heinrich*. Can't Bauer in Göttingen [through Blumenbach, etc.] provide anything medical pertaining to the curing of leprosy through blood, I mean any leads in books?). Jacob and Wilhelm Grimm, *Briefwechsel*, p.308. Although Jacob wished to find specific medical evidence that blood was regarded as a cure for leprosy, Hartmann von Aue likely had only a general understanding of the medical conventions of his time and does not appear to have relied on specific medical practice in crafting his story. Melitta Weiss Adamson has observed that the medical content of *Der arme Heinrich* is "always subservient to his narrative strategies," and the condition that the blood be the heartblood of a virgin is most likely an example "of Hartmann adjusting medical conventions to serve his plot development" (p.138). See Melitta Weiss Adamson,"Illness and Cure in Hartmann von Aue's *Arme Heinrich* and *Iwein*" (in *A Companion to the Works of Hartmann von Aue*, ed. by Francis G. Gentry (Rochester, NY: Camden House, 2005), pp.125-40 (p.138).

54 Jacob and Wilhelm Grimm, *Briefwechsel*, p.318.

55 Jacob and Wilhelm Grimm, *Hartmann von Aue: Der arme Heinrich*, p.173.

the Grimms' understanding of the interrelationship of science and folk belief in the medieval and early modern period. The Grimms' hopes of finding historical evidence that blood was used by doctors to treat leprosy is nevertheless further of note insofar as it highlights their desire to study not only the Christian symbolism of blood and the heart in *Der arme Heinrich* but also actual medical practice. Moreover, although they do not supply direct evidence of historical medical uses of blood, they attempt to link medical practice with folk belief and religious traditions in other ways, as for example when they suggest that the medical practice of bloodletting to heal the body is analogous to the sacrifice of a single individual to save an entire people, as in the Christian tradition.[56] In addition, the very title of the chapter "Heilung des Aussatzes durch Blut" ("The Curing of Leprosy with Blood") in their afterword to *Der arme Heinrich* reads almost as if it were the title of a medical treatise.

More strikingly, in the afterword to *Der arme Heinrich* the Grimms conceptualize their overall philological endeavor in interpreting Hartmann von Aue's work as akin to anatomical dissection:

> Durch die Zerlegung und Erforschung ihrer Gründe werden Gedichte, gleich lebensvollen Gestalten durch die Anatomie grausam zerstört und aufgehoben. Dieses Eindringen in das innere Heiligthum, in so weit es Menschen vergönnt wird, bleibt ebenfalls menschlich und recht; die Poesie aber kleidet alles Herbe und Entsetzende in vollkommene Schönheit nach ihrer Milde und nach ihrem Mitleiden.[57]
>
> (Like life forms in anatomy, poems are horrifically taken apart and annihilated in the analysis and researching of their depths. This penetration into the inner sanctuary remains human and just, to the extent that it is granted to human beings. But poetry clothes everything that is harsh and horrific in a perfect beauty by means of its gentleness and its compassion.)

The "Milde" (mildness, gentleness) of which the Grimms speak here recalls their designation earlier in the afterword of God's curing of Heinrich as a "mildere" ("gentler") outcome than would have taken place in ancient stories of human sacrifice.[58] The maiden's heart in *Der arme Heinrich* is not actually cut out of her body, since Heinrich refuses at the last minute to allow her to sacrifice herself, after which God cures him. Similarly, the Grimms as interpreters might have figuratively dissected *Der arme Heinrich* in their attempts to arrive at an understanding of the laws according to which

56 Jacob and Wilhelm Grimm, *Hartmann von Aue: Der arme Heinrich*, p.201.

57 Jacob and Wilhelm Grimm, *Hartmann von Aue: Der arme Heinrich*, p.207.

58 Jacob and Wilhelm Grimm, *Hartmann von Aue: Der arme Heinrich*, p.206.

cures can be effected in folk belief, but they did not actually perform surgery on it.

Or did they? The central feature of their edition of *Der arme Heinrich* was Wilhelm Grimm's translation of the Middle High German poem into modern German prose. As the Grimms proclaimed when they first announced their intent to publish an edition of *Der arme Heinrich*, "Eine Uebertragung in die heutige Sprache wird diese altdeutsche Sage zu einem allgemein lesbaren Volksbuch machen" (A translation into today's language will make this ancient German tale into a universally readable *Volksbuch*).[59] The Grimms' edition thus includes a translation of Middle High German into New High German and of poetry into prose. It not only changes the form of *Der arme Heinrich* in this way, but also recasts Hartmann von Aue's courtly poem as folk literature. *Übertragung*, the word the Grimms utilize in their announcement for "translation," is also one of the two terms the Grimms employ when in their afterword they discuss the manner in which vitality is transferred or transmitted via blood or other substances from one individual to another. Analogous to this transmission, Wilhelm's translation is meant to effect a rejuvenation of *Der arme Heinrich*, a work that had long before sunk into obscurity because of its medieval language and because of what was viewed as its antiquated spiritual outlook and its repellant subject matter.[60] The Grimms' compiling of the numerous folk beliefs with which they associate *Der arme Heinrich* in their appendix was integral to their efforts to recast the work as folk literature, but in their foreword to the edition they also appeal to the *Volk* by heralding *Der arme Heinrich* as a work for patriotic Germans in the aftermath of the Napoleonic Wars.[61] There, they speak of Hessians who, departing to fight the French, had held up their swords and proclaimed that Hessian blood should defend the fatherland, since this blood lives forever. Seven years of suffering have now purified ("reinigten") the German lands, and in this context the Grimms explain that they initially conceived of their "Bearbeitung eines alten, in sich deutschen, Gedichts als ein geringes Opfer" (edition of an old, in itself German, poem as a humble

59 Wilhelm Grimm, *Kleinere Schriften*, ed. by Gustav Hinrichs (Berlin: Ferdinand Dümmler, 1882), vol.4, p.504. The announcement also states that there will be a critical edition of the original medieval text for "Gelehrte dieses Fachs" (scholars in this field).

60 See Krohn, p.224.

61 In their afterword the Grimms also relate Heinrich to the many characters in folk literature with the names Heinrich, Hans, Hänsel, and Hanswurst. See Jacob and Wilhelm Grimm, *Hartmann von Aue: Der arme Heinrich*, pp.215-16. For an analysis of their refashioning of Hartmann von Aue's courtly work as a "Volksbuch," see Ursula Rautenberg, *Das 'Volksbuch vom armen Heinrich': Studien zur Rezeption Hartmanns von Aue im 19. Jahrhundert und zur Wirkungsgeschichte der Übersetzung Wilhelm Grimms* (Berlin: Erich Schmidt Verlag, 1985), pp.122-28.

offering) – an allusion to the maiden's willingness to give her life as an "Opfer" (sacrifice) for Heinrich.[62] In a further extension of this analogy, they state, "Jetzt hat sich unser gesammtes Vaterland in seinem Blut von dem französischen Aussatz wieder geheilt und zu Jugend-Leben gestärkt" (Now our entire fatherland has been cured in its blood from the French leprosy and has been invigorated with youthful life).[63] In this manner, Hartmann von Aue's work is assimilated into a nineteenth-century German cultural outlook. The Grimms' afterword might read in large part as a study of folk beliefs concerning the vitality of blood, but their foreword illustrates the vitality not just of blood but also of blood symbolism, since the Grimms reutilize the power of this symbolism to make a patriotic appeal to their readers. The foreword associates blood not only with physiological and spiritual vitality and purity, but with nineteenth-century constructions of ethnic vitality as well.[64]

Rüdiger Krohn has pointed out that Wilhelm Grimm's "trendsetting translation" led to the nineteenth-century "stylistic assimilation" of *Der arme Heinrich*, insofar as the work became "combinable in various ways with other, genuine chapbooks, and its recasting as a primer of an unselfish spirit of sacrifice made Hartmann's tale increasingly suitable for education, especially – but not only – that of a youthful public."[65] But while Wilhelm's translation successfully rejuvenated the work for a nineteenth-century audience, the Grimms themselves were at times divided with regard to the translation of literature. Just a few years before the publication of their edition of *Der arme Heinrich*, Jacob and Wilhelm had quarreled over Wilhelm's translation of *Altdänische Heldenlieder, Balladen und Märchen* (*Old Danish Heroic Songs, Ballads, and Tales*, 1811). Again conceiving of philology in terms of physiology, Jacob charged in a letter to Savigny that poetic goodness is "made lame" ("gelähmt") by translation.[66] Whereas he maintained that Wilhelm should have published a critical edition instead of a

62 Jacob and Wilhelm Grimm, foreword to *Hartmann von Aue: Der arme Heinrich*, n.p.

63 Jacob and Wilhelm Grimm, foreword to *Hartmann von Aue: Der arme Heinrich*, n.p.

64 Similar to the Grimms in their foreword to *Der arme Heinrich*, Achim von Arnim responds to the Napoleonic occupation of Prussia by associating blood with vitality and with individual and collective identity in *Isabella von Ägypten, Kaiser Karl des Fünften erste Jugendliebe* (*Isabella of Egypt, First Love of Emperor Charles the Fifth*, 1812). See Dorothea von Mücke, "Blut und Wunder bei Achim von Arnim," in *Die Macht und das Imaginäre: Eine kulturelle Verwandtschaft in der Literatur zwischen Früher Neuzeit und Moderne*, ed. Rudolf Behrens and Jörn Steigerwald (Würzburg: Königshausen und Neumann, 2005), pp.143-56.

65 Krohn, p.227.

66 Jacob and Wilhelm Grimm, *Briefe der Brüder Grimm an Savigny*, p.105. See also Schmiesing, pp.35 and 44.

translation, Wilhelm insisted that literature in effect ceases to exist if it is not translated and transmitted.

The question of what interventions can be made in a text without adulterating it carries over into the Grimms' editorial practice in the *Kinder- und Hausmärchen*. The first edition of the *Kinder- und Hausmärchen* appeared in two volumes in 1812 and 1815, followed by editions in 1819, 1837, 1840, 1843, 1850, and 1857. In addition to these seven *Große Ausgaben* ("large editions"), the Grimms published ten *Kleine Ausgaben* ("small editions") between 1825 and 1858. These small editions contained a selection of fifty tales from the large edition. Over the course of the seven large and ten small editions, the Grimms added tales to the collection and took others out, such that a total of around 240 different tales appeared in the seven *Große Ausgaben* and ten *Kleine Ausgaben*.[67] The seventh and standard large edition of the *Kinder- und Hausmärchen* contains 201 tales and 10 religious legends.

The Grimms not only added and subtracted tales from the collection as a whole but also significantly changed individual tales. An individual tale in the Grimms' collection typically combines or was influenced by multiple oral and/or written sources. The Grimms' intent in mixing two or more variants of the same tale or choosing one variant over another was to restore to their tales, to the greatest extent possible, what they considered a state of organic wholeness. In the preface to the first volume of the first edition, they appealed to readers to help them improve the collection by sending them authentic passages that would make fragmentary tales in the collection whole. In the preface to the second edition (1819), they reported that many of the tales in the first edition now appeared in better form, in part because tales that had been incomplete in the first edition had now been made whole through the incorporation of new passages. They also noted that they had taken out tales that were too foreign in origin and in their place substituted new tales, including tales from Austria and Bohemia. In describing their editorial practice in making these various changes, the Grimms acknowledge in the preface to the second edition that "in allem lebendigen Gefühl" (in all energetic feeling) for poetry there must be a shaping and reshaping of it lest even a faithful transmission make the story into "etwas Unfruchtbares und Abgestorbenes" (something infertile and dead).[68] But they distinguish sharply between editors who damage the integrity of the text by arbitrarily tying and gluing pieces of text together and those who more wisely regard

---

67 See Rölleke, *Die Märchen der Brüder Grimm: Quellen und Studien; Gesammelte Aufätze* (Trier: Wissenschaftlicher Verlag Trier, 2000), p.37.

68 Jacob and Wilhelm Grimm, *Kinder- und Hausmärchen: Ausgabe letzter Hand mit den Originalanmerkungen der Brüder Grimm*, 7th edn, ed. by Heinz Rölleke, 3 vols (Stuttgart: Reclam, 2004), vol. 1, p.23. See also Schmiesing, pp.43-44.

the text as like a plant that requires only gently nurturing as its leaves quietly and naturally unfold. The editor or author who zealously reworks or polishes a text becomes like Midas, they continue, insofar as his too radical touch results in texts that cannot nourish us because they have been turned into indigestible gold. Wilhelm also explained in a letter to Savigny in 1814 that he and Jacob had avoided cutting into a tale "mit critischem Messer" (with a critical knife) whenever they sensed that the tale was alive and had issued from the collective voice of the people as *Volkspoesie*.[69]

Nevertheless, despite their Romantic preference for organicist as opposed to mechanicist modes of editorial intervention, the Grimms' editing of *Kinder- und Hausmärchen* tales at times reveals more cutting, pasting, and reworking than their preface seems to suggest, and at times this reworking significantly alters a tale's plot motivation, characterization, or themes.[70] Scholars including Maria Tatar, Ruth B. Bottigheimer, and Jack Zipes have in particular shown how the Grimms' amendments to their tales reveal their desire to inculcate proper behavior in children, impart their Christian worldview, and enforce their patriarchal views of women.[71] Since Wilhelm was mainly responsible for editing the *Kinder- und Hausmärchen* from the second volume of the first edition on, these changes are in large part attributable to him. Whereas he was more concerned with increasing the tales' appeal to a bourgeois audience, Jacob was somewhat more scientific in his approach.[72] In any case, while it is certainly true that the Grimms' editorial practice reveals more intervention than one might assume on the basis of the editorial philosophy they set forth in their preface to the *Kinder- und Hausmärchen*, contrary to the claims of scholars such as Alan Dundes and John M. Ellis they did not consciously or deliberately manufacture "fakelore."[73] Instead, they endeavored to restore to the texts what they

---

69 Jacob and Wilhelm Grimm, *Briefe der Brüder Grimm an Savigny*, p.187.

70 A prime example is their reworking of the tale "Das Mädchen ohne Hände" ("The Maiden without Hands"). See Schmiesing, pp.80-98.

71 Maria Tatar, *The Hard Facts of the Grimms' Fairy Tales* (Princeton: Princeton UP, 1987), and *Off with Their Heads! Fairy Tales and the Culture of Childhood* (Princeton: Princeton UP, 1992); Jack Zipes, *The Brothers Grimm: From Enchanted Forests to the Modern World* (New York: Palgrave Macmillan, 2002); and Ruth B. Bottigheimer, *Grimms' Bad Girls and Bold Boys: The Moral and Social Vision of the Tales* (New Haven: Yale UP, 1987).

72 See Zipes, pp.30-32.

73 John M. Ellis, *One Fairy Story Too Many: The Brothers Grimms and Their Tales* (Chicago: U of Chicago Press, 1983), p.100, and Alan Dundes, "Nationalistic Inferiority Complexes and the Fabrication of Fakelore: A Reconsideration of Ossian, the *Kinder- und Hausmärchen*, the *Kalevala*, and Paul Bunyan," in *Papers I-IV [and] Plenary Papers: The 8th Congress for the International Society for Folk Narrative Research, Bergen June 12th-17th 1984*, vol. 1, ed. Reimund Kvideland and Torunn Selberg (Bergen: International Society for Folk Narrative Research, 1984), pp.159-60.

genuinely believed were the tales' original messages, even if their amendments frequently reveal their – and in particular Wilhelm's – nineteenth-century bourgeois outlook. Similarly, the Grimms' cultural outlook is revealed in their reading of Hartmann von Aue's courtly poem *Der arme Heinrich* as a German *Volksbuch* of particular cultural relevance to Germans in the aftermath of the Napoleonic Wars. Like their hope of rejuvenating *Der arme Heinrich* by translating it into modern German, moreover, the Grimms' aspiration in publishing their fairy tales was to safeguard the fairy tale at a time when they believed its survival was threatened. Writing in the preface to the *Kinder- und Hausmärchen* of their decision to collect and publish their tales, they lament the waning custom of orally transmitting tales and draw attention to the mortality of the storytellers who still retain the tales in their memory.[74]

In both the afterword to *Der arme Heinrich* and the preface to the *Kinder- und Hausmärchen*, the Grimms thus describe folk texts as living organisms and reveal their conception of their role as editors as that of in effect safeguarding and sustaining the vitalism of their texts.[75] By framing their editorial policies within the context of the *Lebenskraft*-debate of the era – by emphasizing what nourishes the text as a living organism and what debilitates its life energy – they succeeded in establishing discursive affinities between their own philological endeavors and those of contemporary physiologists and natural philosophers who explored interconnections between body and soul. Similarly, Reil compared natural science to philology when he observed that natural scientists study the forms of nature to arrive at an understanding of the laws according to which they arise, just as philologists study ciphers in order to convey the living spirit of a discourse that has otherwise died out.[76] Not only did the Grimms aim to discern the laws of healing in folklore in their study of the vitalistic role blood plays in *Der arme Heinrich* and elsewhere, they also conceived, in keeping with the Romantic *Zeitgeist*, of their overall philological and editorial project in terms of nourishing and preserving the vitalistic energy of their texts. Their actual editorial practice in works such as *Der arme Heinrich* and the *Kinder- und Hausmärchen* might not be as neatly organicist and restorative as they themselves hoped. But then, the medical regimen of pills, mercury salves, baths, magnetic therapy, and electric shocks Reil prescribed for Wilhelm did

---

74 Jacob and Wilhelm Grimm, *Kinder- und Hausmärchen*, vol. 1, pp.15-16.

75 The Grimms' emphasis on vitalism and a restoration of able-bodiedness can also be seen in many of the amendments they made to tales depicting disabled characters, as I show with reference to disability studies theory in *Disability, Deformity, and Disease in the Grimms' Fairy Tales.*

76 Reil, "Ueber das polarische Auseinanderweichen," p.492.

not have a neat outcome either. Yet, in both instances, the focus was on bringing about a desired equilibrium of forces within the body at the risk of introducing even greater disequilibrium. Analogous to the physiologist studying the processes of *Lebenskraft* in the human body, the Grimms aspired to a philological vitalism that would restore rather than deplete the original life of the text.

Fitness and Fitting In

Fitness, Nerves, the Degenerate Body, and Identity: Radical Reality and Modernity in Max Nordau's Aesthetics and Fiction

Nicholas Saul

Abstract: This paper argues that Max Nordau's monistic Darwinism and concept of the subject as embodied place him in the tradition of radical realism. In this context, Nordau's literary works have hitherto been regarded as heteronomous productions, mere allegories of his Darwinian cultural criticism and so without intrinsic value. Here, however, it is argued that his aesthetic writing, which consistently thematizes life in the body, in some degree represents a counter-discourse to his critical work. First, the novella 'Pas de chance' (1879) is seen to reflect in literature on the relative cognitive performance of literature and scientific discourse, and ultimately to celebrate literature's autonomous cognitive power. Second, contrary to the anti-Naturalistic, anti-neurasthenic prescriptive and proscriptive aesthetic of *Entartung* (1892-1893), both this novella and "Mahâ-Rôg" (1906) are shown to foreground as a kind of Wilhelminian Gothic (often extreme) physical and social ugliness. Thus paradoxically they both legitimate its representation and their characters' attempts to come to terms with their negative embodiment. Nordau's literary work thus includes what his theory demands to exclude and so trangresses the scientistic and medicalized normativity there propounded.

Max Nordau, born Max Simon Südfeld in the Jewish community of Pest, has an undesirable popular reputation and, perhaps on that account, has generally been little read or studied. On the one hand he is known as the author of the study *Entartung* (1892-1893),[1] an impressive, acute, in part amusing but also depressing, abusive and deliberately divisive diagnosis of everything that was wrong with modern art, which is condemned in biologistic terms unacceptable today as degenerate. Worse, and despite Nordau's unsecret Jewishness, this notorious analysis paradoxically became linked with the dismal *Entartete Kunst* travelling exhibition of 1933, in which National Socialist cultural bosses publicly denounced what they considered to be 'unhealthy' art.[2] On the other hand Nordau, the quondam renegade Jew,

---

1 Max Nordau, *Entartung*, 2 vols (Berlin: Carl Duncker, 1892-1893).

2 See Thomas Anz, "'Gesund' und 'krank.' Kriterien der Kritik im Kampf gegen die literarische Moderne um 1900," in *Ethische contra ästhetische Legitimation von Literatur*, ed. by Walter Haug and Wilfried Barner (Tübingen: Niemeyer, 1986), pp. 240-250; T.Anz, *Gesund oder krank? Medizin, Moral und Ästhetik in der deutschen Gegenwartsliteratur* (Stuttgart: Metzler, 1989), pp. 33-52; Jens-Malte Fischer, "Dekadenz und Entartung – Max Nordau als Kulturkritiker des Fin de Siècle," in *Fin de Siècle. Zu Literatur und Kunst der Jahhrundertwende*, ed. by Roger Bauer and Eckhard Heftrich (Frankfurt a.M.: Klostermann, 1997), pp.93-111.

was of course converted to Zionism by Theodor Herzl, and is today revered as one of the founding fathers of the state of Israel. It is hard to know what to make of such apparent contradictions, so that the obvious move, if one seeks the truth about Nordau, is to return to his texts. In the last fifteen years that is precisely what scholars have done, and there has been a mini-renaissance (or nascence) of Nordau studies associated initially with the names Schulte, Söder and Stanislawski[3] and most recently Zudrell, Person and Kottow.[4] These, to bring us up to date, have done away with the myths. Nordau is for example not a racist, or at least only a racist insofar as almost every Western intellectual around 1900 factually and unreflectingly subscribed to the commonplace, then-authoritative post-Darwinian anthropology of opposed natural and cultural peoples, at either end of a hierarchy of sophistication. Nor was Nordau an ideological Zionist. Rather, he seems to have been persuaded into that cause on ethical and cosmopolitan grounds. Indeed, he was no kind of nationalist, believing national consciousness to be a mere construct, and remained all his life a vehement critic of monarchism and militarism. But this is not our focus here. That is Darwinism and literature. Nordau was a qualified and practising medical doctor, and as such a convinced biological and cultural Darwinist of the first wave and strictest observance.[5]

It is his unapologetic Darwinian anthropology which links him with the cognitive interest of this volume, Howard N. Tuttle's concept of monistic radical reality, its tradition in nineteenth-century positivistic science (and earlier epistemes), and their common relation to literary production.[6] Now

---

3 See on this Michael Stanislawski, *Zionism and the Fin de Siècle. Cosmopolitanism and Nationalism from Nordau to Jabotinsky* (Berkeley, Los Angeles, London: University of California Press, 2001), esp. pp. xv-xvi, 26-27, 74-97; also Hans-Peter Söder, "Disease and Health as Contexts of Modernity: Max Nordau as Critic of Fin-de-Siècle Modernism," in *GSR*, 14 (1991), 473-487; and Christoph Schulte, *Psychopathologie des Fin de Siècle. Der Kulturkritiker, Arzt und Zionist Max Nordau* (Frankfurt a.M.: Fischer 1997).

4 Petra Zudrell, *Der Kulturkritiker und Schriftsteller Max Nordau. Zwischen Zionismus, Deutschtum und Judentum* (Würzburg: Königshausen & Neumann, 2003); Jutta Person, *Der pathographische Blick. Physiognomik, Atavismustheorien und Kulturkritik 1870-1930* (Würzburg: Königshausen & Neumann, 2005); Andrea Kottow, *Der kranke Mann. Medizin und Geschlecht in der Literatur um 1900* (Frankfurt a.M., New York: Campus, 2006), pp.127-173.

5 See Hedvig Ujvári, "Entartung und konventionelle Lügen. Zum Zivilisationsbegriff von Max Nordau," in *Trans. Internet-Zeitschrift für Kulturwissenschaften* 14 (February 2003) on Nordau's early *Pester Lloyd* essays, 1873-1874: http://www.inst.at/trans/14Nr/ujvari14.htm#t100. Accessed 3 April 2013.

6 Howard N. Tuttle, *Human Life as Radical Reality. An Idea Developed from the Conceptions of Dilthey, Heidegger and Ortega y Gasset* (New York and Washington: Peter Lang, 2005).

Darwin's is in intention nothing but a conventionally value-free and falsifiable scientific theory, admittedly of almost unchallenged authority up to the present day. Hence Darwin himself painstakingly avoided deriving any moral, philosophical or theological meanings from his theory. It is thus at best only partially co-extensive with any variety of "Darwinism," the philosophical or other interpretation of what Darwin's original theory might be taken to mean.[7] As for Germany, German "Darwinism" indeed really means German "Darwinisms," of which tradition for our purpose there can probably be said to exist in the nineteenth century two major strands. Both are Spencerian in their orientation, conceiving of evolution as a teleological or progressive process of development analogous to the growth of a tree, and viewing the evolution of life as a positive movement from less to more complex species, less to more sophisticated and valuable organisation. Karl Eibl sees the dominant strand of German Darwinism after Ludwig Büchner as being idealistically filtered, open to devotional and aesthetic neo-Romantic discourses.[8] Philip Ajouri suggested recently that there was however, notably in figures like Vischer and Keller (and I would add Raabe and Jensen), a more materialist strand, more sceptical of religion and progress, and more akin to the hard-bitten English mainstream tradition.[9] Nordau, I would suggest, belongs in the latter tradition of appropriated Darwinism. He believes that human nature is exhaustively explicable in physical and physiological terms derived from the newly dominant discourse of Darwinian positivistic natural science. The content of consciousness is wholly causally determined by material factors: the body, inheritance, environment (and of course chance). In this sense, then, Nordau's psychophysical Darwinist an-

7 See on this discrepancy Peter J. Bowler, *The Eclipse of Darwinism: Anti-Darwinian Evolution Theories in the Decades around 1900* (Baltimore: Johns Hopkins UP, 1983); *The Non-Darwinian Revolution: Reinterpreting a Historical Myth* (Baltimore: Johns Hopkins UP, 1988); "Herbert Spencers Idee der Evolution und ihre Rezeption," in *Die Rezeption von Evolutionstheorien im 19. Jahrhundert*, ed. by Eva-Marie Engels (Frankfurt a.M.: Suhrkamp, 1995) pp.309-25.

8 Karl Eibl, "Darwin, Haeckel, Nietzsche: Der idealistisch gefilterte Darwin in der deutschen Dichtung und Poetologie des 19. Jahrhunderts. Mit einer Hypothese zum biologischen Ursprung der Kunst," in *Fritz Mauthner: Sprache, Literatur, Kritik*, ed. by Helmut Henne and Christine Kaiser (Tübingen: Niemeyer, 2000) pp.87-108.

9 Philip Ajouri, *Erzählen nach Darwin. Die Krise der Teleologie im literarischen Realismus: Friedrich Theodor Vischer und Gottfried Keller* (Berlin: de Gruyter, 2007); also Nicholas Saul, "Raabes Geschichte(n): Realismus, Erzählen, Historie, Prähistorie," in *Jahrbuch der Internationalen Raabe-Gesellschaft* (2009), 22-39; and N. Saul, "'Once in Human Nature, a Thing Cannot be Driven Out': Evolutionary Aesthetics in Wilhelm Jensen's *The Legacy of Blood* (1869). An Early Response to Darwin," in *The Evolution of Literature. Legacies of Darwin in European Cultures*, ed. by Nicholas Saul and Simon J. James (Amsterdam and New York: Rodopi, 2011), pp. 239-253.

thropology can be captured by Tuttle's broadly-framed notion of radical reality, which defines human life and the body as the same, and sees human life as circumscribed by the holistic and monistic experience of an embodied subject (Tuttle, p. 79) – an experience which precludes any aprioristic mode of knowledge. Nordau's critical and aesthetic works are in this sense his pioneering explorations of the consequences of new, monistic Darwinian rules of the game of life for modern humanity. His project, however, has had even worse consequences for his reputation in one area – aesthetics – than some of the things previously outlined. For like his near-contemporary, Wilhelm Bölsche,[10] Nordau rejects the chief aesthetic axiom of the grand idealist tradition – aesthetic autonomy – and welcomes positivistic natural science and its philosophical derivatives as the dominant providers of truth.[11] Thus aesthetic discourse, rather than being creative of meaning, can, so it seems, only be a parasite on or a servant of the prior truth of science.

All that suggests two chief cognitive interests of this essay. First, I shall ask how far, as some critics of Nordau's almost unstudied yet beststelling literary works have suggested,[12] those works really can be regarded as mere allegories of his philosophical and scientific position. For if that were so, his literary works would indeed be bereft of autonomous dignity and any "Leistung der Form,"[13] that is, the semantic or cognitive difference conferred by aesthetic shaping. Indeed, they could hardly be classified as works of art at all. Second, having rejected that criticism, I shall explore the significance in those works of the most radical reality, the body, for his Darwinist anthropology. To these ends I will investigate two literary texts from either end of Nordau's career, both novellas: "Pas de chance!" from Nordau's first collection *Seifenblasen* (1879) and "Mahâ-Rôg" (1906) from the collection of the same name.[14] The first, published before virtually all of Nordau's theo-

---

10 See Wilhelm Bölsche, *Die naturwissenschaftlichen Grundlagen der Poesie. Prolegomena zu einer realistischen Ästhetik. Mit zeitgenössischen Rezensionen und einer Bibliographie der Schriften Wilhelm Bölsches*, ed. by Johannes J. Braakenburg (Munich: dtv, 1976), pp.1-65 (p.25).

11 See Max Nordau, "Mene, Tekel, Upharsin," in M.Nordau, *Die conventionellen Lügen der Kulturmenschheit* (Leipzig: Schlicke, 1883), pp.1-35 (pp.30-32); also "Evolutionistische Ästhetik," in M.Nordau, *Paradoxe* (Leipzig: Elsaesser, 1885), pp.280-310 (pp.291, 309).

12 See Stanislawski, pp.23, 30-31, Zudrell, pp.15, 273-275, and Melanie A. Murphy, *Max Nordau's Fin-de-Siècle Romance of Race* (New York: Peter Lang, 2007), p.130.

13 The term belongs to Peter Pütz, who applies it to drama in his *Die Leistung der Form. Lessings Dramen* (Frankfurt a.M., Suhrkamp, 1986).

14 *Seifenblasen. Federzeichnungen und Geschichten* (Leipzig: Reclam, 1879), pp.82-92; *Mahâ-Rôg und andere Novellen* (Berlin: Alfred Scholl, 1906), pp.3-88. Nordau's collection is not to be confused with the later collection of the same name by a rather different kind of Darwinian writer, Kurd Laßwitz, *Seifenblasen. Moderne Märchen* (Hamburg: Voss, 1890).

retical or publicistic writings, is in my view not at all a soap bubble, more a programmatic exploration of the relative truth content of literary and scientific (medical) discourse. The second is an example of Nordau's characteristic deployment of the theme of the meaning of our bodies in the post-Darwinian world for the way we live our lives. I also include references to *Entartung* and some shorter essays of Nordau to contextualise the points.

Those who have looked at *Seifenblasen*[15] tend to take the title at its word – that is, without registering the foregrounded hint of defensive irony by a new author longing despite his fears for recognition. I think this is emblematic of the way all Nordau's literary work is read. Yes, this volume contains several potboilers and space fillers in imitation of other writers, for example, Karl May. But "Pas de chance!" does not belong in this category, in particular its concluding tale. However, it does foreground many of Nordau's life-long thematic obsessions. The central protagonist and first-person narrator is a Parisian doctor who possesses a more than passing acquaintance with the dissection theatre. Then there are the obsessive themes: the body, the intellectual pursuit of the "Mysterium des Lebens" ("mystery of life", *Seifenblasen*, p. 84), its corollaries death and love, and their corollary, woman. The narrative is situated in the dissection theatre of the Hôtel Dieu in May 1877, tellingly in the depths of the ancient hospital beside the Seine and below the water-level, lit dimly by a few tiny high, porthole-like windows which are regularly submerged under the murky water. This macabre environment is nineteenth-century positivistic science made literary symbol: decor-free, dominated by dissection tables, devoted wholly to the pursuit of the mystery of life through the analysis of its lifeless remains. The subterranean, or submarine, dissection theatre, then, figures the deep structures of the psychophysis.

Doctors habitually drop in to see whether any interesting corpses have turned up, and this is the occasion of the narrative. Our doctor happens upon one as yet undissected, fresh cadaver, which grips his whole attention: a strikingly beautiful, evidently proletarian, young woman with no outward, visible mark of the cause of death. Her skin is flawless and peachy, her features regular, she is indeed a classical beauty (*Seifenblasen*, p. 87). Her face bears only a diminished trace of the agony of death, her still open eyes reveal nothing of what she has seen in the past or beyond the grave (p. 86). Of course we recognise here something like the primal scene of Gabriel von Max's painting *Der Anatom* (1869), and this tale is surely in part an ekphrasis

15 See Zudrell, pp.262-263; Murphy, pp. 49-50.

of that image.[16] But by contrast to the Max painting, the body does exhibit one curious sign for a nineteenth-century woman: a tattoo.[17] Her name, evidently, "Marie Balok," a year, "1876," and the further words, "Pas de chance!" (p. 87).

So how and why did she die? Faced with this riddle, the dissection room becomes the site of what Jürgen Link identified as the major function of aesthetic discourse: its ability to cite other discourses and reflect on them at a metalevel.[18] Hence we find a kind of cognitive competition in truth-finding occurring between the spontaneous narrative production of our writing doctor, and the immediately subsequent actual autopsy of the professional pathologists, both forensic in intention: between aesthetic and natural-scientific discourses, then. The result, significantly, is a draw. The narrator finds himself extrapolating metonymically from the text on Marie's body *qua* "Inhaltsverzeichniß des unglückseeligen Menschenlebens" (contents list of an unhappy human life, p. 87) a retrospective biographical dream (pp. 87-88). She is, he imagines, a typical Parisian, first-generation immigrant *grisette*, familiar from many French Naturalist experimental novels, like Nana or Lulu no better than she should be and using what weapons nature has given her to fight in the metropolitan struggle for existence (p. 88), encountering poverty, poorly-paid work seamstressing or some such, violent lovers in Montmartre, alcohol, and so forth, and above all a yearning for something undefined but better (p. 89).

The crux of the tale is reached on a picnic excursion to St Germain, the lung of Paris (p. 90), when Marie experiences a sudden impulse to be tattooed. Now if we follow Lombroso, tattooing in the nineteenth century is the prerogative of sailors, male proletarians, and criminals – all of them, to him, degenerates. But the impulse comes to Marie under the decentring influence of alcohol, when she experiences in a micro-epiphany the primal desire for self-understanding and self-expression. She herself composes this cryptic arabesque of her identity, which is crudely tattooed by her lover onto her right upper arm. That this primitive cryptogram is the essence of her be-

16 Compare Zudrell, pp. 262-262 (who is content with a bare, functionally allegorical ekphrastic reading); also Elisabeth Bronfen, *Nur über ihre Leiche. Tod, Weiblichkeit und Ästhetik* (Munich: Kunstmann, 1994), p.89.

17 See the commentary on tattoos as proper to degenerates and primitives by Nordau's guru, Cesare Lombroso, *Der Verbrecher (homo delinquens) in anthropologischer, ärztlicher und juristischer Beziehung*, 3 vols (Hamburg: Verlagsanstalt/Königl. Buchhandlung, 1894) [1st edn. 1876], I, pp.253-269.

18 Jürgen Link, "Literaturanalyse als Interdiskursanalyse. Am Beispiel des Ursprungs literarischer Symbolik in der Kollektivsymbolik," in *Diskurstheorien und Literaturwissenschaft*, ed. by Jürgen Fohrmann and Harro Müller (Frankfurt a. M.: Fischer, 1988), pp.284-307.

ing is indicated by the narrator's observation that tiny drops of blood ooze from every puncture, to merge with the indigo forming the very text (p. 91). Her identity is thus merged with her body, her body sublated through her lifeblood into the words and numbers. No luck, obviously, in attaining her heart's desire, is the motto, the Alpha and Omega of her existence, written on her body for those with eyes to see. This, he concludes, was not only her motto but also a prophecy.

Now this immanent reflection in creative writing of creative writing is not presented as a certain historical truth. But it is confirmed by the autopsy results (p. 92). Marie, it turns out, had a weak heart valve, and she died of a heart attack in a moment of passionate excitement. The exact science of dissection, then, like the reconstruction of the imagination, discloses a fundamental heart problem, so that the one discourse is demonstrated by the tale to complement the other. In Dilthey's contemporary language the aesthetic experience of a "Geisteswissenschaft" (i.e., a matter of empathetic understanding) complements the empirical explanation of a "Naturwissenschaft" (i.e., a matter of logical explanation).[19] There are, then, for Nordau on this construction of the tale not one but two routes of the mind into the mystery of life (and death), each of equal dignity and with its own characteristic cognitive performance: art and science. But the ability to say this, the "Leistung der Form," (the achievement of literary form) is the unique prerogative of literary discourse.

In this apology of aesthetic autonomy, Nordau seems to re-affirm Max's notorious affirmation of patriarchal gender discourse as disclosed by Bronfen. In Gabriel von Max's painting the male gaze of the anatomist upon the beautiful cadaver is suggested to be transmediated into patriarchal scientific discourse on the body. In Nordau's story the disempowered and intoxicated woman is in some sense at least the author of discourse on her identity, and the lover merely her amenuensis. Then, however, the narrating doctor overwrites it with his version of the truth. There is also an irony here in this would-be irenic reconciliation of the two cultures. Nordau, a lifelong venerator of Goethe, is perhaps unaware of the rupture with the classical aesthetic tradition which his literary thought-experiment to reconcile imaginative synthesis and anatomical dissection entails.[20] He seems oblivious of

---

19 Compare Safia Azzouni on Laßwitz and the two cultures read through Dilthey: S.A., Azzouni, Safia: "How Wilhelm Dilthey Influenced Popular Science Writing: Kurd Laßwitz's *Homchen. Ein Tiermärchen aus der Kreide*," in *Historical Perspectives on Erklären and Verstehen*, ed. by U. Feest, *Archimedes*, 21 (2010), 61-79.

20 Goethe is of course the model of 'healthy' literature in *Entartung*. See *Entartung*, I, p.26., II, p.462.

Goethe's extremely negative view of the cognitive achievement of the dissection theatre (indeed any analysis of the living) in seeking human truth.[21]

That said, we have at least established that there is more to Nordau's writing than has hitherto been recognised. Aesthetic discourse does, despite Nordau's views expressed elsewhere, possess some autonomy, even if it does not challenge the master discourse of science, and is still legitimated by it.

The mature Nordau moved on to establish himself as the cultural policeman of the fin de siècle, continuing to write fiction but seemingly unable to resist offering ever fiercer publicistic attacks on any contrary tendencies. Thus he famously earnt Karl Kraus's disapproval for aesthetically urinating on almost every feature he encounters in the literary landscape.[22] This critical strategy has psychophysical roots in Nordau's rigorous insistence on reducing cultural history to the principles of Darwinian biological and social evolution. *Entartung* medicalizes and pathologizes the entire cultural domain, arguing that there are actual psychophysical causes of all negatively regarded cultural phenomena, such as Naturalism, Impressionism, Wagner, mysticism, and so forth (*Entartung*, p. viii). The external causes Nordau seems mainly to locate in the modern way of life in the big city (*Conventionelle Lügen*, p. 17-18), which, apart from restricting access to natural requirements such as exercise and fresh air, subjects individuals to a massive excess of stimulating information (*Entartung*, pp. 71-73), and also exposes them to poisons such as alcohol, tobacco, and other narcotics (*Entartung*, pp. 63-65). The external causes manifest themselves however also internally in the psychophysis as stigmatic maladjustments or deformations of the nervous system (*Entartung*, p. 32), whereby our nerves are either excessively receptive or productive in transmitting sense information for processing in the brain. This in turn results in distortion of the way neural pathways are formed and linked by association and habituation in order adequately to model self and world in the brain. The resultant severe perceptual distortions of self and world ensure that all adaptive behaviour through the foundational activities

21 The paradigmata of this attitude are Goethe's well-known attack on Newton's light-splitting *experimentum crucis* and his perhaps lesser-known advocacy of the use of models of human limbs in preference to dissection practice for medical students in *Wilhelm Meisters Wanderjahre*. See Johann Wolfgang Goethe, *Sämtliche Werke, Briefe, Tagebücher und Gespräche*, ed. by Hendrik Birus, Dieter Borchmeyer et al, 40 vols (Frankfurt a.M.: Deutscher Klassiker Verlag,), 1. Abteilung, vol. 10: *Wilhelm Meisters Wanderjahre*, ed. by Gerhard Neumann and Hans-Georg Dewitz, pp.602-610. On this theme Jürgen Barkhoff, "Goethes Ehrfurchtsgebärden in den *Wanderjahren*," in *Anthropologie und Literatur um 1800*, ed. by J.B. and Eda Sagarra (Munich: iudicium, 1992), pp.161-186 (165-166).

22 Karl Kraus, "Max Nordau," in *Die Fackel*, No. 200, VII. Jahr, 3 April 1906.

of attention, judgement and will is misemphasized and misplaced. Hence, we might see here a tendency to lassitude and pessimism or to mysticism, thanks to the failure to establish normal patterns of causal inference.[23] These literally, psychophysically degenerate, world-views are then compulsively expressed by artists in the form of Impressionism (*Entartung*, pp. 51-52 pp. 88-89), Wagnerianism, etc. All modern art seeks thus by definition merely to excite our hungry or satiated nerves in untoward ways (*Entartung*, p. 21). Worst: the stigmata or symptoms of this inner, rather than outward degeneration (*Entartung*, p. 34). This is where Nordau thinks to add to Lombroso's theory: they are first acquired and then passed on in Lamarckian fashion to the next generation en masse (*Entartung*, p. 74).[24]

Above all, intervention in media is significant, since for Nordau all communication functions by suggestion, so that what matters in this culture of weak-willed and neurasthenic mass city-dwellers is to generate the strongest possible suggestions in the receptive psyches of the mass. In communicative acts, just as in hypnotic suggestion, the recipient, so Nordau argues, is habituated to reproduce exactly what the transmitter sends. The transmitter sends an exact reproduction of his mental state encoded in natural or conventional signs, in which new sensual data experienced are interpreted by being matched with memory data of analogous states already established through association and habit. This judgement is then passed on through the medial connection, just as yawning in one subject encourages yawning in third parties, or as in Darwin's notion that the formation of an expression by one person encourages the inner formation of a concomitant emotion in another.[25] In each case the communication stimulates a sense-impression and an inner, nervous response by association which precisely reproduces the sender's state of mind in the receiver, even down to the molecular level.[26] The role of the will is crucial in this culture of neurasthenia and pessimism.[27] The stronger the mental activity and in particular the stronger the will in the recipient, the greater the resistance to this suggestive mode of communication (*Paradoxe,* pp. 231-32). Conversely, the weaker the will, the lower the resistance. Only the genius is proof against such effects. The geni-

23 See *Entartung*, esp. pp.97-110.

24 On this see Person, pp.125-142. Nordau rejects Lombroso's view of the potential genius of "degenerate" artists (*Entartung*, p.45).

25 See Charles Darwin, *The Expression of the Emotions in Man and Animals*, ed. by Francis Darwin (London: Murray, 1904) [1st edn. 1872], pp.25-82.

26 See Max Nordau, "Suggestion," in Nordau,, *Paradoxe*, pp.222-242, esp. pp.224-230; also *Entartung*, p.61.

27 On the collective phenomenon of nervosity and the move to the will as therapy see Joachim Radkau, *Das Zeitalter der Nervosität. Deutschland zwischen Bismarck und Hitler* (Munich: Propyläen, 2000) [1st edn. 1998], esp. pp.385-485.

us, in fact, as possessor of the strongest mind and intellect positively forms the consciousness of the mass in this way, and is the motor of cultural evolution (*Paradoxe*, pp. 236-37).

And so Nordau condemns himself to his lonely fate, a creative lifetime of forcefully aggressive diagnostic polemics. Fascinatingly, given the medical analogy which is the *dispositif* of his life, Nordau seems to offer little actual therapy. Healthy, rather than sick books should be published and publicized, although this sounds more like prevention than cure. Otherwise, he recommends setting up ethical societies and suchlike. In any case the degenerate will die out by definition thanks to their lack of adaptive qualities (*Paradoxe*, p. 270). And indeed it is hard to see how the psychophysical damage from which the degenerates suffer could be repaired. But that said, if we know that Nordau continued apace to write critical polemics against the sickness of the age, what becomes of his literature? Surely he should avoid producing yet another "Buch [...] vom Kranken" (book [...] of a sick man) and offer another unfashionable and rare yet precious "Buch vom gesunden Menschen" (book of a healthy person) *Paradoxe*, p. 272)?

Turning now to "Mahâ-Rôg" (1906) and Nordau's later writings, we discover however that he does not follow his own precept. Here one might have thought that as one who attacked as degenerate the Naturalist foregrounding of pessimism, determinism, the squalid and the ugly, equally as one who derived the experience of aesthetic beauty like Darwin and Spencer from the display entailed in sexual selection, Nordau would attempt suggestively and with maximum willpower to propagate some positive images and norms.[28] Not a bit of it. At the level of both content and form the bad and the ugly are not merely excluded but positively included, indeed sometimes they predominate in a kind of (if not Naturalist) then at least late Wilhelmine social Gothic. Typical of this Gothicization, and despite his move to internalize the physiognomic symptoms of degeneration by comparison with Lombroso, is Nordau's continued fascination with squalid and ugly physicality of all kinds. There are in his tales some enthusiastic allusions to and descriptions of post mortems, of course.[29] But I am thinking here particularly of descriptions of faces, which almost always are ugly faces juxtaposed to telling effect with beauty. In Nordau's second story collection, *Seelenanalysen* (1892) the tale "Panna" recalls for example the willed contrast of Marie Balok's classical beauty and crude tattoo in *Seifenblasen*. Here, the beautiful, if slightly low-browed Panna marries Pista out of compensatory guilt. Pista

---

28 See Nordau, "Evolutionistische Ästhetik," in Nordau, *Paradoxe*, pp.290-291.

29 See "Ein Sommernachtstraum" in *Seelenanalysen* (Berlin: Verein der Bücherfreunde, 1892), pp. 1-29 (p.14); and especially "Panna," in *Seelenanalysen*, pp.63-170 (pp.111-112).

was formerly a handsome man. However Panna, in a moment of fury prompted by her inherited choleric disposition, has gravely disfigured him by smashing half a brick in his face. One eye is put out, the nose crushed, the upper lip split, left scarred and partly hairless, and several front teeth are missing (*Seelenanalysen*, pp. 63-64). Another example is offered by "Wie Frauen lieben"[30] in the form of the painstakingly captured ekphrasis of the 1848 revolutionary Baudin's memorial in Montmartre. Here, Baudin's quotidien features are made noble and transfigured into sublimity, so Nordau's narrator seems to think, by the rendition of the large bullet hole in his forehead from which blood and brain-tissue are spilling (*Seelenanalysen*, pp. 220-221), frozen by the artist for the spectator's and reader's eternal edification. Or, finally, there is the carefully-described, grotesquely lopsided disposition of Geheimrat Behr's face in "Auf Abbruch" from the collection *Mahâ-Rôg*[31] after his stroke (*Mahâ-Rôg*, pp. 266-267). Nordau is otherwise keen to show how faces decay over time.[32] But it is the story "Mahâ-Rôg" which takes the palm here.

For this Hindu phrase, as the narrator tells us, means the great sickness: leprosy, known since 1873 to be caused by bacterial infection, but then still beyond cure or even palliative care. In this tale, set in contemporary colonial India, three generations of Indian women in the same family, all (needless to say) spectacularly beautiful, are threatened by the bacillus, which of course excludes them both from life in general and from the generation of life which is their biological destiny. Here the eldest, Rani, nobly chooses the living death (*Mahâ-Rôg*, pp. 42, 50) of the colony within the colony, the leper colony, in order to minimize the risk of contagion for her young daughter, the extremely beautiful, almost white (!) Udschli, who grows up, for a time, until her impending marriage to the handsome Dasa, in ignorance of her mother's fate and the absent but ever-present danger of her own latent infection. Again, Nordau does not shirk description of the horrors leprosy inflicts on a woman's face (*Mahâ-Rôg*, pp. 17, 21, 39), even if he restricts the detail to the revealed face of Rani's companion, and leaves us to imagine what lies behind Rani's veil. And when he is not emphasising the death in life that is the leper's fate, he compares their status to that of animals (p. 21). Thus it is no surprise when Udschli, mindful of the danger to her own equally beautiful (and white; pp. 2, 28, 56-57, 60, 70) daughter Tschandni, mindful that even the living death of the colony is no necessary protection for others, and – almost equally important – unable to accept the leper colony as her own fate (p. 42), rejects this solution. When the illness, following

30 *Seelenanalysen*, pp.213-253.

31 *Mahâ-Rôg*, pp.217-273.

32 See "Die Kunst zu altern," in *Seelenanalysen*, pp.171-212.

a long false dawn, at last breaks out, she learns from her Indian doctor (as ever in Nordau, a raisonneur figure) of an alternative, mystical cure. If she will bury herself alive, her sacrifice will purify her daughter; a life freely offered to death buys a life (p. 43). Just as the terrible symptoms begin to appear she takes this horror upon herself (p. 68). And indeed Tschandni *is* saved from the disease, perhaps thanks to the sacrifice, perhaps because she was never infected.

Here, however, comes a surprising twist. Tschandni is of course crushed with gratitude that her life, and above all her beauty, have been saved, and venerates her mother. But she does not follow the path to marriage and to life which her redeemed beauty would seem to pre-ordain. Rather, she chooses to live in seclusion at the temple, becomes a scholar and eventually trains as a Vedic doctor. All these are no doubt intended to signify her choice of a different mode of thanksgiving and an unheard-of – non-Darwinian – autonomy and sovereignty over life. And yet her beauty, saved at such a price, eventually causes her death. She turns out to be a kind of *femme fatale* for the local Rajahs. One forces her hand in marriage, his brother fights over her, so that at last Tschandni, having escaped poisoning by jealous first wives, herself commits suicide with mandrake root. In a dreamlike confrontation with her mother, she reproaches her for saving her beauty, which in truth was her and her mother's curse (p. 88).

The meaning of the twist seems clear. Beauty may be an evolutionary adaptation and aesthetic beauty may derive from that. But the function of beauty in the cultural domain, as alpha males compete to pass on their hereditary material, is deeply ambiguous, certainly for a woman who, as is the case with Tschandni (like a more fortunate version of Marie), wants to write her own destiny. Indeed, the most powerful and perhaps intentionally significant scene in this polemic *against* the cultural use of aesthetic beauty is in conventional terms its most horrific and unbeautiful, when Udschli, having learnt of her fate but still living in the world, visits her mother Rani in the leper colony, the domain of the excluded, and at last, rediscovering their common humanity, includes this horrifying ruin of a woman in the human gesture of the embrace:

> In diesem Augenblick hatte sich Udschli der ganzen Länge nach auf das schmale harte Lager geworfen, das arme Menschengetrümmer in ihre Arme genommen und an sich gezogen und ihr, am ganzen Leibe bebend, ins Ohr gerufen: 'Dein Kind, deine Udschli' [...] die Kranke [...] überließ sich mit einer Wonne, die das Zittern ihres Unterkiefers und das Zucken ihrer weißen Lippen verrieten, der unbekannten Empfindung des Gehätscheltwerdens und sie suchte mit ihren jämmerlichen Armstummeln in schwachen, unsicheren Streichelbewegungen Udschlis Körper zu betasten. So ruhten Mutter und Tochter Brust an Brust und wurden nach lebelanger Trennung wieder eins. (*Mahâ-Rôg*, pp. 51-52)

> (At that moment Udschli had thrown herself upon the narrow hard bed, drawn up and taken into her arms the poor ruin of a human being, and, trembling in her whole body, called into her ear: 'Your child, your Udschli' [...] in an ecstasy betrayed by the quaking of her lower jaw and twitching of her white lips the sick woman [...] abandoned herself to the unknown experience of being caressed; with the pitiful stumps of her arms she attempted to caress Udschli's body in feeble and uncertain strokings. Thus mother and daughter rested in each other's bosom and became one again after a lifetime of separation.)

Not perhaps what one might have expected from the apostle of Goethean classicism, the ostensible enemy of sick art and of Naturalism. It is just this moment of the representative inclusion of the excluded, it seems to me, which suggests that Nordau's later literary works in many ways offer a counter-discourse to the aesthetic normativity so ruthlessly propagated in his cultural criticism.[33] Emerging as they do from the non-idealistic strand of German Darwin reception, they propagate a hard doctrine of existential ateleological choice, a doctrine of affirmed authenticity, a tough love of acceptance of the radical reality of our given or imposed physical shape and state. In propagating fitness of an unexpected, sometimes counter-selective variety, they expressly confront the nerves of the degenerate reader and argue a moral choice *for* the apparently unfitted. In this they both undermine conventional gender and aesthetic normativity and hesitate to denounce, qualities which are all too lacking in the critical works. These cognitive qualities come demonstrably from their aesthetic form. Nordau's literary *œuvre* functions, then, to this extent as an indicator of the indispensible nature of the *Geisteswissenschaften* (humanistic studies) as our moral guide through radical reality in the age of science. Is aesthetic experience, then, as a consummate trait of the *Geisteswissenschaften*, a translation of *Lebenskraft* into a new medium?

---

33 See the programmatic essay of Caroline Pross, in which she both reconstructs the emphatically foregrounded presentation of the deviant-degenerate-apathetic Eynhardt in *Die Krankheit des Jahrhunderts* through Agamben's inclusive-exclusive biopolitical concept of banishment and notes how the text simultaneously subverts that categorisation: "Die Kunst der Unterscheidung. Zur Darstellung der Anormalität in der Erzähliteratur des 19. Jahrhunderts," in *Bann der Gewalt. Studien zur Literatur und Wissensgeschichte*, ed. by Maximilian Bergengruen, Roland Borgards (Göttingen: Wallstein, 2009), pp. 361-386 (esp. pp.375-386).

No Body?
Radical Gender in *Memoirs of a Man's Maiden Years* (1907)

Stephanie M. Hilger

Abstract: Published under the pseudonym N.O. Body, *Aus eines Mannes Mädchenjahren* (1907) is the autobiographical account of Martha or, as she was later called Karl, Baer. Born with ambiguous genitalia in 1885, Martha was raised as a girl. After experiencing masculinizing development during puberty, Martha consulted medical professionals, among them Berlin sexologist Magnus Hirschfeld, and eventually decided to continue life as a man. Baer's anxiety regarding his ambiguous body converged with the fear of anti-Semitism that he experienced as a Jew living in Imperial Germany. The popularity of Baer's memoir indicates the curiosity about non-normative gender identities in the context of the two-sex model in the early twentieth century. Positioned at the intersection of literary and medical discourses about intersexuality – until recently called hermaphroditism – Baer's case highlights the connections between constructivist and materialist accounts of the human body and the necessity of establishing a bridge between scientific research and humanistic inquiry.

In *Human Life is Radical Reality* (2005), Howard Tuttle argues that "the radical fact of our experience" is "the life of each one of us."[1] At the same time that an individual's life is the most immediate and therefore most radical reality to that person, it is also the most obscure entity because it is enmeshed in a complex web of surrounding factors:

> The 'I' and its circumstances are so closely interdependent, so presuppositional of each other, that they cannot be separated in the radical reality of human life. Even the body must appear within 'my life.' The fingers and toes of a baby must first appear in and take meaning from the primary life of the child as part of its circumstances.[2]

Tuttle argues that, in human life, "even the body" needs a social and physical context to acquire meaning. The newborn child's body awaits signification and thereby underlines the corporeal dimension of human life. The role of the body in shaping human subjectivity has occupied philosophical thinkers since the inception of the discipline. The Western philosophical tradition has, according to Elizabeth Grosz, been marked by "somatophobia," a sus-

---

1 Howard N. Tuttle, *Human Life is Radical Reality: An Idea Developed from the Conceptions of Dilthey, Heidegger, and Ortega y Gasset* (New York: Peter Lang, 2005), p.1.

2 Tuttle, p.172.

picion of the body as unruly and unreasonable.[3] Grosz traces this distrust of corporeality to the Platonic and Aristotelian binary of mind versus body. She outlines the process by which this model was first translated into the Christian dualism between body and soul and then into the Cartesian opposition between mind and matter. Grosz argues that this somatophobia permeated thinking from Antiquity to the Early Modern period and that it caused the subordination of the body to the *cogito* in modern and contemporary philosophy.

Twentieth-century feminist philosophy moved the body from its previously subordinate position to the center of attention by focusing on its sexed nature. As a result, corporeality ceased being a lesser aspect of human subjectivity unworthy of philosophical inquiry. In addition, it became possible to reconsider the body, and the female body in particular, outside of patriarchal and universalizing modes of interpretation. The focus on the body posed the danger of absolute materialism and its separation of the body not only from the mind but also from its surroundings. New materialist thinkers such as Grosz reacted to this threat by highlighting the dangers of replacing the dualism between mind and matter with the binary of culture versus nature: "The body is not opposed to culture, a resistant throwback to a natural past; it is itself a cultural, *the* cultural, product."[4] Grosz' writings exemplify new materialism's focus on the cultural dimension of corporeality.[5] In its anti-essentialism, therefore, new materialism shares social constructivism's main concern. Conversely, social constructivist thinkers not only explore the body as a discursive product but also investigate its materiality. In her *Bodies that Matter*, Judith Butler, for example, performs a Foucauldian analysis of the body's socially determined nature while examining its physical dimension.[6] Like Grosz, Butler questions the nature/culture dualism through an investigation of the sex/gender binary by postulating that there is no body outside of culture. Gender and sex are both culturally constructed and, therefore, nature itself is always already part of culture.

Concrete illustrations for the intersection between new materialist and social constructivist thinking with respect to the gendered body can be

---

3 Elizabeth Grosz, *Volatile Bodies: Toward a Corporeal Feminism* (Bloomington: Indiana UP, 1994), p.5.

4 Grosz, p.23.

5 For a recent collection of essays by new materialist thinkers, see Diana Coole and Samantha Frost, *New Materialisms: Ontology, Agency, and Politics* (Durham: Duke UP, 2010). *New Materialisms* is informed by the "claim that foregrounding material factors and reconfiguring our very understanding of matter are prerequisites for any plausible account of coexistence and its conditions in the twenty-first century" (p.2).

6 Judith Butler, *Bodies that Matter: On the Discursive Limits of "Sex"* (New York and London: Routledge, 1993).

found in the work of cultural historians such as Thomas Laqueur and Londa Schiebinger, who both outline the changes in the representation of the body in anatomy and medicine throughout the ages. Laqueur argues that, over the course of the eighteenth and nineteenth centuries, a gradual shift occurred from the one-sex to the two-sex model.[7] Whereas the one-sex model had posited a continuum on which individuals were positioned depending on their proximity or distance to the male ideal – with the most feminine women being the most removed – the two sex-model presented men and women as incommensurably different and therefore diametrically opposed creatures. Laqueur discusses this shift mainly in the representation of human reproductive organs, while Schiebinger observes it in the representation of the entire body.[8] In the eighteenth century, sexual difference began to be perceived as permeating every aspect of the human physique, from the nervous system to its skeletal structures. Illustrations of the female body for medical textbooks illustrate this shift and render the confluence of gender and sex (and, by extension, culture and nature) particularly visible. Anatomists of the period did not randomly select a female skeleton for their drawings. They either picked a woman who most perfectly represented the feminine ideal or, like the ancient Greek painter Xeuxis, assembled a skeleton from the parts of several women's bodies and exaggerated their relative proportions. In the eighteenth and nineteenth centuries, the ideal woman was portrayed as the Rousseauian ideal, that is a mother with wide hips and a narrow ribcage. This portrayal underlined women's reproductive mission and undermined their claims for equality at a time when the Revolutionary ideal of *fraternité* was interpreted literally as solidarity between just men. Cultural ideals of femininity, however, not only shaped images of the body in medical textbooks. They also affected the materiality of the body itself. Schiebinger suggests that anatomists might not have manipulated reality when they exaggerated the contrast between the hips and the ribcage to emphasize women's childbearing mission. They could simply have selected the skeleton of a woman whose torso had been deformed by decades of wearing corsets made of rigid whalebone stays. Medical textbooks literally illustrate the fact that the body is both matter and culture. The permeable boundary between nature and culture is reflected in the shifting line between sex and gender.

Eighteenth- and nineteenth-century anatomical textbook illustrations emphasize the centrality of the body's reproductive dimension in representations of its materiality. As Judith Butler reminds us, etymologically, "matter

7 Thomas Laqueur, *Making Sex: Body and Gender from the Greeks to Freud* (Cambridge: Harvard UP, 1990).

8 Londa Schiebinger, *The Mind Has No Sex?: Women in the Origins of Modern Science* (Cambridge: Harvard UP, 1989).

[is linked] with mater and matrix (or the womb) and hence, the problematic of reproduction."[9] A body that is unable to reproduce complicates the issue of materiality and forces a reconsideration of the male-female binary, which informs all inquiries into the corporeal dimension of human life. One such body is that of the intersex individual, commonly called hermaphrodite in reference to the Greek myth of Hermes' and Aphrodite's two-sexed child. Intersex bodies do not conform to either male or female standard anatomy but are situated in an intermediate space. These divergences from the norm take different forms. They affect the reproductive organs, genitals, chromosomes, hormones, or all of them, and do so with different degrees of severity.[10] Because intersexuality manifests physiologically, it appears simply to be a matter of matter (that is biology and nature) and yet it is not. If the eighteenth century really witnessed the shift from the one-sex to the two-sex model, as Laqueur and Schiebinger suggest, then it appears logical that Enlightenment scientists and their successors strove to understand the intersex body because of its potential to destabilize the new system, in which men and women were presented as polar opposites.

In Antiquity, the fusion of Hermaphroditus' male form with that of the nymph Salmacis resulted in a body that was viewed as the incarnation of human perfection because it was both male and female, yet neither one completely. The hermaphrodite continued to occupy the middle position on the one-sex continuum until, with the gradual shift to the binary system of gender, it became viewed first as a grotesque monster and, then, beginning in the eighteenth century, to a mistake of nature in need of classification and, ultimately, medical intervention. Rather than considering the hermaphrodite as a figure of sexual ambiguity, Enlightenment thinkers focused on the literal meaning of the term "hermaphrodite" as an organism that could be both male and female in the reproductive process. Such bodies exist in

---

9 Butler, p.31.

10 Estimates of the prevalence of intersexuality differ widely, ranging from 0.02% to 2% of all live births. Since the term intersexuality encompasses a variety of conditions, the number varies depending on which manifestations are included. If only individuals with both ovarian and testicular tissue are counted, the number drops sharply, to 12 in 1 million. For an overview of the variety of conditions grouped under the heading "intersexuality" and their occurrence, see the following works: Anne Fausto-Sterling, *Sexing the Body: Gender Politics and the Construction of Sexuality* (New York: Basic Books, 2000); Catherine Harper, *Intersex* (Oxford and New York: Berg, 2007); Morgan Holmes, *Intersex: A Perilous Difference* (Selinsgrove: Susquehanna UP, 2008); Katrina Karkazis, *Fixing Sex: Intersex, Medical Authority, and Lived Experience* (Durham and London: Duke UP, 2008); Sharon Preves, *Intersex and Identity: The Contested Self* (New Brunswick and London: Rutgers UP, 2005); and Vernon A Rosario, "Quantum Sex: Intersex and the Molecular Deconstruction of Sex," *GLQ: A Journal of Lesbian and Gay Studies*, 15.2 (2009), 267-284.

the animal kingdom – slugs and snails are examples – and the plant world, yet there is no recorded case in the human realm. Whereas animal hermaphrodites are characterized by a doubled reproductive capacity, intersex humans are generally affected by reduced fertility and even infertility. Enlightenment science's literalization of human hermaphroditism served an ideological goal. By demonstrating that humans, unlike animals, could not be literal hermaphrodites, Enlightenment scientist positioned humans higher on the evolutionary scale. They underlined humans' intellectual superiority and thereby justified their domination of the animal and plant world. These scientists argued that humans could only be *pseudo*hermaphrodites that, upon closer inspection, could be identified as either male or female and classified accordingly in the two-sex model. They thereby also undermined the possibility of hermaphrodites in a more extended, pre-eighteenth century sense, that is as individuals with ambiguously sexed bodies. The eighteenth-century debate regarding terminology in encyclopedia articles and medical treatises exemplifies the replacement of the gender continuum with the gender binary, which the intersex individual cemented precisely because he/she was excluded from it. Clearly, the intersex body mattered, if only through its negation. Anne Fausto-Sterling calls this the "hermaphrodite vanishing act:"[11] "People of mixed sex all but disappeared, not because they had become rarer but because scientific methods classified them out of existence."[12]

Nineteenth-century anatomists expanded on the Enlightenment effort to classify the intersex body as either male or female by performing autopsies. Tissue samples were examined under the microscope in search of either ovarian or testicular markers. In this context, Alice Dreger posits that, during the "Age of Gonads," "the only true hermaphrodite was a dead hermaphrodite."[13] Dreger's statement regarding the gonads, the embryological structures that give rise to either the ovaries or the testicles, is more than just a provocative comment; it, instead, summarizes a historical reality. Since there were few non-emergency exploratory surgeries and biopsies in the nineteenth century, the "correct" classification could only be established post-mortem. Such autopsies resulted in the publication of a scientific report, sometimes called *mémoire* in a French-language context. Another type of *mémoire* often accompanied this medical document, a legal report arguing

11 Fausto-Sterling, p.37.

12 Fausto-Sterling, p.39.

13 Alice Dreger, "A History of Intersexuality: From the Age of Gonads to the Age of Consent," *The Journal of Clinical Ethics* 9.4 (Winter 1998), 345-355 (p.346). For a detailed discussion of the nineteenth-century approach to ambiguous genitalia, also see Alice Dreger, *Hermaphrodites and the Medical Invention of Sex* (Cambridge: Harvard UP, 1998).

for a change in gender with city or county officials. With the gradual popularization of the autobiographical genre, yet another type of memoir emerged, that written by the hermaphrodite him/herself. This text enabled the intersex individual to transform him/herself from a silent object of medical science and legal discourse to a speaking subject in charge of his/her self-representation.

In the early twentieth-century, with the development of the field of sexology, the discovery of hormones, and the momentous advances in surgical techniques, the intersex body gradually changed from an object of post-mortem examination to a patient in need of surgical correction. As Sander Gilman has observed, the medical advances of this period converged with an explosion of published memoirs by individuals with non-normative gender identities and sexual orientation.[14] One of these accounts is the German intersex autobiography *Aus eines Mannes Mädchenjahren* [*Memoirs of a Man's Maiden Years*] (1907), written by Martha or, as she was later called Karl, Baer. Born with ambiguous genitalia in 1885, Martha was raised as a girl. After experiencing masculinizing development during puberty, Martha consults medical professionals, among them Berlin sexologist Magnus Hirschfeld, and eventually decides to continue life as Karl. In 1907, after being legally reassigned as male,[15] Karl Baer published his autobiography under the pseudonym N.O. Body in Berlin.[16] The author's choice of the English pseudonym N.O. Body engages medicine's construction of the ambiguously sexed body as a *pseudo*hermaphrodite and its implied negation of the hermaphrodite in the broader sense. Baer, being interpellated by the hermaphrodite vanishing act, seemingly accepts it with a pseudonym that can be read as either "no body" or "nobody." And yet, Baer's pseudonym also questions the perception of the intersex individual as a human non-entity by presenting the "no" as two capitalized initials, that is N.O., which indicate a person's individual identity.[17] Therefore, even though Baer's "N.O." can be read as "no," it also undermines this reading and thereby asserts the personhood of

14 Sander Gilman, "Preface: Whose Body Is It, Anyway? Hermaphrodites, Gays, and Jews in N. O. Body's Germany," in N.O. Body, *Memoirs of a Man's Maiden Years*, trans. by Deborah Simon (Philadelphia: U of Pennsylvania P, 2006), pp.vii-xxiv (p.xi).

15 Hermann Simon, "Wer War N.O. Body?," in N.O. Body, *Aus eines Mannes Mädchenjahren*. Reprint (Berlin: Edition Hentrich, 1993), pp.167-246 (p.178).

16 In the following, for reasons of readability, I will refer to the author of *Aus eines Mannes Mädchenjahren* as "Karl Baer" and "he" since this is the identity he chose for himself.

17 In the text of the memoir itself, the identity of Karl/Martha is hidden behind the name Norbert/Nora, which establishes a fictional name for the capitalized initials N.O. in the pseudonym N.O. Body.

the individual who seemingly denies his body and thereby his identity. And yet, as Helga Thorson argues, "the body is very much present in Body's autobiography" in all its ambiguity.[18] The pseudonym "N.O. Body" contains an implicit question mark by the author who simultaneously hides and reveals his liminal body precisely because it troubles the border between sex and gender.

N.O. Body's *Memoirs of a Man's Maiden Years* were marketed for a mass audience and at least six editions were published over the course of two years.[19] Book reviews assured readers that the "autobiography [was] legitimate reading material – even for middle-class women"[20] by highlighting the author's "keusche Zurückhaltung" (chaste restraint)[21] and emphasizing Baer's suffering.[22] The popularity of his memoir indicates the curiosity about non-normative gender identities in the context of the two-sex model in the early twentieth century. [23] At the same time that this model seemed to have consolidated, it was already being questioned by the realities that it could not accommodate. In addition to the popularity of autobiographies by individuals with non-normative gender identities, this period witnessed a marked increase in scientific writings on sexuality in general and the intersex condition in particular by such scientists as Magnus Hirschfeld, Baer's physician. Hirschfeld, founder of the field of sexology, opposed the notion that there were only pseudo-hermaphrodites who could, upon closer inspection, be categorized as either male or female.[24] In Volume II of his medical textbook *Sexualpathologie*, which carries the title *Sexuelle Zwischenstufen* [Intermediate Sexual Stages], Hirschfeld argues that, in many cases, "selbst mit Zuhilfenahme des Mikroskops weder in vivo, noch bei einer Operation, und auch nicht einmal post mortem die Möglichkeit besteht, eine sichere

18 Helga Thorson, "Masking/Unmasking Identity in Early Twentieth-Century Germany: The Importance of N.O. Body," *Women in German Yearbook*, 25 (2009), 149-173 (p.150).

19 David Brenner, *German-Jewish Popular Culture Before the Holocaust: Kafka's Kitsch* (London and New York: Routledge, 2008), pp.47-48.

20 Thorson, p.152.

21 In Simon, p.193.

22 Except for Deborah Simon's published English translation of N.O. Body's *Aus eines Mannes Mädchenjahren,* Rudolf Presber's foreword, and Hirschfeld's epilogue, all translations are mine.

23 For a detailed discussion of the early twentieth-century reception of Baer's memoir, see Simon, p.191-196.

24 Detailed accounts of Hirschfeld's life and work can be found in the following studies: Manfred Herzer, *Magnus Hirschfeld: Leben und Werk eines jüdischen, schwulen und sozialistischen Sexologen* (Frankfurt/M. and New York: Campus, 1992) and Elena Mancini, *Magnus Hirschfeld and the Quest for Sexual Freedom: A History of the First International Sexual Freedom Movement* (New York: Palgrave Macmillan, 2010).

Geschlechtsdiagnose zu stellen" (it is not possible to establish a clear diagnosis of sex, even with the help of a microscope, either in vivo or during surgery, and not even post mortem).[25] At the transition point from the Age of Gonads to the Age of Surgery, Hirschfeld posits the possibility of truly ambiguous bodies that do not fit into the two-sex model. He resists the binary system of his predecessors and instead highlights the "Fülle von Vermischungsarten" (abundance of possible combinations),[26] thereby allowing for the possibility of a human hermaphrodite in the broader sense and referencing the earlier one-sex model of continuity. One of Hirschfeld's case studies in *Sexuelle Zwischenstufen* is that of N.O. Body/Karl Baer, who consulted him as a young adult before transitioning to a male identity and writing his memoir.

The radical reality of the intersex individual is constituted through the dialogue that occurs between patient/Baer and practitioner/Hirschfeld. The materiality of Baer's body is embedded in and literally shaped by the surrounding circumstances. Baer consulted Hirschfeld for a diagnosis that would clarify the future course of his life. The physician, however, did not make the medical decision regarding Baer's body unilaterally. It ultimately emerged from what Ronald Carson calls the hyphenated space between doctor and patient. Carson uses the image of the hyphen because it "calls attention to the distance between parties to the clinical encounter" but also functions as "a bridge across the divide."[27] The hyphen's double function as a divider and a connector is visible in the interaction between Hirschfeld and Baer. Hirschfeld's case report on Baer and the memoir written by Baer-as-N.O. Body enter into dialogue with each other explicitly and implicitly. They are connected through a hyphen that draws attention to the permeability of both genres. Hirschfeld's three-page medical report, composed several years before the publication of *Aus seines Mannes Mädchenjahren* in 1907, is collated into his textbook *Sexuelle Zwischenstufen*, which was published a decade after Baer's autobiography, in 1918. Hirschfeld's original report is inserted into his textbook and set off typographically to signal its status as a preexisting document; it appears in a smaller font. The prefatory material that introduces this report in the textbook and the concluding section were written in 1918, after the publication of Baer's memoir. In his preface to the report, Hirschfeld explicitly refers to Baer's autobiography:

---

25 Magnus Hirschfeld, *Sexuelle Zwischenstufen. Das männliche Weib und der weibliche Mann* (Bonn: A. Marcus und E. Webers Verlag, 1918), p.78.

26 Hirschfeld, *Sexuelle Zwischenstufen*, p.89.

27 Ronald Carson, "The Hyphenated Space: Liminality in the Doctor-Patient Relationship," in *Stories Matter: The Role of Narrative in Medical Ethics*, ed. by Rita Charon and Martha Montello (New York and London: Routledge, 2002), pp.171-182 (p.171).

> Der nächste Fall betrifft einen als Mädchen verkannten Mann, der unter dem Namen N. O. Body (nobody = niemand) und dem Titel: "Aus seines Mannes Mädchenjahren" vor einigen Jahren seine aufsehenerregende und in der Tat recht lehrreiche Lebensbeschreibung veröffentlicht hat.[28]

> (The next case concerns a man who was mistaken for a girl and who, a few years ago, using the name N.O. Body ["nobody" = nobody], published a spectacular and indeed rather instructive life description with the title "Of a Man's Maiden Years.")

In the same way that Baer's memoir is referenced in Hirschfeld's *Sexuelle Zwischenstufen*, Hirschfeld's medical views are included in Baer's memoir. *Aus eines Mannes Mädchenjahren* was published with an epilogue written by Hirschfeld, in which he describes Baer as an example of the intermediate sexual stages discussed in his *Sexuelle Zwischenstufen*. The publication chronology of these texts highlights the multiple connections between the documents produced by doctor and patient. The body of each text – consisting of different parts produced at different moments and then collated for publication – is as complex as the body of the historical subject who is reconstructed through these narratives.

Hirschfeld's medical report consists of four parts: the patient's past (individual and family history), present (physical appearance and interests), a separate section devoted to Baer's genitals and sexual drive, and finally Baer's future (prognosis). Baer's memoir can be read either as an explicit response to Hirschfeld's written report, which he in all likelihood read because it was part of the legal file that he forwarded to various offices when he petitioned for his gender reassignment as male, or as an implicit reaction to Hirschfeld's medical exam of his ambiguous body.

Like Hirschfeld's report, Baer's autobiographical narrative is chronological. It is, however, unlike the medical document, not linear and devotes a different proportion of narrative space to each life period. Baer's narrative consists of key memories of his childhood and adolescence that are narrated as chronologically arranged traumatic flashbacks:

> Mein Gedächtnis reicht bis ins dritte Lebensjahr zurück. Aus dieser Zeit stammt ein Erinnerungsbild. [...] Ein Maimorgen in meiner kleinen Heimatstadt. [...] Alle Kinder sind froh und lustig – bis auf eines, das abseits an der Dachtraufe steht [...]. Nun kommen die andern Kleinen herbeigelaufen; sie sammeln sich um dieses kleine Mädchen – wenigstens trug auch dieses Kind Mädchenkleider – und schreien: 'Junge, böser Junge!' Da schreckt das Kind zusammen und läuft weinend ins Haus. Dieses Kind war ich. So lange ist das her, und dennoch hallen das

[28] Hirschfeld, *Sexuelle Zwischenstufen*, p.44.

> brennende Weh und der Kindertrotz noch in mir nach, wenn ich an diese Szene denke.[29]

> (My memory reaches back to my third year. From that time comes a scene that stands as clearly before my eyes as if I had seen myself in it only yesterday. A May morning in my small hometown. […] All the children are merry and cheerful, except one who is standing to one side beside the rain gutter. […] Now the other children come running and gather around this little girl, this child is wearing girls' clothing in any case, and shout, 'Boy, boy, bad boy!' The startled child then runs into the house, crying. That child was me. It was so long ago, and yet the burning pain and childish defiance still resound within me when I recall this scene.)[30]

Baer narrates this event from a third-person perspective. He both distances himself from this traumatic childhood memory and recreates its immediacy through the use of the present tense before solving the puzzle for the reader. Baer's recollection echoes Hirschfeld's description of his patient's interaction with her peers: "Auch äußerlich sah sie mehr knabenhaft aus und litt schon damals unter diesbezüglichen Bemerkungen und Neckereien ihrer Mitschülerinnen" (Even her outward appearance looked more like that of a boy and she already, at that time, suffered from her fellow female students' remarks and teasing on this account).[31] Hirschfeld refers to Baer's suffering in one sentence, whereas Baer describes his peers' teasing in great detail. This rhetorical difference is not surprising considering the respective genre, the detached scientific report and the personal life narrative. What is significant to note, however, is that Baer, like Hirschfeld, resorts to a third-person perspective to describe his own history; he describes himself as "the child." The autobiographical subject can access painful experiences only through the indirection and opaqueness typical for the recovery of traumatic moments. The third-person narrative perspective echoes the split between Baer's past identity as Martha and his present existence as Karl; Baer's trauma stems from his sexually ambiguous body and his surroundings' reaction to it. Geertje Mak points out that the process of "writing works as therapy […] more than just writing down an already existing memory, the reader seems to witness his *process* of remembering."[32] During this process of re-

29 N.O. Body, *Aus eines Mannes Mädchenjahren* (Berlin: Gustav Rieckes Buchhandlung, 1907), p.18-19.

30 N.O. Body, *Memoirs of a Man's Maiden Years*. Trans. Deborah Simon (Philadelphia: U of Pennsylvania P, 2006), p.8-9.

31 Hirschfeld, *Sexuelle Zwischenstufen*, p.45.

32 Geertje Mak, *Doubting Sex: Inscriptions, Bodies and Selves in Nineteenth-Century Hermaphrodite Case Histories* (Manchester and New York: Manchester UP, 2012), p.220.

covery, Baer accesses particularly traumatic moments from his past and externalizes them through writing.

Baer's account, especially of the early years, consists of a montage of such flashbacks. His presentation of these memories engages another section of Hirschfeld's report, his detailed description of Baer's ambiguous genitals, of what would be called a hypospadic micropenis with one partly undescended testicle in present-day medical terminology. Again, Baer accesses painful memories related to his anatomy by presenting a snapshot that depicts the behavior of the people surrounding him:

> Nur ein Bild ist mir aus jener Zeit geblieben, eine flüchtige, blitzartige Erinnerung. Ich muß damals noch sehr klein gewesen sein. Eine große, blonde Frau zog mir unter Scherzen das Nachtröckchen aus und setzte mich in eine Wanne. Ich sah aber, daß ihr trotz der frohen Worte Tränen in die Augen traten. [...] Heute kann ich mir das Bild erklären, das ich lange im Gedächtis bewahrte. Es war meine Mutter, die mich auskleidete [...] [die] Frau, die weinen mußte, wenn sie den Körper ihres Kindes sah.
>
> (I have retained only one image from that period, a fleeting flash of memory. I must have been very small when a big blond woman took off my little nightgown, joking all the while, and placed me into a tub. But I saw that, despite her cheerful words, tears came to her eyes. [...] Today I understand the image, which long remained in my memory. It was my mother who undressed me [...] the woman who was moved to tears by the sight of her child's body.)[33]

In contrast to Hirschfeld's detailed presentation and measurements of Baer's genitals in his medical report, Baer remains indirect in his description of his anatomy. It is always presented through the eyes of others, in this case the mother who cries while bathing her baby or, later in the memoir, Baer's female childhood friends: "Du, guck mal, [sie] ist ja ganz anders, wie wir" ("Look here, [she] is very different from us").[34] The closest Baer comes to directly describing this difference is by referring to the "kleinen Vorzug, den meine kleinen Freundinnen so sehr bewunderten" ("small advantage that my little girlfriends admired so much").[35] Later in the text, Baer reports his desire to remove this marker of difference by describing his childhood castration fantasies. These fantasies become especially pronounced once little Martha begins her schooling: "Mit einem Anflug heroischer Größe säbelte ich in einem unbewachten Augenblick auf diesen armen Finger los. Ob das sehr weh tat, weiß ich nicht mehr. Aber der Schnitt muß wohl recht tief gewesen sein, denn die Narbe ist heute noch sichtbar" ("With a burst of heroic

33 Body, *Aus eines Mannes*, pp.22-23 and Body, *Memoirs*, p.10.

34 Body, *Aus eines Mannes*, p.38 and Body, *Memoirs*, p.18.

35 Body, *Aus eines Mannes*, p.39 and Body, *Memoirs*, p.19.

strength, in an unguarded moment, I sawed away on that poor finger. I no longer remember whether it was very painful. But the cut must have been fairly deep, because the scar is visible to this day").[36] This act not only expresses Baer's resentment of learning to write, but also functions as a displaced desire to remove that part of herself that makes little Martha different from her classmates. Baer's indirection stands in stark contrast to Hirschfeld's scientifically explicit description of his patient's genitals in a section that is separated from the other three parts discussing Baer's past, present, and future. Neither Baer's name nor any personal pronouns are ever used in this section. His body parts are described as "*die* Geschlechtsteile" (the genitals), "*die* Urethra" (the urethra), "*die* Prostata" (the prostate) etc.[37] The medical narrative also performs a kind of castration, albeit a different one; it removes Baer's non-normative genitals textually from the different stages of his life in order to reattach them in the concluding section, the prognosis.

In the autobiography and the medical report, the reattachment of the metaphorically and textually severed member, that is Baer's reassignment as male, occurs through the determination of his sexual orientation. Both narratives devote a significant amount of space to this aspect. Hirschfeld highlights Baer's attraction to women, as does Baer himself. But they differ on one important aspect. Whereas Hirschfeld states with scientific determination that "[d]ie ersten geschlechtlichen Regungen traten im 25. Lebensjahr auf" (the first sexual urges emerged in the twenty-fifth year),[38] Baer paints a different picture. He describes his sexual childhood play with his female friends "um Eltern und Erzieher darauf aufmerksam zu machen, wie früh sich das Sexualempfinden in manchen Kindern regen kann" ("to draw the attention of parents and teachers to how early sexual feelings may arise in some children").[39] Baer directly opposes Hirschfeld's one-line statement that locates his patient's sexual awakening at the age of twenty-five. This contradiction not only poses questions about the relation between doctor and patient and the problematic correlation between sexual orientation and gender identity but also highlights the issue of audience for Hirschfeld's report.[40]

---

36 Body, *Aus eines Mannes*, p.50 and Body, *Memoirs*, p.25.

37 Hirschfeld, *Sexuelle Zwischenstufen*, p.46-47.

38 Hirschfeld, *Sexuelle Zwischenstufen*, p.47.

39 Body, *Aus seines Mannes*, p.39 and Body, *Memoirs*, p.19.

40 For a discussion of the connection established in nineteenth- and early twentieth-century medical studies between hermaphroditism and homosexuality, see Gilman, "Whose Body Is It Anyway: Hermaphrodites, Gays, and Jews in N.O. Body's Germany," in *Jewish Masculinities: German Jews, Gender, and History*, ed. by Benjamin Maria Baader, Sharon Gillerman, and Paul Lerner (Bloomington and Indianapolis: Indiana UP, 2012), pp.138-150 (pp.142-145).

Did Baer not tell Hirschfeld about his childhood feelings or did the doctor not report these episodes for strategic reasons? Not discussing Baer's sexual childhood play in his report, which was part of Baer's legal petition for gender reassignment, allows the physician to focus on the mature adult's decision to embark on a heterosexual lifestyle. Hirschfeld states that "Als höchstes Ideal steht ihr eine dauernde eheliche Verbindung vor Augen" ([h]er highest ideal is a permanent marital relation).[41] He also outlines the perils of not directing Baer's sexual drive into a socially sanctioned outlet by mentioning that Baer procured cyanide to commit suicide because, as Martha, he could not get married to the woman he loved. This fact is mentioned twice, in the report itself and also in the prefatory material that was added later to introduce the report in *Sexuelle Zwischenstufen.* Hirschfeld's conclusion provides the basis for Baer's legal case: "Es kann nach allem nicht dem geringsten Zweifel unterliegen, daß es sich [hier] um einen Fall von irrtümlicher Geschlechtsbestimmung handelt. [...] [Baer] [ist] in Wirklichkeit ein Mann" (After all this there cannot be the slightest doubt that [this case] is one of erroneous sex determination. [...] [Baer] is in reality a man).[42] This conclusion seems surprising after reading the report, which highlights Baer's biological ambiguity by pointing at the absence of a prostate, testicle, and spermatozoa. All indications are that biologically Baer is intersex but, as Hirschfeld argues in his epilogue to Baer's memoir – entering into dialogue with Baer's version of his own life – "*[d]as Geschlecht des Menschen ruht viel mehr in seiner Seele als in seinem Körper*, oder, um mich einer mehr medizinischen Ausdrucksweise zu bedienen, viel mehr im Gehirn als in den Genitalien."[43] ("*[t]he sex of a person lies more in his mind than in his body*, or to express myself in more medical terms, it lies more in the brain than in the genitals").[44] Hirschfeld's didactic statement in his epilogue to Baer's memoir appears as an attempt to silence the ambivalences in his own writings about Baer. Yet these ambivalences constantly reappear as contradictions in Hirschfeld's writings. In Hirschfeld's epilogue, for example, the body does not matter (because of the supremacy of the brain as *cogito* rather than body part) and yet it does when he highlights its importance through another contradiction. Hirschfeld, on the one hand, argues that Baer is an example of "irrtümlich[e]

41 Hirschfeld, *Sexuelle Zwischenstufen*, p.47.

42 Hirschfeld, *Sexuelle Zwischenstufen*, p.47.

43 Magnus Hirschfeld, "Nachwort," in *Aus eines Mannes Mädchenjahren*, by N.O. Body (Berlin: Gustav Rieckes Buchhandlung, 1907), pp.213-218 (p.214).

44 Magnus Hirschfeld, "Epilogue," in *Memoirs of a Man's Maiden Years*, by N.O. Body. Trans. Deborah Simon (Philadelphia: U of Pennsylvania P, 2006), pp.109-111 (p.110).

Geschlechtsbestimmung" ("erroneous sex determination")[45] and thereby claims that he is biologically male, while, on the other, describing him as intersex by stating that

> Fälle, wie der vorliegende, nur extreme Formen sexueller Zwischenstufen sind, von denen leichtere Grade auf körperlichem und seelischem Gebiete in großer Mannigfaltigkeit vorkommen. Ja, es scheint sogar fraglich, ob die moderne Ansicht richtig ist, welche in diesen Abweichungen von der Norm schlankweg Störungen und Mißbildungen erblickt oder ob nicht die antike Auffassung mehr für sich hat, die hier nur eigentümliche Spielarten, Varietäten und Besonderheiten der Gattung Mensch sah. In der Kunst des Altertums, sowohl der redenden, wie der bildenden, ist ja bekanntlich der Hermaphrodit eines der beliebtesten Darstellungsobjekte.
>
> (cases like the one described here are only extreme forms of intermediate sexual stages which, to a lesser extent, occur in a great variety of physical and mental forms. Indeed, it even seems questionable whether the modern view is correct, a view that sees in these deviations from the norm outright dysfunctions and deformities, or whether there is not more to be said for the view held in antiquity, which regarded [these deviations from the norm] merely as curious varieties and special individuals within the human race. It is well known that in the art of antiquity, both in its spoken and visual forms, the hermaphrodite is one of the most popular figures.)[46]

By referencing the figure of the hermaphrodite, Hirschfeld posits the possibility that Baer's case is not one of erroneous sex determination but, instead, an example of an intermediate stage on a sexual continuum.

The contradictions present in Hirschfeld's epilogue to Baer's memoir also permeate the medical report itself, both with respect to its content and its position in Hirschfeld's medical textbook. In it, Hirschfeld states that Baer is "in Wirklichkeit ein Mann" (in reality a man).[47] This seemingly definitive statement clashes with the fact that Hirschfeld included Baer's case in the second volume of his textbook. The volume's title "Sexuelle Zwischenstufen" (Intermediate Sexual Stages) echoes Hirschfeld's reference to hermaphrodites in the epilogue to Baer's memoir. In this volume, the report is included in the chapter "Hermaphroditism" and is presented in-between two studies of intersex patients. Also, each page of the case study bears the header "I. Kapitel: Hermaphroditismus" (Chapter I: Hermaphroditism), which – along with the detailed description of Baer' ambiguous genitals in the report itself – causes a dissonance with Hirschfeld's assertion that Baer suffered from erroneous sex determination at birth. The inclusion of

45 Hirschfeld, "Nachwort," p.213 and Hirschfeld, "Epilogue," p.109.

46 Hirschfeld, "Nachwort," p.215 and Hirschfeld, "Epilogue," p.110.

47 Hirschfeld, *Sexuelle Zwischenstufen*, p.47.

this report, which served as a petition to city officials for Baer's gender reassignment, in his medical textbook causes a dissonance that points at Hirschfeld's ambivalence with respect to Baer's case.[48] Hirschfeld's writings do not resolve his patient's gender trouble; in fact, they generate more questions about what really is the matter with his body. Hirschfeld, like Baer, does not manage to silence the ambiguities in his writings. Unlike Thorson, who highlights the ambiguities in Baer's writing and opposes them to Hirschfeld's narration, which "tends to clarify the corporeal confusion by situating the patient's body back within the framework of dichotomous sex classification based primarily on secondary and tertiary sex characteristics and heterosexual desire,"[49] I would argue that, like Baer's autobiography, Hirschfeld's writings raise more questions than they provide answers. Thorson's tentative "tends to clarify" indicates this possibility, yet refrains from reading medical discourse with the same close reading strategies as autobiographical writing. Despite his own assertions to the contrary, Hirschfeld is not able to contain Baer's body within the binary system. The medical *mémoire* is marked by the same kind of signifying tensions as the autobiographical memoir. Reading medical writing against the grain, that is not merely as a record of facts but as a discursive product itself, reveals the multiple connections between the body as matter and the body as social construct.

In light of the contradictions in Hirschfeld's writings on Baer, the absence of a gonadal biopsy is striking, in particular because the doctor references such a procedure in his report: "eine genaue Untersuchung [...], die

---

48 Hirschfeld's report is also included in his published lecture, "Drei Fälle von irrtümlicher Geschlechtsbestimmung," *Medizinische Reform: Wochenschrift für soziale Medizin, Hygiene und Medizinalstatistik*, 51 (1906), 614-617. In the introduction to the lecture, Hirschfeld argues that Baer belongs to a "Gruppe von Personen zweifelhaften Geschlechts, nämlich diejenigen deren Geschlechtszugehörigkeit bei der Geburt unsicher, sich erst im postpubischen Alter nach einer oder der anderen Richtung entscheidet" (group of people of doubtful sex, namely those whose sex is unclear at birth and develops in one or the other direction only in postpuberty) ("Drei Fälle," p.614). In his lecture, Hirschfeld contextualizes Baer's case differently than in his textbook. His lecture highlights the role of puberty in clarifying the biological sex of individuals with ambiguous genitalia. However, this categorization aimed at proving Baer's male biology also raises questions because it clashes with a statement in the report itself: "Die Geschlechtsteile [...] zeigen [...] zur Zeit einen durchaus männlichen Typus" (The genitals presently show a male type indeed) ("Drei Fälle," p.615). The mention that Baer's genitals "presently" appear as male raises the question whether they might change in the future. Hirschfeld's lecture is permeated by similar ambiguities as his other writings on Baer's case. Hirschfeld strives to present Baer as unambiguously male, yet, by doing so, he highlights his intersexuality.

49 Thorson, p.155.

natürlich nur operativ vorgenommen werden könnte, [wäre] für die genaue Feststellung der Keimdrüsen sehr wesentlich" (a detailed exam [...], which can naturally only be performed operatively, would be very crucial in exactly determining the gonads).[50] Unlike their nineteenth-century predecessors, Hirschfeld and his colleagues were no longer confined to removing tissue samples during an autopsy. The development of operative techniques in the early twentieth century made biopsies on living persons possible. In Baer's case, either such an examination was never performed or, if it was performed, the results were not reported. Perhaps, as Ulrike Klöppel argues, the primordial importance of the gonads in sex determination began to be questioned during this period.[51] This could explain why Hirschfeld, on the one hand, referenced this procedure yet, on the other, did not perform it. Or, assuming that Hirschfeld still considered the gonads to be a major factor, he might have feared a discovery that contradicted both his patient's wishes and his own claim that Baer was in reality a man suffering from erroneous sex determination at birth.[52] A biopsy could have transformed the hyphen between doctor and patient from a connecting bridge to a separating barrier, causing Baer to act on his threat of suicide. Hirschfeld highlights the fact that this most negative of all outcomes did not occur. In the postscript to his medical report in *Sexuelle Zwischenstufen*, he underlines his patient's successful life as a heterosexual male by emphasizing that he was both married and employed.

At the same time that Baer's wishes might have driven Hirschfeld's "rectification" of Baer's sex determination at birth, Hirschfeld's own position in early-twentieth century Germany could also have shaped his decision. Hirschfeld possibly transferred his own anxieties as a gay man living in Berlin under the threat of paragraph 175 – which prohibited homosexuality – onto Baer. Hirschfeld might have acquiesced to his patient's wish to become

---

50 Hirschfeld, *Sexuelle Zwischenstufen*, p.46.

51 Looking at German medical discourse in particular, Ulrike Klöppel disagrees with Dreger's theory that the nineteenth century was the "Age of Gonads." She argues that, "in the last third of the nineteenth century the importance of the gonads was again scrutinized [...] well-known physicians were of the opinion that the differentiation of sexual characteristics was not primarily dependent on the gonads" (Ulrike Klöppel, "Who Has the Right to Change Gender Status? Drawing Boundaries between Inter- and Transsexuality," in *Critical Intersex*, ed. by Morgan Holmes (Farnham: Ashgate, 2009), pp.171-187 (p.173). This revision in thinking on the importance of the gonads could have influenced Hirschfeld, who wrote his report of Baer at the beginning of the twentieth century.

52 In one of his published lectures, Hirschfeld posits the existence of humans with gonadal tissue containing both testicular and ovarian markers. He references his own research and that of a colleague. In this light, his reticence to perform a gonadal biopsy on Baer becomes significant. See Hirschfeld, "Drei Fälle."

a heterosexual man rather than remain a homosexual woman because he himself knew the challenges of openly living homosexuality. In addition, reassigning Baer as male provided his patient with greater freedom to explore his interests in education and travel. By becoming Karl, Baer also circumvented the danger of being arrested for cross-dressing, which he faced as Martha because of his masculinized appearance.[53] The medical decision to reassign Baer as male occurs in the space where doctor and patient meet. The cultural and historical context that informs this decision directly affects the matter of Baer's body. When Baer begins his life as a man, he comments on the resulting physical changes:

> Mein Körper, von keinem Mieder und keinem anderen einengenden Kleidungsstück mehr eingeschnürt, entfaltete sich freier und kräftiger. Ich turne und treibe jede Art von Gymnastik. [...] Ich bin kräftiger und breiter geworden, meine Haltung freier, und körperlich dürfte aus meinen Mädchenjahren wohl nichts zurückbleiben, als eine leichte Schnürfurche.
>
> (My body, which was no longer constricted by bodices and other tight articles of clothing, developed freely and became stronger. I now do gymnastics and other kinds of exercise. [...] I have become stronger and broader, my posture is freer, and physically nothing is likely to remain of my girlhood years other than a slight furrow left behind from tight lacing.)[54]

By liberating his patient from the feminizing mold of his childhood, Hirschfeld provides the starting point for the literal reshaping of Baer's body. He reverses the feminizing process initiated by the constraints of fashion based on the two-sex model. At the same time, however, he reinforces this same binary by categorizing Baer as male rather than intersex, thereby providing the basis for Baer's desire to achieve a more masculine physical appearance. Hirschfeld's ambiguous statements regarding Baer's case highlight the reality that medicine is not a science per se but rather, as Kathryn Montgomery-Hunter argues, an interpretive practice, in which the doctor's task is to collect as much information as possible, organize it, interpret it, and then establish a plan of action.[55] Because medical practitioners are engaged in an interpretive act, they are not completely disengaged from their objects of study: human bodies. The liminal bodies of intersex individuals such as Baer highlight the liminality in the doctor-patient relationship particularly forcefully.

---

53 Gilman, "Whose Body," p.144.

54 Body, *Aus eines Mannes*, p.204 and Body, *Memoirs*, p.105.

55 Kathryn Montgomery-Hunter, *Doctors' Stories: The Narrative Structure of Medical Knowledge* (Princeton: Princeton UP, 1991)

The relationship between Hirschfeld and Baer is rendered more complex by the fact that Hirschfeld was not only Baer's physician but also served as the intermediary between Baer and Rudolf Presber, a central figure on the early twentieth-century German publishing scene. He introduced Baer to Presber, thereby serving as the conduit for the composition and eventual publication of Baer's life story.[56] Hirschfeld's motivations for this introduction are debatable and perhaps even questionable. Yet, Hirschfeld's function as a bridge between Baer and Presber allowed the patient to transform himself from the silent object of medical discourse into the speaking autobiographical subject. The question of agency in Baer's own description of his gender trouble is complex, especially if we consider the voracious appetite with which Baer's memoirs were consumed, threatening to transform him from a medical case study into an object of popular voyeurism. The complexity of these issues makes the juxtaposition of the doctor's *mémoire* with the patient's memoir a productive endeavor; it allows us to explore the signifying ambiguities in this encounter with respect to the shifting barrier between sex and gender in the early twentieth century.

Hirschfeld possibly transferred his own anxieties regarding not only his sexual orientation but also his ethnicity onto Baer. Hirschfeld and Baer were both Jews living in early twentieth-century Germany. They therefore experienced the double threat of paragraph 175 and anti-Semitism. Fear of prejudice and discrimination permeates Baer's autobiography, which performs a complex process of identity concealment. In addition to his choice of pseudonyms, Baer engages in a "durchgängig verwendetes Verfremdungsprinzip" (consistently employed principle of de-familiarization),[57] which translates the identifiers of Baer's Jewish identity, such as holiday celebrations, into Christian ones. Most significantly, Baer presents his Jewish identity as French ancestry:

> Wir sind kein deutsches Geschlecht. Unsere Vorfahren kamen aus Frankreich. [...] Wir führen unser Geschlecht auf einen Franziskaner-Mönch zurück, der zur Zeit der Reformation seinen Orden verließ und eine Familie gründete [...] Schon äußerlich sind wir von den andern Bergheimern leicht zu unterscheiden: Schwarze oder braune Augen, braunes, welliges Haar, scharf geschnittene südländische Züge findet man nicht oft unter den Sachsen und Franken jener Bergtäler.

---

56 Rudolf Presber wrote a foreword to *Aus eines Mannes Mädchenjahren*, which describes their initial encounter and his encouragement of Baer to write his life story: Rudolf Presber, "Vorwort," in N.O. Body, *Aus eines Mannes Mädchenjahren.* (Berlin: Gustav Rieckes Buchhandlung, 1907), pp.7-13.

57 Simon, p.180.

> (Our lineage is not German. Our forefathers came from France. [...] We trace our lineage back to a Franciscan monk who left his order at the time of the Reformation and founded a family [...] Our outward appearance alone is enough to distinguish us from the other inhabitants of Bergheim: black or brown eyes, brown wavy hair, and sharply defined southern European features are seldom found among the Saxons and Franconians of those mountain valleys.)[58]

Baer translates being Jewish as being French, which, as Gilman points out, was "viewed in Baer's time as a racial category as well as a political one."[59] Baer hints at the religious dimension to his identity through his reference to the non-conforming Franciscan monk. This monk, who forsakes celibacy and becomes a father, is positioned at the origin of Baer's invented family lineage. This fictional reproductive illegitimacy is reflected back onto Baer in the form of a variant body that cannot reproduce. His playmates' statement that Baer is different therefore applies not only to his genitals but also to the remainder of his physical appearance. Baer's body threatens the gender binary and the early twentieth-century body politic within which it is embedded. As Shelley Baranowski has argued, the comparatively late creation of a unified Germany in 1871 resulted in the exclusion of all types of others, racial, religious, and sexual, in order to create the illusion of a common identity shared by all Germans.[60] The discrimination of those who were perceived as non-Germans was enmeshed with intolerance towards those whose bodies and behaviors questioned the two-sex model of gender. Elena Mancini points out that, after German unification, the "clause of the Prussian penal code [Paragraph 143, which prohibited same-sex acts between men] would be extended throughout the German kingdoms and be renamed Paragraph 175."[61] German unification was accompanied by the exclusion of difference, sexual, racial, and ethnic. Hirschfeld's decision to reassign Baer as male and therefore transform his attraction to women into heterosexual desire normalizes one of Baer's divergent subject positions in early twentieth-century Imperial Germany. After becoming Karl, Baer continues his life as a heterosexual Jewish man in Berlin. While Baer concealed his Jewish identity in his writing, he did not in his life. He eventually became director of the Berlin B'nai B'rith lodges, a Jewish fraternal organization. He emigrated to Palestine in 1938 and died near Tel Aviv in 1956, at age seventy-one. Similarly, Hirschfeld, who had embarked on a lecture tour in the early thirties, would never return to Germany and died in exile in France in 1935. Both of their

58 Body, *Aus eines Mannes*, p.29-30 and Body, *Memoirs*, p.13-14.

59 Gilman, "Preface," p.xix.

60 Shelley Baranowski, *Nazi Empire: German Colonialism and Imperialism from Bismarck to Hitler* (Cambridge: Cambridge UP, 2011), pp.9-66.

61 Mancini, p.154.

bodies were excluded from the German body politic of the Wilhelmine era because of their threat to its ethnic constitution and gendered organization.

All indications are that Baer's gender reassignment occurred only in writing, that is in Hirschfeld's legal petition to city officials that argued for the change of "Martha" to "Karl." Hirschfeld, however, not only shaped bodies on paper; he also pioneered gender reassignment operations in the early 1930s, thereby ushering in what Alice Dreger has called the "Age of Surgery."[62] As both a proponent of sexual intermediate stages and a leading figure for such surgeries, which were not only performed on individuals who felt that they lived in the wrong body but were also used to normalize variant genital morphology, Hirschfeld's legacy is complex. Hirschfeld has been seen as both a sexual reformer and as an implementer of sexual hygiene, that is as simultaneously resisting and reinforcing his time's restrictive gender discourses.[63] The persecution of the members of the "Institut für Sexualwissenschaft" and its eventual closure in 1933 make it clear that normalizing surgeries, like other cosmetic surgeries, were seen as subversive acts by the Nazis since they erased those markers of sexual and ethnic difference that the totalitarian regime used for purposes of discrimination and persecution.[64]

The complicated legacy of gender reassignment surgeries pioneered by Hirschfeld became clear in the sixties and seventies, when such operations became the first recourse for babies born with non-standard genital anatomy, in an attempt to spare them the identity challenges faced by Baer. Surgeons mainly decided according to cosmetic considerations and surgical feasibility, sometimes in consultation with pediatric endocrinologists. Psychologists and psychiatrists encouraged parents to perform the operation, which strove to create either male or female credible external genitalia, before the age of two so as to enable the child's formation of a stable gender identity.

When these children became adults in the 1990s, the negative outcomes of such normalizing surgeries began to surface. The most vocal patients and medical practitioners opposed to these operations formed the activist organization, the ISNA, the Intersex Society of North America, in 1993.[65] Draw-

---

62 Dreger, "A History," p.349.

63 Herzer in particular discusses Hirschfeld's ambivalent legacy.

64 As Gilman points out, the year 1933 also saw the closing of the public clinic for cosmetic surgery in Germany: Gilman, "Preface," p.xxiv.

65 See Milton Diamond's work on medical research cautioning against normalizing surgeries in infancy: Milton Diamond, "Pediatric Management of Ambiguous and Traumatized Genitalia," *The Journal of Urology*, 162 (1999), 1020-1028; Milton Diamond and H. K. Sigmundson, "Sex Reassignment at Birth: Long Term Review and Clinical Implications," *Archives of Pediatrics and Adolescent Medicine*, 151 (March 1997),

ing on "lessons from the intersexed,"[66] the ISNA criticized corrective genital surgeries in general but also, more specifically, questioned the practice of performing such invasive procedures on individuals who, due to their age, were not able to give informed consent and thereby broached the question of medical ethics.[67] The ISNA's efforts lead to the 2006 publication of a consensus paper, which established new treatment guidelines emphasizing patient-centered approaches and the importance of informed consent. The implementation of these new guidelines has been so slow that the new standard of care exists mainly on paper rather than in medical practice. This situation has led to an impasse, resulting in the disbanding of the ISNA and the creation of a new organization, the Accord Alliance, in 2008.[68]

The Accord Alliance has proposed a change in terminology from "intersexuality" to "DSD/disorders of sex development," which it defines as "congenital conditions in which development of chromosomal, gonadal or anatomic sex is atypical." The rationale for this change is that, "[b]y contrast to [...] older terms, the term DSD refers to a condition that a person *has*, not who a person *is*."[69] At the same time that the new definition distances itself from what the Accord Alliance considers to be stigmatizing terms such as hermaphroditism and even intersexuality, the choice of the term "disorder" also pathologizes variations in physical appearance that do not necessarily require medical intervention because they do not involve malignant tissue changes. The concept of "disorders of sex development" has, therefore, come under scrutiny by activist groups, medical practitioners, and cultural critics alike.[70] The Accord Alliance's focus on the body rather than on identity, however, ultimately leads to the same impasse encountered in earli-

---

298-304; Milton Diamond and Hazel Glenn Beh, "Surgical Treatment of Infants with Ambiguous Genitalia: Deficiencies in the Standard of Care and Informed Consent," *Michigan Journal of Gender & Law*, 7.1 (2000), 1-63; and Milton Diamond and Hazel Glenn Beh, "An Emerging Ethical and Medical Dilemma: Should Physicians Perform Sex Assignment on Infants with Ambiguous Genitalia?," *Michigan Journal of Gender & Law*, 7.1 (2000), 1-63.

66 Suzanne Kessler, *Lessons from the Intersexed* (New Brunswick: Rutgers UP, 1998).

67 For discussions of the ethical dimension of medical treatments for intersexuality, see Sharon Systma, ed., *Ethics and Intersex* (Dordrecht: Springer, 2006).

68 The website of the ISNA at http://www.isna.org provides a description of the transition from the ISNA to the Accord Alliance.

69 http://www.accordalliance.org/learn-about-dsd/faqs/

70 In *Bodies in Doubt: An American History of Intersex* (Baltimore: Johns Hopkins UP, 2009), cultural critic Elizabeth Reis proposes a "new term, still DSD, but standing for divergence of sex development" rather than "disorder" (xv). Medical researcher Milton Diamond similarly suggests replacing the term "disorder" with "variation" ("Variations of Sex Development Instead of Disorders of Sex Development," in *Archives of Disease in Childhood*, 27 July 2006, electronic letter, n.p.).

er terminologies, namely the question of what is considered normal and who answers this question. The definition of DSD mentions "chromosomal, gonadal or anatomic" characteristics, thereby highlighting the body's materiality, yet its focus on these traits' "atypical" appearance implies value judgments that highlight the body's cultural dimension. Nature and culture merge even in definitions that explicitly present and conceive of the body as pure matter.

The order of the physiological elements in the new definition of DSD as atypical conditions of "chromosomal, gonadal or anatomic sex" is significant. This definition lists historical diagnostic markers, starting with the current emphasis on the chromosomes. Whereas, up to the eighteenth century, the general anatomical appearance tended to be the decisive indicator, in the nineteenth century the emphasis gradually moved to the gonads. In the twentieth century, attention shifted to the chromosomes when scientists began to search for a testis-determining gene on the Y chromosome. They discovered it in 1990 and called it *SRY*. However, as Vernon Rosario points out, this chromosome-related discovery did not resolve all ambiguities; the diverse manifestations of intersexuality limited the implications of the identification of *SRY*.[71] When, additionally, scientists realized that testes could form in the absence of *SRY*, a Pandora's box was opened that has not been closed again. In addition to outlining the limitations of the chromosomal model, Rosario's discussion of its testocentric bias uncovers the lack of scientific interest in ovarian development, thereby highlighting biomedical research's focus on male bodies. Looking beyond the testocentric chromosomal model, Rosario highlights the possibilities generated by molecular genetics for understanding sex determination not as a binary system but as "quantum sex," a complex process in which multiple genes interact with a variety of environmental factors, each contributing a small part to the eventual biological make-up of the individual.[72] This emerging scientific research agenda might trigger a change from the binary model of sex to one that emphasizes the diversity in sexual anatomy by investigating its various components.

The quantum model of sex echoes the pre-Enlightenment gender continuum that was based on the one-sex model. Both models consider humans not as exclusively male or female but as a mixture of male and female characteristics in various proportions, thereby allowing for non-pathologizing approaches to intersexuality. In addition to outlining a non-dualistic view of gender and sexuality, the quantum model proposes a dynamic process of interaction between body and environment. It emphasizes the fact that

---

71 Rosario, p.274.

72 Rosario, p.279.

genes are not expressed in isolation but through their interaction with the environment, thereby creating a link to new materialist and social constructivist humanistic inquiry. The environment consists not only of material factors but also of the cultural context that shapes these material elements. The body and the environment acquire meaning only through their interaction with the culture in which they are embedded.

The bridge between scientific research and humanistic inquiry allows for the investigation of the complex intersections between nature/sexuality and culture/gender. Howard Tuttle's observation that a baby's body acquires meaning from its engagement with surrounding circumstances expresses the interactionist perspective and thereby supplements Vernon Rosario's argument from a humanistic vantage point. Every individual is a product of its interaction with the environment and culture. Through their liminality, intersex bodies such as Karl Baer's crystallize the complex nature of these interactions. Their existence poses questions regarding the ethics of physical interventions that are geared at normalizing and thereby erasing their signifying difference at various moments in history. The burgeoning field of bioethics emphasizes the fact that bodies matter and they do so through their interaction with their surroundings.[73] This is what constitutes the radical reality of human life.

[73] A discussion of the challenges currently facing bioethics, in particular of the need to problematize normative binaries, can be found in Margrit Shildrick and Roxanne Mykitiuk, eds., *Ethics of the Body: Postconventional Challenges* (Cambridge and London: MIT Press, 2005).

Naturphilosophie and Murder: The Limits of Scientific Explanation in Döblin's *Die beiden Freundinnen*

Cate I. Reilly

> Man resembles a cauldron, a reservoir of energy [...]. He is composed of the most remarkable substance. It insatiably demands new material, new water, new salts, new sugar. It is, by nature, characterized by endless fluctuation, desire, the obligation of need, by struggle, inequality, new constellations, equalizations, levelings-off. This is its metabolism and its inner life.[1]

Abstract: When Alfred Döblin penned *Die beiden Freundinnen und ihr Giftmord* (Two Girlfriends Commit Murder by Poisoning) in 1924, the story already bore all the hallmarks of sensationalist true-crime fiction: a working-class lesbian affair, an epistolary romance, and a premeditated murder by poisoning. It was based on the Klein/Nebbe murder trial that had then recently taken the Berlin press by storm. Unlike the extensive press coverage of the case, Döblin's book-length work deliberately avoided causal explanations for the murder. Instead he explored multiple unconscious drives and the metabolic exchange of energy that motivated the protagonists' actions. Written at a key moment in Döblin's development as a psychiatrist and psychoanalyst, *Die beiden Freundinnen* is part of his long-term project to return the *Naturwissenschaften* to a study of the organic interconnections and mutual dependency he perceived as linking all natural phenomena. In this essay I examine *Die beiden Freundinnen* in light of Döblin's early reflections on Nietzsche's *Will to Power,* his emerging *Naturphilosophie* in the 1920s, and his commitment to investigating the hidden motors of human consciousness.

The Klein/Nebbe Case

In February of 1922, Ella (Elli) Klein, a 22-year-old hairdresser in Berlin, began mixing large doses of arsenic into her husband's evening meals. She and her married lover, Margarete Nebbe, who shared the same apartment building in Berlin-Lichterfelde, had plotted to murder both of their abusive spouses and begin life anew together. Elli's husband, Willi Klein, grew increasingly ill as he was simultaneously poisoned and nursed by his wife during the subsequent four months. He died on April 1 in Berlin's Lichtenberg Hospital, allegedly from alcohol poisoning. An autopsy requested by his

1 Alfred Döblin, "Das Wasser," *Die Neue Rundshau,* 33.2 (1922), 853-858 (p.853): "Der Mensch sieht aus wie ein Kessel, ein Kraftreservoir [...]. Die Stoffmasse, aus der er selbst besteht, ist von der sonderbarsten Art. Sie verlangt unersättlich neue Stoffe, neues Wasser, neue Salze, neuen Zucker. Sie hat in sich die Eigenart und es ist ihre Natur, unaufhörlich zu schwanken, wünschen, begehren zu müssen, der Kampf, die Ungleichheiten, die neuen Konstellationen, Ausgleichungen, Umebnungen. Das ist ihr Stoffwechsel und ihr Seelenleben." Unless otherwise noted, all translations are my own.

mother and a subsequent search of Elli's apartment (which revealed the nearly 600 letters she and Margarete had exchanged during the course of their affair) exposed the homicide. In March of 1923, the two women were tried in a highly publicized case that drew together Weimar Berlin's increasing fascination with sexual deviance, a widespread belief in a new, urban criminality and a sense that the war had produced large-scale forms of political, moral and social disorder that were wholly antithetical to traditional, Wilhelmine values.[2]

Major Berlin newspapers featured articles on the case by authors ranging from anonymous reviewers to Klein and Nebbe's defense attorney, and literary figures such as Josef Roth and Robert Musil. In August of 1924, Rudolf Leonhard, a communist literary-political activist and the newly named editor of The Forge press, wrote to leftist journalist Alfons Paquet to announce the forthcoming publication of his series, *Aussenseiter der Gesellschaft: Die Verbrechen der Gegenwart* (*Outsiders of Society: The Crimes of Today*). Elli Klein and Margarete Nebbe's story, as retold by Alfred Döblin, would be the first book in the series.

Döblin's modification of the real-life protagonists' names (Elli Klein became Elli Link, Margarete Nebbe became Margarete Bende) for his novel *Die beiden Freundinnen und ihr Giftmord* (*Two Girlfriends Commit Murder by Poisoning*, 1924) represents but one of the myriad ways in which he dramatically reconfigured the story of the homicide for his contribution to the series. Published between 1924 and 1925, *Aussenseiter* enlisted modern authors including Theodor Lessing, Ernst Weiß and Iwan Goll to write dramatized versions of criminal cases that had caused major press sensations in Weimar Germany.[3] The 14 volumes represented a radical departure from previous German collections of *causes célèbres,* such as the *Neue Pitaval* (1842-1890). Viewing the justice system as a potentially culpable party, the works in Leonhard's series set out to narrate the collapse of normative distinctions between "criminal" and "non-criminal" by pointing out the justice system's failure to conclusively establish causality, assign blame, and judge expertise.[4]

---

2 For a discussion of the representational functions and social effects of murder in the Weimar Republic, see Sace Elder, *Murder Scenes. Normality, Deviance and Criminal Violence in Weimar Berlin* (Ann Arbor: University of Michigan Press, 2013).

3 Die Schmiede is also responsible for publishing Kafka's *Ein Hungerkünstler* (1924), and – posthumously – *Der Prozess* (1925). A full list of publications can be found in Frank Hermann and Heinke Schmitz, *Der Verlag die Schmiede 1921-1929. Eine kommentierte Bibliographie* (Morsum/Sylt: Cicero Press, 1996).

4 An overview of the Pitaval tradition in Germany and elsewhere can be found in: *Literatur und Kriminalität. Die gesellschaftliche Erfahrung von Verbrechen und Strafverfolgung als Gegenstand des Erzählens. Deutschland, England und Frankreich 1850–1880,* ed. by Jörg Schönert (Tübingen: Max Niemeyer, 1983); *Erzählte Kriminalität. Zur Typologie und*

While the Berlin press battled over competing explanations for Elli and Margarete's actions (underdeveloped reproductive organs? sexual pathology? a structural crisis in Weimar society? the advent of the "New Woman?"), Döblin's *Die beiden Freundinnen und ihr Giftmord,* hereafter *Freundinnen,* consciously avoided any single explanatory mechanism or organizing principle. "Sein ganzes Buch ist nichts andres als ein grosses Fragezeichen hinter diese und hinter alle 'Tatsachen,'" (His entire book is nothing other than an immense question mark after these and after all 'facts'), Hans Siemsen remarked in a 1925 review for the Berlin weekly, *Die Weltbühne.*[5] Like the subsequent works in Leonhard's series, *Freundinnen* upset the Pitaval's neat causal linkages and offered a surfeit of motivating factors in place of a tidy plot. On the one hand, *Freundinnen*'s emphasis on arbitrariness and plurality is a paradigmatic example of Expressionist prose, which highlighted unpredictability. Döblin even concluded *Freundinnen* with the observation that, "Es hat so sich ereignet. [...] Aber es hat sich auch nicht so ereignet" (It [the case] happened this way. [...], but it also didn't happen this way).[6] On the

---

Funktion von narrativen Darstellungen in Strafrechtspflege, Publizistik und Literatur zwischen 1770 und 1920, ed. by Jörg Schönert in collaboration with Konstantin Imm and Joachim Linder (Tübingen: Max Niemeyer, 1991). Particularly relevant for *Aussenseiter* is Alexander Košenina's "Juristische Fallgeschichte: Theodor Lessings Haarmann. Die Geschichte eines Werwolfs" (1925), *Non Fiktion,* 3 (2008), H. 1/2, p.83-94. Walter Benjamin, Bertolt Brecht and Siegfried Kracauer all took up the related question of the *Kriminalroman* in short essays during the late 1920s and early 1930s. See Walter Benjamin, "Kriminalromane, auf Reisen," in *Gesammelte Schriften,* ed. by Rudolf Tiedemann and Hermann Schweppenhäuser, 7 vols (Frankfurt am Main: Suhrkamp, 1972-1989), vol. 4.1, pp.381-382; Bertolt Brecht, "Über die Popularität des Kriminalromans" in *Gesammelte Werke,* ed. by Elisabeth Hauptmann in collaboration with Rosmarie Hill, 20 vols (Frankfurt: Suhrkamp, 1967), vol. 19, pp.450-458; Siegfried Kracauer, "Mordprozeß und Gesellschaft," *Die Neue Rundschau* 42 (March 1931), pp.431-432. Lynn M. Kutch and Todd Herzog explore German crime fiction's twenty-first-century legacy in their edited volume, *Tatort Germany: The Curious Case of German-Language Crime Fiction,* ed. by Lynn M Kutch and Todd Herzog (Rochester, NY: Camden House, 2014).

5 Hans Siemsen, review of *Die beiden Freundinnen,* by Alfred Döblin, *Die Weltbühne* 21, I (1925), 360-361, reprinted in *Alfred Döblin im Spiegel der zeitgenössischen Kritik,* ed. by Ingrid Schuster and Ingrid Bode (Bern: Franke Verlag, 1973), pp.157-158 (p.157).

6 Alfred Döblin, *Die Beiden Freundinnen und ihr Giftmord,* (Berlin: Die Schmiede, 1924), p.112. This echoes Carl Einstein's comment in "Über den Roman: Anmerkungen" (1912) that the plot "kann auch anders endigen," as well as Robert Musil's idea of *Parallelaktion* in *Der Mann ohne Eigenschaften* The arbitrariness of Expressionist prose has been pointed out by Walter H. Sokel in, "The Prose of German Expressionism." Sokel quotes both Einstein and Musil but does not invoke *Freundinnen.* See Walter H. Sokel, "The Prose of German Expressionism," in *Expressionism as an International Literary Phenomenon,* ed. by Ulrich Weisstein (Amsterdam: John Benjamins Publishing Company, 2011), pp.69-88.

other hand, Döblin's exposé of the multiple moving parts and conflicting drives in the Klein/Nebbe case is unique to his developing *Naturphilosophie* (philosophy of nature), which substituted bundles of organic interconnections in place of the more reductive, linear alternatives he saw in the *Naturwissenschaften* (natural sciences).

Döblin's short essays from the 1920s provide key insights into a transitional period for his *Naturphilosophie*, as it emerged between his early, formative exposure to Nietzsche and the publication of *Das Ich über der Natur* (The Ego over Nature, 1927).[7] *Freundinnen* falls in the center of these reflections and contributes to them in its own right. Despite its relative brevity and its place well outside the list of Döblin's standard works, it was composed at a key moment in the production of lengthier, more familiar, texts such as *Berge Meere und Giganten* (Mountains, Seas, and Giants, 1924); *Manas*, 1928; *Berlin Alexanderplatz*, 1929. In concert with these longer works, and in conversation with the shorter essays, it sketches a landscape of metabolic exchange.

The energy transactions that *Freundinnen* describes are at once anticipated by Nietzsche's *Will to Power*, indebted to Döblin's psychiatric training and in dialogue with Freud's drive-discharge model. In this essay, I explore *Freundinnen* as part of, rather than an exception to, Döblin's evolving vision of the interconnections joining all natural phenomena and binding inorganic matter into dynamic organic wholes. I begin by examining Döblin's Nietzsche reception alongside mathematician Felix Hausdorff's philosophical writings. Turning to Döblin's 1923 essay "Blick auf die Naturwissenschaft" (A View of Natural Science), I look at Döblin's construction of an interconnected "I" by rejecting formulaic, empiricist reasoning. Finally, I read *Freundinnen* in light of the Freudian hydraulic metaphors Döblin employs to describe the changes in Elli and Margarete's soul-states, prior to, during and after the murder. These sources (Nietzsche, Freud, the essays) though never mentioned explicitly in *Freundinnen,* are nevertheless important for understanding the text's contribution to reimagining the relationship between body and psyche.

Overall, Döblin's attention to metabolic processes would seem to place *Freundinnen* squarely on the side of the anti-vitalists in the *Lebenskraft*-debate outlined by Alfred Noll in 1914.[8] Yet, *Freundinnen* draws freely on ideas pro-

---

7 This book-length meditation on the individual as an inseparable component of an organic totality is one of Döblin's most robust theoretical formulations of the position. It draws directly from material he had published in earlier essays of the 1920s and often lifts passages wholesale from them.

8 "Die 'Lebenskraft' in den Schriften der Vitalisten und ihrer Gegner," ed. by Alfred Noll, Voigtländers Quellenbücher 69, 4th ed. (Leipzig: R. Voigtländer, 1914). For a full discussion of Noll's text, see the introduction to this volume.

posed by thinkers from both sides of the discussion Noll cites, leaving the question of a separate life force open in the text. In Döblin's retelling, the Klein/Nebbe case is presented as the seemingly unmediated account of interactions between multiple drives (childishness, sexuality, perversion, etc.). Elli, Margarete and Willi are described as victims of their own unwieldy vital impulses. These impulses alternately open and block regions of their souls, which are themselves discussed by Döblin in material terms. Elli and Margarete often appear uncontrollably propelled to action in the text, as though the *vim vital* both arose from them and independently stimulated their behavior from without.

Much of the text also hinges on the inseparability of these drives from one another and from the surrounding environment. *Freundinnen*'s emphasis on the interplay between concrete matter and more abstract soul-states recalls Johann Christian Reil's anti-vitalist contention that matter is a form of *Kraft* associated with growth and movement as well as Ludwig Büchner's belief that motion results from the net effect of matter's constant, dynamic metamorphosis.[9] By the time Döblin wrote *Das Ich über der Natur*, his position was more clearly on the side of Noll's anti-vitalists. In *Freundinnen*, however, Döblin (like Nietzsche) focuses predominantly on the release, absorption and uptake of energy. His characters are "Kessel, Kraftreservoir(s)" (cauldrons, reservoirs of energy), a phrase he had used in an earlier essay for *Die Neue Rundschau* and that Leo Greiner echoed in a 1925 review of *Freundinnen*, stating: "(d)er ganze Ablauf wird durch eine geheime Energie sichtbar" (the entire process [of the case] becomes visible through the workings of a secret energy).[10] By placing *Freundinnen* in conversation with such earlier essays as "Blick auf die Naturwissenschaft" and Döblin's Nietzsche texts, it becomes possible to resituate the Klein/Nebbe case within Döblin's *Naturphilosophie* and to illuminate its contribution to thinking life – in Howard Tuttle's words – as the metabolism of circumstances.[11]

9 Johann Christoph Reil, *Von der Lebenskraft*, ed. Karl Sudhoff (Leipzig: Johann Ambrosius Barth, 1910). Originally published in *Archiv für Physiologie*, 1 (1796) and Ludwig Büchner, *Kraft und Stoff. Empirisch-Naturphilosophische Studien* (New York: E. Steiger, 1872). Both are discussed at greater length by John A. McCarthy in the introduction to this volume.

10 Leo Greiner, review of *Die beiden Freundinnen,* by Alfred Döblin, *Berlin Börsen-Courier (Morgenausgabe),* 29 March 1925, reprinted in *Alfred Döblin im Spiegel der zeitgenössischen Kritik*, ed. by Ingrid Schuster and Ingrid Bode (Bern: Franke Verlag, 1973), pp.155-156 (p.156).

11 Howard N. Tuttle, *Human Life is Radical Reality: An Idea Developed from the Conceptions of Dilthey, Heidegger, and Ortega y Gasset* (New York: Peter Lang, 2005), pp.97-100.

Döblin's *Naturphilosophie* between Romanticism and Psychiatry

Döblin's *Naturphilosophie* in the 1920s was characterized both by a Romantic belief in the fundamental unity of nature (a nature imbued with animate qualities that could be revealed through hidden signs) and by his vision of an organic cosmos held together through a "universelle Zusammenhang" (universal interconnection).[12] Roland Dollinger has identified five key themes in Döblin's *Naturphilosophie* between the 1920s and the 1933 publication of *Unser Dasein* (Our Existence). These include: the animation (*Beseeltheit*) of all natural phenomena, the interrelatedness of these phenomena, a continuity between organic and inorganic forms, the concept of human consciousness as an expression of a spiritual faculty residing in all things, and the belief in an anonymous force that both lends particularity to various elements in nature and places its divisibility as a whole into question.[13] Of these, *Freundinnen* highlights interrelatedness of natural phenomena and an innate dynamism between particularity and unity, while downplaying *Beseeltheit* and the spiritual elements of consciousness.

German Romanticism's exploration of the primordial elements of consciousness and formulation of transcendental categories of the mind was also complemented by advances in psychiatry that emerged at the turn of the century.[14] *Freundinnen* was an attempt to escape from more reductive, empiricist explanations for human behavior (like those offered by the 21 experts testifying at the trial) by offering an alternative model for the invisible workings of the human soul. Unlike models associated with a particular school of thought (whether sexology, criminology or pathological anatomy), Döblin's representation of the Klein/Nebbe case drew piecemeal from his philosophical training, medical knowledge and psychiatric expertise.

Taken together, *Freundinnen* and the short essays published in *Die Neue Rundschau* (The New Review) and the *Vossische Zeitung*, illuminate how Döblin's critique of modern reductive empiricsm and biologism simultaneously looked back toward a Romantic philosophy of wholeness and leapt dramatically forward to imagine a new type of fiction as equally conversant

---

12 Alfred Döblin, *Das Ich über der Natur* (Berlin: Fischer, 1928), p.11. Döblin's early university years had involved exposure to Spinoza, Hegel, Schopenhauer and the burgeoning Neo-Kantian movement (through lectures by Max Dessoir in Berlin), alongside his medical training.

13 Roland Dollinger, "Alfred Döblins Naturphilosophie in den Zwanziger Jahren," in *Philosophia naturalis: Beiträge zu einer zeitgemässen Naturphilosophie*, ed. by Thomas Arzt, Roland Dollinger and Maria Hippius (Würzberg: Königshausen und Neumann, 1996), pp.135-150 (p.136).

14 Andreas Killen, *Berlin Electropolis: Shock, Nerves and German Modernity* (Berkeley: University of California Press, 2006). Killen argues that the etiological shift from somatic explanations to psychogenic ones in Germany dates from well before Freud.

in psychoanalysis as in its psychiatric competitors and predecessors. This is particularly borne out in *Freundinnen*'s paratext. The extensive 1924 appendix – lengthier than in any subsequent volume in Leonard's series – diagrammed changes in the women's psyches over the course of the murder and offered a graphological analysis (with facsimile samples) of their handwriting. [15]

These diagrams have been rightfully interpreted as the juncture of the competing and complementary influences of clinical psychiatry and psychoanalysis both in Döblin's medical training specifically and in Germany more generally. Döblin received his doctoral degree in clinical psychiatry from the University of Freiburg in 1905, following the completion of a general medical degree at Berlin's Friedrich Wilhelm University (1900-1904). After several years of private practice (and time as a military doctor), he began training at the Berlin Psychoanalytic Institute, under Ernst Simmel.[16] In the early 1920s, psychiatry and psychoanalysis were most valuable to Döblin as part of a foundational metapsychology linked to Romantic notions of the hidden, irrational forces buried deep within the human personality. Instead of the neatly packaged explanations advanced by experts at the trial, Döblin's epilogue to *Freundinnen* advocated seeing the relationship between body and psyche in terms of the "Gesamtorganismus" (collective organism) and "Seelenmasse" (literarally "soul mass"; i.e., the mesh of interconnections binding the soul to its surrounding environment).[17]

In short, *Freundinnen* did not simply fictionalize the by-then familiar Klein/Nebbe crime narrative, or supplement existing psychiatric speculations on the cause of the murder. Rather, it supplied a narrative landscape for Döblin's *Naturphilosophie*, in much the same manner (at least structurally speaking) as had his science fiction epic, *Berge Meere und Giganten* (Mountains Seas and Giants), published the same year. The landscape of *Freundinnen* is populated with psychiatric and psychoanalytic terminology and reference points, but is not limited to them. Much of the language of *Freundinnen*, with its repeated emphases on somatic and psychological swellings, distensions, and expulsions, on fecal matter and "Unrat" (garbage) of the soul, stands at the crossroads of Nietzsche's ideas on the will to power and Freud's drive-discharge theory.[18] Döblin's membership in the Berlin Psychoanalytic Insti-

15 Alfred Döblin, *Die beiden Freundinnen und ihr Giftmord* (Berlin: Die Schmiede, 1924), p.114. The 1924 edition is the only one to include the full appendix of materials and diagrams.

16 Veronika Fuechtner, *Berlin Psychoanalytic: Psychoanalysis and Culture in Weimar Republic Germany and Beyond* (Berkeley: University of California Press, 2011), pp.19-21.

17 Both terms are Döblin's own invention but also owe a debt to the writings of Ernst Simmel. For further information on Döblin's connections to Simmel, see Veronika Fuechtner, *Berlin Psychoanalytic*, Chapter 1.

18 "Wie denkt man Menschen zu sanieren, die ihren inneren Unrat nicht loswerden?"

tute during this time and his work in the associated polyclinic, make it tempting to read *Freundinnen* exclusively as a work of psychoanalytic true crime fiction. These hydraulic metaphors, however, extend beyond the ideas Freud articulated in *Jenseits des Lustprinzips* (*Beyond the Pleasure Principle*, 1920) or *Das Ich und das Es* (*The Ego and the Id*, 1923). They are connected with Döblin's long-term project to wrest the *Naturwissenschaften* from the grips of an empty biologism and mathematical empiricism. The metabolic landscape that Döblin constructs in *Freundinnen* is criss-crossed by the intakes and expulsions that undermine the distinction between inside and outside, a goal common to other volumes in the *Aussenseiter* series. Foreshadowings of ideas that would emerge more robustly in *Das Ich über der Natur*, are already present in *Freundinnen*: e.g., an omnipresent spiritual principle (*Urgeist*), death as reabsorption into an anonymous mass, the cradling of individuals in the innumerable manifestations of matter.

Despite Döblin's bid for unity, English and German scholars have largely seen *Freundinnen* in terms of a thematic fracturing. By selecting one element from its montage-like composition, scholars have looked to draw out a single premise in the work as a whole. Veronika Fuechtner's lucid unpacking of Döblin's connections to the Berlin Psychoanalytic Institute and *Freundinnen*'s debt to war neuroses and Devin Fore's exposition of the text's struggle with the legacy of realist reportage have gone furthest in attending to its polyvalence.[19] Scholars such as Walter Müller-Seidel and Manfred Maiwald have attended to the juridical implications of *Freundinnen* and Todd Herzog has explored *Freundinnen* in terms of the advent of the *Kriminalroman*, its place in Rudolf Leonhard's 1924-1925 *Outsiders* series and Foucaultian power discourse.[20] Work on the implications for gender and sexuality, moti-

---

(How can one imagine the rejuvenation of man without cleaning out his inner filth?). Alfred Döblin, "Praxis der Psychoanalyse," *Vossische Zeitung*, 301 (28.6.1923), pp.2-3. Reprinted in Alfred Döblin, *Kleine Schriften, Band 2: 1922-1924*, ed. by Anthony W. Riley (Olten, Freiburg im Breisgau: Walter Verlag, 1985), pp.270-274 (p.272). I cite from the reprint edition throughout.

19 See Devin Fore, "Döblin's Epic: Sense, Document and the Verbal World Picture," *New German Critique,* 99 (Fall 2006), 171-207; Fuechtner, *Berlin Psychoanalytic* and Fuechtner, "Döblin's Medical, Psychiatric and Psychoanalytical Work," in *A Companion to the Works of Alfred Döblin,* ed. by Roland Dollinger, Wulf Koepke, and Heidi Tomann Tewarson (Rochester, NY: Camden House, 2004), pp.111-139.

20 See Todd Herzog, *Crime Stories: Criminalistic Fantasy and the Culture of Crisis in Weimar Germany* (New York: Berghahn Books, 2009); Walter Müller-Seidel, "Alfred Döblin, Die Beiden Freundinnen und ihr Giftmord. Psychiatrie, Strafrecht und moderne Literatur," in *Literatur und Recht: Literarische Rechtsfälle von der Antike bis in die Gegenwart*, ed. by Ulrich Mölk. (Göttingen, Wallstein Verlag, 1996), pp.356-369. Manfred Maiwald, "Die beiden Freundinnen und ihr Giftmord. Juristische Betrachtungen zu einem lite-

vated by Maria Tartar's *Lustmord: Sexual Murder in Weimar Germany* (1995) has produced studies by scholars on both sides of the Atlantic.[21] In contrast to these later scholars, Döblin's contemporaries often complained of the text's failure to adhere to a standard genre classification, a critique also leveled at *Berlin Alexanderplatz* in early reviews. Preoccupied by the problem of genre, these reviewers generally missed the work's explosion of classification as a whole.

The Klein/Nebbe trial created a sufficient sensation in 1923 to warrant its subsequent inclusion in studies of the Weimar era female poisoner and criminal anthropology in the 1920s.[22] Books such as the 1920 *Bilder-Lexikon der Sexualwissenschaften* (Visual Dictionary of the Sexual Sciences) and Erich Wulffen's 1923 *Das Weib als Sexualverbrecherin* (Woman as Sexual Criminal) encouraged the conflation of femininity and the nature of the poisoner by building on the stereotype of the *Giftmischerin* (female poisoner).[23] Scholarship on the juridical elements in Döblin's rendering of the case has used these studies as a historical context for examining differences between the real events of the trial and their artistic instantiation, looking at how both Döblin and *Outsiders* transformed contemporaneous views of the female murderer.

Overall, these types of assessments benefit from isolating an individual theme and generating a picture of relative consistency. Far from being reductive, they help make the text's plurality manageable and construct meaningful links with other major discourses in Weimar Studies: changing sexual

---

rarischen Prozessbericht," in *Literatur und Recht: Literarische Rechtsfälle von der Antike bis in die Gegenwart*, ed. by Ulrich Mölk (Göttingen: Wallstein Verlag, 1996), pp.370-382.

21 Maria Tartar, *Lustmord: Sexual Murder in Weimar Germany* (Princeton: Princeton UP, 1995). Tartar's book has inspired far more general studies on sexual murder in the Weimar Republic than can be listed here. To name only those that focus specifically on *Freundinnen*, see Laura Leann Wilson, *The Portrayal of the Female Figure in Alfred Döblin's Die beiden Freundinnen und ihr Giftmord* (Lafayette, Indiana: Diss., Purdue University, 2001) and Ute Karlavaris-Bremer, "Außenseiterinnen der Gesellschaft. Alfred Döblins Erzählung Die beiden Freundinnen und ihr Giftmord aus interdisziplinärer Perspektive," in *Tatsachenphantasie - Alfred Döblins Poetik des Wissens im Kontext der Moderne / Internationales Alfred-Döblin-Kolloquium Emmendingen 2007,* ed. by Sabina Becker and Robert Krause (Bern: Peter Lang AG, 2008), pp.265-278.

22 Isabella Claßsen, *Darstellung von Kriminalität in der detuschen Literatur, Presse und Wissenschaft 1900-1930* (Frankfurt: Land, 1988); Hania Siebenpfeiffer, *Böse Lust. Gewaltverbrechen in Diskursen der Weimarer Republik* (Köln: Böhlau Verlag, 2005); Inge Weiler, *Giftmordwissen und Giftmörderinnen : Eine diskursgeschichtliche Studie* (Tübingen: M. Niemeyer, 1998).

23 The famous case of Nannette Zwanziger, née Schönleben, who poisoned multiple persons between 1801 and 1811 is an essential cultural reference point *Freundinnen.* See "Anna Margareta Zwanziger, die deutsche Brinvillier," in *Aktenmäßige Darstellung merkwürdiger Verbrechen*, ed. Johann Paul Anselm von Feuerbach (Gießen, 1828), pp.1-53.

politics and gender paradigms, perceptions of crime, the rise of competing psychodiagnostic models and treatments, and retrospective evocations of the "glitter and doom" preceding National Socialism.[24] The disadvantage, however, is a corresponding tendency to slot *Freundinnen* into a paradigm that ignores the work's own (very deliberate) inconsistencies. *Freundinnen* is a story about un-storying, a quasi-Tolstoyan unraveling of meaningful causal logic. It privileges disorientation, but not along the lines of a Joycean stream-of-consciousness. This is disorientation from without. "Mit dem Kausalitätsprinzip frisiert man" (It is easy to be manipulative using the principle of causality), Döblin asserted in the epilogue.[25] The "characters" – Elli, Margarete, the two husbands – turn out to be subsidiary to a polemical critique of factual representation and a study of the multiple internal motors driving human action. Anti-mimesis trumps the specifics of the case.

While this approach is, of course, subject to critique (how far should one follow Döblin in his overtures to disorder? where do these give way to a morass of contingencies? is this type of reading a political or an aesthetic position, or some combination of the two?), it begs the question of what a reading might look like that takes Döblin on his own terms. By attending to the plurality of texts at work in the background, I take a more holistic approach to examining what these terms might be and how they are constructed. Focusing on Döblin's alternating presencing and concealment of the several major philosophical, medical and psychoanalytic concerns in *Freundinnen* helps unpack the epistemological basis for placing the "Fragezeichen hinter alle 'Tatsachen,'" (question mark after all facts).[26] Writing, after all, was not "Nägelkauen und Zahnstochern" (nail-biting and picking one's teeth).[27] It was, for Döblin, a public duty as intimately tied to self-renunciation as to the restoration of the missing physiognomic, spatial and temporal elements so violently isolated by psychological prose: "ein amputierter Arm; Atem, ohne den Menschen der atmet; Blicke ohne Augen" (an amputated arm; breath without the man who breathes; gazes without eyes).[28]

Döblin's early (1902/1903) response to Nietzsche is an important benchmark in his metaphysical education. It lays the groundwork for his re-

24 Jochen Hung, Godela Weiss-Sussex, Geoff Wilkes, eds. *Beyond Glitter and Doom. The Contingency of the Weimar Republic* (Munich: Iudicium, 2012).

25 Döblin, *Freundinnen*, p.114. There is presumably also a word play here on "frisieren," which can mean either "to coif hair" (connected to Elli's profession as a hairdresser) or "to manipulate something."

26 Siemsen, review of *Die beiden Freundinnen*, p.157.

27 Döblin, "An Romanautoren," p.17.

28 Döblin, "An Romanautoren," p.18.

jection of biologism (a point that repeatedly reappears in *Freundinnen*) and marks one of the earliest instances that he identifies empiricism's misappropriation of the sciences. It is also key for his conception of causality. Like Nietzsche, Döblin believed that considering the unconscious drastically complicated establishing even basic sequences of cause and effect, since cause could only be identified retrospectively.[29] "So können Kugeln aus dem Unsichtbaren auf uns treffen, uns verändern, und wir merken nur die Veränderung" (Thus bullets are able to hit us from the invisible, to change us, and we note only the change), he remarked on the final page of *Freundinnen*.[30] Nietzsche saw the destruction of neat linearity as a return to reality, not a departure from it. Likewise, Döblin found that the impossibility of pinning down a reliable causal chain had deep narrative resonance. It was the point at which the dismembered limbs scattered by realism's naive quest for order could be restored to a more organic unity. Döblin's reading of Nietzsche helped him envision alternative forms of interconnectedness and an "I" that was organically embedded in its surroundings, ideas he would later draw on in his retelling of the Klein/Nebbe case.

Ball, Chain and Straightjacket: Thoughts from the Landscape of Zarathustra

In the seven months between August 1902 and March 1903, while still a medical student in Berlin, Alfred Döblin composed two essays on Nietzsche, whom he had been reading extensively. Not published until 1978, both essays explore Nietzsche's work in the context of his intellectual development and a twentieth-century scientific genealogy.[31] The texts are split between a critical review of Nietzsche's *Will to Power* and elements of his moral doctrine, and Döblin's immense regard for Nietzsche as the "Philosoph des Entwicklungsbegriff" (philosopher of the concept of growth).[32] Döblin would return periodically to Nietzsche throughout his career, and as Döblin biographer Wilfred Schoeller has remarked, no other philosopher – neither Freud nor Marx – is as frequently cited by him.[33] In

29 Fuechtner, *Berlin Psychoanalytic*, p.34.

30 Döblin, *Freundinnen*, p.117.

31 See John A. McCarthy, "Die Nietzsche-Rezeption in der Literatur 1890-1918," in *Hansers Sozialgeschichte der deutschen Literatur*, vol. 7, ed. by York-Gothart Mix. (Munich, Carl Hanser Verlag: 2000), pp.192-206 (p.205).

32 Döblin, "Zu Nietzsches Morallehren" (1903). First printed in *Nietzsche und die deutsche Literatur*, ed. by Bruno Hillebrand, pp.331-358. Reprinted in Alfred Döblin, *Kleine Schriften, Band I*, pp.29-55 (p.35).

33 Quoted in Michaela Spreng, *Alfred Döblins Nietzsche-Verständnis vor dem Hintergrund der philosophischen Schriften Felix Hausdorffs* (Munich: Herbert Utz Verlag, 2014), p.7.

the course of his life, Döblin's relationship to Nietzsche was inflected by far more intricate currents and countercurrents than these initial texts reveal. Although only giving occasional glimpses of Döblin's own cosmology, the two essays nonetheless sketch the formative lines that Döblin would return to in his later work.

The essays underscore Döblin's skepticism of any philosophy modeled on scientific positivism, which he sees as harkening back to the original emptiness of the natural sciences' grounding in philosophy.[34] The two texts also expose important links to the writings of mathematician and philosopher Felix Hausdorff, whose books published under the pseudonym Paul Mongré acted as a guiding force behind Döblin's initial encounters with Nietzsche and introduced essential ideas on chaos to Döblin's thinking. Taken together, the essays bring to light Döblin's belief in the simultaneous existence of inexorable natural forces, a magnificently impersonal collective and an individual consciousness able to navigate both. The experts who testify at Elli and Margarete's trial are wholly insensitive to these concerns and only see the women as either discrete individuals moved by personal, situation-specific desires or as manifestations of a cultural crisis. Döblin's reaction to these experts' misreading and his depiction of it in *Freundinnen* are shaped by the same critique of aggressive individualism and the same dismissal of human actions' basic transparency that appears in his Nietzsche essays.

Döblin read Nietzsche in 1902 and 1903 through the dual lens of his psychiatric training and Hausdorff's ideas on chaos. He responded positively to Nietzsche's vision of anarchic energy and flux, though more critically to his biologically and psychologically based "Einheitlichkeit" (unity). He was made uneasy by Nietzsche's use of medical language to present himself as the "philosophical doctor of an ailing Western culture."[35] Döblin pondered

---

34 See John A. McCarthy, *Remapping Reality: Chaos and Creativity in Science and Literature* (Amsterdam: Rodopi, 2006), p.115 and McCarthy, "Die Nietzsche-Rezeption in der Literatur 1890-1918," p.205.

35 As described by Daniel Ahern, *Nietzsche as Cultural Physician* (University Park, PA: Pennsylvania State UP, 1995). The full passage in Döblin reads: "[D]as medizinische ‚Bild' ist nicht absichtslos hier gewählt. Hier ist ganz unbildlich von der Medizin als Wissenschaft und Heilkunde die Rede –, und man sieht, wie sich auf diesem Wege, von der Biologie ausgehend, Nietzsche zum Moralisten entwickeln musste" (The medical "picture" is in no way unintentional here. This is a wholly literal interpretation of medicine as a science and healing art – and one sees in this way, that on the basis of biology, Nietzsche must develop into a moralist.) See "Der Wille zur Macht," in Alfred Döblin, *Kleine Schriften, Band I: 1902-1921*, ed. by Anthony W. Riley (Olten, Freiburg im Breisgau: Walter Verlag, 1985), p.16-17. David Midgley reviews Döblin's early philosophical thinking in "Metaphysical Speculation and the Fascination of the Real," in *Alfred Döblin's Paradigms of Modernism*, ed. by Steffan Davies and Ernest Schonfield (Berlin: Walter de Gruyter 2009), pp.7-27.

whether elements of Nietzschean thinking reduced the rich "Räumlichkeit" (spatiality) and "Dinglichkeit" (materiality) of the world to little more than "Erhaltungsbedingungen" (conditions for survival). [36] Under the tyranny of this biological *Weltanschauung*, the will to power became an all-unifying cosmic principle.[37] Yet, for Döblin, no anthropomorphic view of human cognition's development could ever be equal to the task of generating a full-blown cosmology. Nietzsche could not become a wholly successful spokesman for the *Naturwissenschaften*, Döblin argued, because his claims were still indebted to "ein bis zum Exceß naturwissenschaftliches Zeitalter" (an excessively scientifically-minded age).[38]

"Zu Nietzsches Morallehren" (On Nietzsche's Moral Doctrine, 1903) extended the same critique to moral judgments. In Döblin's reading, Nietzsche's attempt to treat morality as a psychological phenomenon was a further variation on shunting morality under the heading of biology. "Der Standpunkt der empirischen Weltanschauung, auf die wir Nietzsche stoßen sahen, ist ein subjektiver Idealismus auf biologisch sensualistischer Grundlage" (The standpoint of the empirical world-view, which we saw Nietzsche encounter, is a subjective idealism built on biological sensualistic foundations).[39] To Döblin, Nietzsche's moral doctrine ran aground by returning to these biological justifications. Yet Döblin was also determined to rehabilitate a spirit of metaphysical inquiry and redeem it from the legacy of empiricism. If Nietzsche had called for a return to frenetic knots of energy and a consideration of the body as a social structure forged from many souls, Döblin adopted a parallel position: the human body and its attendant psychological processes should be treated as an integrated unit engaged in cyclically merging with and differentiating itself from the surrounding environment.

Döblin's Nietzsche-reception should also be placed amidst psychiatry's shifting terrain between the 1890s and 1900s. The somatically oriented un-

---

36 "Wie auf den 'Verstand,' so wird auch auf die Sinnlichkeit das Wort in Kantischen Sinne genommen, das biologische Leitprinzip angewandt: Räumlichkeit, Dinglichkeit erweisen als tierische Erhaltungsbedingungen" (The guiding biological principle is applied, just as in the case of the 'intellect,' to sensibility in the Kantian sense of the word: spatiality and materiality prove to be animalistic conditions for survival). Alfred Döblin, "Der Wille zur Macht als Erkenntnis bei Friedrich Nietzsche" (1902) in Döblin, *Kleine Schriften, Band*, vol. 1, pp.13-29 (p.17).

37 Döblin's title is taken from the 1901 posthumous publication of the same name, but the text also indicates that he was familiar with the will to power as it appears in texts such as *Jenseits von Gute und Böse* and *Also Sprach Zarathustra.*

38 Döblin, "Der Wille zur Macht," p.19.

39 Döblin, "Zu Nietzsches Morallehren" (1903), in Döblin, *Kleine Schriften, Band*, vol. 1, pp.29-55 (p.33).

derstandings of the mind advocated by neuropathology and psychiatric autopsy had waned since the height of their popularity in the 1860s and 1870s. By the time Emil Kraepelin, the founder of scientific psychiatry, delivered his inaugural lecture at the university of Dorpat in 1887, he could state with assurance that "pathological anatomy could not fill the nosological vacuum created by the disintegration of the concept of unitary psychosis."[40] Unitary psychosis dated back to Romantic psychology of the 1830s. It rejected the notion that there were different forms of mental disease. The *seeming* varieties were actually only various stages of a single disease process. Pathological anatomy, on the other hand, constituted a major advantage for early psychiatry by stalwartly defending disease differentiation. By the 1890s, however, the medical community had grown frustrated with the obvious limitations of psychiatric autopsy, neuropathology, and localization theories.[41]

Beginning in the 1890s, German psychiatrists placed increasing trust in psychophysics, the use of "experiments designed to measure psychological reaction times to external stimuli, as well as various mental functions and capacities such as memory, decision making, attention span, etc."[42] Psychophysicists attempted to lay bare the invisible components of human cognition that had eluded their more literally minded peers.[43] Döblin's 1905 dissertation, "Gedächnisstörungen bei der Korsakoff'schen Psychose" (Memory Impairment in Korsakoff's Psychosis), emphasized precisely "das Unkörperliche im Körper" (immaterial elements in the material body), in keeping with the turn toward psychophysics and away from pathological anatomy.[44]

---

40 Quoted in Eric Engstrom, *Clinical Psychiatry in Imperial Germany: A History of Psychiatric Practice* (Ithaca: Cornell UP, 2003), p.127.

41 See Engstrom, *Clinical Psychiatry*, Chapters 4-5. The epilogue to *Freundinnen* even paid tacit reference to the waning popularity of assigning physiological coordinates to the mind's invisible processes: "Darüber hinaus liegen sehr entfernte und unkenntliche Motore. Mann kann manche menschliche Organe schneiden, ohne dass wir es merken; diese Organe sind empfindungslos" (Above all, there are very remote and unrecognizable motors [for human actions]. One can slice into many a human organ, without perceiving them; these organs are insensate). Döblin, *Freundinnen*, p.116-117.

42 Engstrom, *Clinical Psychiatry*, p.131.

43 Engstrom, *Clinical Psychiatry*, p.125.

44 Alfred Döblin, *Gedächnisstörungen bei der Korsakoffschen Psychose.* Inaugural-Dissertation zur Erlangung der medizinische Doktorwürde, Medizinische Fakultät der Albert-Ludwig-Universität zu Freiburg i.B., 1905, pp.14-15. The full passage reads: "Muss man es doch als ein ewig erstaunliches Geheimnis hinstellen, wie in einem Körper oder an ihm sich etwas Unkörperliches bindet, wie dieses Unkörperliche sich mit den Körperzuständen wandelt und sich, wie der Psychiater täglich sieht, verwirrt bei gleichzeitiger vielfacher Verwirrung von Körperfunktionen und Körperelementen" (Surely it is an eternal, wonderful mystery: how certain immaterial and incorporeal

Döblin's response to Nietzsche was therefore framed by a shift from localizable aberrations in psychiatry to a less visible mental cartography. His understanding of Nietzsche in terms of cognitive biologism is tied, in part, to an overall change in the methodology for identifying mental aberrations. A new intellectual culture interested in rejecting the "mechanistic thinking in the 19th century sciences" was on the rise.[45] Psychosomatic understandings were giving way to psychogenic ones. Döblin was not opposed to the *Naturwissenschaften* per se, but simply to what he perceived as their misappropriation: "Man misverstehe mich nicht: hier ist nicht die Rede an dieser Stelle, von dem Bankrott der naturwissenschaftlichen Philosophie oder der philosophischen Naturwissenschaft" (Do not misunderstand me: this is not a discussion about the bankruptcy of scientific philosophy or the philosophical sciences). His objection was in another key entirely. "Queen" philosophy had been bound with "Kette, Kugel und Zwangsjacke" (ball, chain and straightjacket) by her rightful servant, empiricism.[46] Nietzsche, Döblin wrote, was "die liebende Mörderin Dejanira" (loving murderer Dejanira) who had thrown a Nessus-shirt over the limbs of philosophy, and, by proxy, science.[47]

Nietzsche's radical materialism and dismissal of atomism – both of which Döblin supported – still had to be considered in light of the empiricist and positivist trends of the late nineteenth century. The case of empiricism was by far the most egregious because it had so dramatically misunderstood the true and proper aim of the *Naturwissenschaften*: to yield insight into human life as part of its experienced embeddedness in the sensible world. Chaos and fragmentation were to be embraced, as Döblin learned from his simultaneous reading of Felix Hausdorff's philosophical works, since they contained the possibility of an abiding skepticism in empirically established givens and linear thinking.

In this regard, Michaela Spreng has pointed out Döblin's large (and largely overlooked) debt to Nietzsche scholar and mathematician Felix Hausdorff.[48] Hausdorff, a professor at the University of Greifswald between 1913 and 1933, is best known for his 1914 textbook *Grundzüge der Mengenlehre* (Principles of Set Theory), which served as the fundamental work on set

---

elements are attached within and to a human body, how they transform in accordance with the bodily state, and how – as the psychiatrist witnesses daily – they entangle themselves with the simultaneous, composite disorder in the body's functions and parts).

45 Midgley, "Metaphysical Speculation," p.18.

46 Döblin, "Der Wille zur Macht," p.26.

47 Döblin, "Der Wille zur Macht," p.26.

48 Spreng, *Döblins Nietzsche-Verständnis*, pp.14-22.

theory for much of the twentieth century and contributed to the development of topology. Hausdorff produced eighteen works under the pseudonym Mongré between 1897 and 1904.[49] The first of these, *Sant'Ilario: Gedanken aus der Landschaft Zarathustras* (Saint Illario: Thoughts from the Landscape of Zarathustra, 1897), a collection of aphorisms, was composed in the style of *Zarathustra*, but with key differences in content. Hausdorff had no intention of copying Nietzsche wholesale or even mimicking him. The aphorisms allowed Hausdorff to "position himself [...] next to Nietzsche in an effort to release his individual thoughts and gain the freedom needed to question conventional norms."[50]

The positioning "next to Nietzsche" on Hausdorff's part was as much topographic as metaphorical. Sant'Ilario refers to the small coastal town near Genoa where Hausdorff composed the *Gedanken*, not far from the site where Nietzsche wrote sections of *Zarathustra.* Döblin was familiar with *Sant'Ilario*, and quotes from it directly in the opening pages of "Der Wille zur Macht als Erkenntnis."[51] Far more of Döblin's essay than is actually cited relies on formulations lifted or paraphrased directly from *Sant'Ilario*; Döblin even replicates some of Hausdorff's stylistic choices.[52] Hausdorff thus becomes a vehicle for Döblin to position himself "next to" Nietzsche, critically assessing the latter's philosophical contributions while simultaneously relying on those contributions as a platform for his own thought: an opportunity to grasp hold of their shared metaphysical commitments while dismissing the biological formulations with which he took issue.

Döblin and Hausdorff also shared a belief in the basic indeterminacy of the world. For Hausdorff, the world was "empörend viel Unsinn, Sprung, Zerrissenheit, Chaos, 'Willensfreiheit'" (full of outrageous nonsense, cracks, fragmentation, chaos, "free will").[53] Hausdorff's 1898 work, *Das Chaos in kosmischer Auslese* (Chaos in Cosmic Selection), which freely played off the idea of Darwinian selection in its title, expanded on Kant's transcendental critique. Hausdorff suggested that the true nature of the universe was inherently unknowable. In the face of manifest chaos, each individual simply selected an interpretable version for herself. Condemning the naïve realism of transcendental positivism, Hausdorff developed a "transcendenter

49 Walter Purkert, "The Double Life of Felix Hausdorff/Paul Mongré," *The Mathematical Intelligencer*, 30.4 (2008), 36-50.

50 Purkert, "Double Life," p.39.

51 Döblin, "Der Wille zur Macht," p.14.

52 Anthony Riley, the editor of Döblin's *Kleine Schriften* was the first to point out that much of Döblin's essay had connections to Hausdorff's work. For specific examples, see Alfred Döblin, *Kleine Schriften, Band I*, p.376, Note 14 as well as Spreng, *Döblins Nietzsche-Verständnis.*

53 Translation from Purkert, "Double Life," p.40.

Nihilismus" (transcendental nihilism), which called even the linear progression of time into question.[54]

Döblin hardly embraced "transcendental nihilism," but his early Nietzsche readings – Nietzsche through Hausdorff, Nietzsche next to Hausdorff – raised the possibility of a structurally chaotic cosmology as the "proper" heritage of the *Naturwissenschaften* and the need to question steadfast empirical givens. Whether these were an incipient biologism, a belief in man's evolutionary *Höherentwicklung* (higher development), or an anthropomorphically driven view of human cognition, the facts produced by each needed to be subjected to rigorous critique. The *Tatsachenphantasie* (fantasy of fact) Döblin named in the 1913 Berliner Programm would indeed carry literature away from men and toward the "kinetisch[e] Phantasie und [das] Erkennen der unglaublichen realen Konturen" (kinetic imagination and the recognition of [the world's] incredible, real contours).[55]

Yet not all facts in the fantasy of fact were equal. These "real contours" had to be differentiated from empiricist imposters, and the poisonous Nessus-shirt lifted. By reworking the causally-minded press coverage of the trial and incorporating documentation lifted directly from the evidence that had been presented there (the women's letters, transcripts of their dreams, handwriting samples), *Freundinnen* experimented both with debunking the tradition of realist reportage and with defying the conventions of deductive reasoning. The same evidence that had so easily assigned the women to neat typologies could, Döblin felt, also be used to arrive at an entirely open-ended, richly plural, understanding of the case that would take into account its proper status as "ein massives Stück Leben" (a massive slice of life).[56]

A Zero Sum Game

In May of 1926, Döblin delivered a speech to the Berlin Psychoanalytic Institute marking the occasion of Sigmund Freud's 70th birthday. He noted that one could point the sharpest telescope into the heavens without seeing God, and the most delicate microscope into the cerebral cortex without ob-

54 "Transcendental nihilism" is Hausdorff's own term for his philosophical project in *Das Chaos in kosmischer Auslese*. Further explication of Hausdorff's understanding and use of this term can be found in Werner Stegmaier, "Ein Mathematiker in der Landschaft Zarathustras. Felix Hausdorff als Philosoph," *Nietzsche-Studien*, 31 (2002), 195-240. As Paul Gray has pointed out, it was here, rather than in his academic work, that Hausdorff's ideas on set theory first developed. See Paul Gray, *Plato's Ghost: The Modernist Transformation of Mathematics* (Princeton: Princeton UP, 2008).

55 Döblin, "An Romanautoren," p.18.

56 Döblin, *Freundinnen*, p.117.

serving anything beyond cells and fibers. There was a hole in this sort of positivism. "Welches Instrument sollte man gebrauchen in dieser instrumentwütigen Zeit?" (What type of instrument should we use in this instrument-mad age?), he asked. "Keins. Nur den einfachen ruhigen und undogmatischen Blick" (None. Only the simple, dispassionate and unprejudiced gaze).[57] Döblin's comment is far from being a technophobic critique of life in the urban metropolis. Instead, it points out that with psychoanalysis, he believed, Freud has discovered the appropriate viewing apparatus for gazing into the psyche's depths.

The question of the proper type of perception – the correct *Blick* – for the *Naturwissenschaften* appeared in Döblin's writing throughout the 1920s, both in his medical texts and feuilleton-style essays. In February of 1923, just one month prior to the start of the Klein/Nebbe trial, he published "Blick auf die Naturwissenschaft" (A View of Science) in *Die Neue Rundschau.* Like many of Döblin's texts in the early twenties, it is at once a review of contemporary literature, an anti-empiricist indictment and a call to a new, foundational metaphysics. "Blick auf die Naturwissenschaft" is, however, unique in establishing an analogy between the emptiness of mathematical logic and the emptiness of the individuated, psychic "I." The dry, linear empiricism of faddish numbers is a type of false religion. Döblin resurrects the sovereign/servant confusion from his Nietzsche essays, claiming that modern physics' elevation of formulae is comparable to the medieval Church's presentation of its priests as God.[58] Mathematical reasoning was increasingly inaccessible to laymen, despite also being increasingly likely to make sweeping claims about the nature of the world.[59]

On the second page of "Blick auf die Naturwissenschaft," Döblin sardonically pointed out the impotence of neat mathematical formulations in

---

57 Alfred Döblin, "Zum siebzigsten Geburtstag Sigmund Freuds," *Vossische Zeitung*, 104 (5.5.1926), reprinted in Alfred Döblin, *Kleine Schriften, Band 3: 1925-1933*, ed. by Anthony W. Riley (Olten, Freiburg im Breisgau: Walter Verlag, 1985), pp.44-55 (p.45).

58 Alfred Döblin, "Blick auf die Naturwissenschaft," *Die Neue Rundschau*, 34.2 (1923), 1132-1138 (p.1133).

59 Döblin did not settle on physics by chance. He had encountered Einstein's relativity theory in 1917, while serving as a wartime physician. Excited by its claim to an "allgemein wissenschaftlichen Standpunkt" (general scientific point of view), his dozens of readings (and conversations about it) nevertheless failed to make the text comprehensible to him. Döblin recounts this story in *Das Ich über der Natur,* in a section entitled, "Die Verirrung der mathematischen Naturwissenschaft" (The Aberration of Mathematical Science) (p.18). This is particularly important because Döblin reproduces the equation exactly as it appears in "Blick" in this section (p.17), along which much of the surrounding text (p.16-17). The duplication establishes an important continuity between his early essays and his more robust metaphysical claims in *Das Ich über der Natur.*

the face of the pressing ontological questions, a subject he returned to in *Freundinnen*:

> Man fragt: was ist das Leben, die Welt, und bekommt die Antwort
> (One asks: what is life, the world, and receives the answer):
>
> $$\sum_{ik} \frac{\delta\delta \times \kappa}{} \left(\sqrt{-D}\ \sigma ei\Theta_{ik}\right) - \frac{1}{2}\sum_{ik} \sqrt{-D}\ \frac{\delta\sigma ik}{\delta \times e}\Theta_{ik} = 0$$
>
> Der wirklich schauende Anblick eines vertrockneten Blattes ist mehr wert, als eine Bibliothek babylonischer oder moderner Formeln.
> (The truly perceptive gaze on a withered leaf has greater value than a library of Babylonian or modern formulae).[60]

The statement deliberately sets up several types of interpretive confusion. On the most obvious level, it pokes fun at a massive discontinuity between question and answer: the first, a query into the nature of life and the world, and the second a reply, given in terms of an esoteric formula, inaccessible to much of the *Neue Rundschau's* readership (and today's). The example showcases a visual clash between the Latin alphabet (in which the question is asked) and the mathematical notation in which the answer is given. Actually reading the equation for understanding is beside the point. Any attempt at looking for an answer to life's meaning in the formula would be a naively literalist misinterpretation. The equation is a token and gesture - insignificant in itself, only the sign of an enigma.[61] Furthermore, the equation is deliberately nonsensical. While it would have been easy for Döblin to lift a "real" formula from a physics text, he instead selects an arbitrary expression.

Yet even readers with little or no understanding of mathematical notation would still have been able to see that this was a case of basic subtraction. Half of one quantity removed from another results in "0." The startling discontinuity between the complicated string of mathematical symbols to the left of the equals sign, and the (literal) nothingness to the right ("0"), smacks of translational inaccuracy, as though a lengthy statement in a foreign language had been rendered with a simple "no." The elaborate bulk of the material to the left is also visually disproportionate to the highly legible

60 Döblin, "Blick," p.1133.

61 With that in mind, however, the notational choice is not arbitrary. Those who were able to decipher it would have observed its use of tensors, which the 1915 publication of Einstein's paper on general relativity had introduced to physics. The presence of tensors (which help to define laws that remain valid regardless of changes in the coordinate systems describing them), implicitly reinforces Döblin's critique of physics' inability to adequately explain the workings (and meanings) of the world.

zero-quantity to the right. The equation not only writes its own emptiness, but affirms Döblin's overall point that mathematical constructions can give no ("0") answer to significant ontological questions, despite their pretension to universal applicability. Indeed, these formulae are themselves void of content, yielding nothing.

In addition, the formula is unreflective, unable to take account of itself or to pay tribute to the major representational shift it puts into play. This is linked to Döblin's earlier point in "Blick" that:

> Man hat die Pflicht nachzudenken, was es heißt, dass Dinge, Vorgänge sich zahlenmässig verhalten. Die Bedeutung der arithmetischen Anordnung, die Möglichkeit mathematisch zu formulieren ist ein Problem. Ja, ein Geheimnis.
>
> (One has a duty to consider what it means that physical objects and actions behave numerically. The meaning of mathematical order, the possibility of formulating the world mathematically is a problem. Indeed, a mystery.)[62]

It is disturbing to Döblin that, under the recent blossoming of a "mathematischer Behandlung der Naturwissenschaften" (mathematical treatment of the sciences), it is possible to transition between representational systems with such ease and nonchalance.[63] Far from being a universal language, mathematical representation inserts cryptic abstractions, accessible to only a select few. "Wird etwas deutlicher dadurch? Und wem?" (What is made clearer this way? And for whom?), he asks.[64] The purported universalism of these formulae turns out to be little more than an easy cover for an insidious depersonalization, a wrenching out of the world's concrete existence. The immediately perceptible components of the material world are transformed beyond recognition. Döblin contrasts the "blossoming" of mathematical treatment in the sciences with the paradoxically greater value of a "withered leaf."

The rest of the essay is devoted to reviewing alternative approaches. Döblin moves rapidly through reflections on Fritz Mauthner's *Geschichte des Atheismus im Abendlande (History of Atheism in the West,* 1923); Fritz Kahn's *Das Leben des Menschen* (*The Life of Man*, 1922); Hans Wolfgang Behm's *Entwicklungsgeschichte des Weltalls, des Lebens und des Menschen* (*History of the Development of the Universe, Life and Men*, 1923); Hermann Schulte-Vaertings *Soziologische Abstammungslehre (Sociological Theory of Evolution*, 1923) and Freud's

---

62 Döblin, "Blick," p.1132.
63 Döblin, "Blick," p.1133.
64 Döblin, "Blick," p.1133.

Das Ich und das Es (1923) among others. Connecting their shared interest in a commonly accessible scientific language, a new naturalism, and the inherent multiplicity of the "I," Döblin positions these writers as viable options to the abstruse, rarified language of the contemporary sciences and to the belief in individual psychology, to which realism alone would have access.

Toward the end of the essay, Döblin observes that the individual psychological realm is a "Hohlraum in der Natur" (lacuna in nature) and can yield nothing.[65] Like the equation introduced at the beginning of the piece, the individualized "I" turns out to be empty, a nil quantity or 0. And, as in the case of the equation, it produces a lacuna. Like the equation, the lone "I" is a self-contained, disconnected, finite statement, without any correlation to the experience of the sensate world. Neither the "I" nor the equation can supply answers to questions about the meaning of life, though both – the essay hints – are all too readily offered as solutions. This "I," Döblin seems to say in "Blick," is as inscrutable as the mass of mathematical symbols, despite its apparent legibility.

The "Hohlraum in der Natur" is at once the metaphorical blank cast open by the hollow "I," and the more literal, geographical opening in the landscape; "Hohlraum" is also intended in a geological sense, as the "hohler, zerklüfteter Raum, namentlich in Gebirgszügen" (an empty, jagged cavity, namely in mountain ranges).[66] The "Blick" of the sciences is a gaze in the same way one might think of a "Blick auf den See" (view of the sea) or "auf die Stadt" (of the city), as though the sciences were a transversable location. The essay aims to restore the working of physical, chemical and astronomical processes to their rightful temporal and spatial coordinates, rescuing them from the timeless, dimensionless domain of formulae.[67]

This restoration takes place both through Döblin's figurative language and through the structure of the essay. The essay is set up to mimic the equation it contains: it is balanced on either side by nil quantities (the zero of the formula and the "nichts" of the singular "I"). The many additional works cited (Freud, Kahn, etc.) form a deliberate excess within the equational logic. The citations are not only counterexamples in the sense that they provide alternatives to empirical thought, but in that they operate beyond the essay's parallelism and so exceed the claim that 0 = 0. Far from being a marker of bad faith, the final line, which stands alone at the bottom of the page, "Dies sage ich nur andeutend und vorläufig" (I say this all hint-

65 Döblin, "Blick," p.1138.

66 Jacob Grimm and Wilhelm Grimm, *Deutsches Wörterbuch*, vol. 4 (Leipzig: S. Hirzel, 1877), p.1720.

67 Döblin, "Blick," p.1133: "bestimmte physikalische, chemische, astronomische Abläufe."

ingly and preliminarily), should be read as an intentional refusal of the finality of mathematical proof.[68] This final line of the essay rejects a tight summary or a recapitulation of earlier points, mediating against the notion that the piece has been organized to logically culminate at a particular juncture or to supply an answer. Metapsychology (embodied in Freud, Kahn, Behm, etc.) is the counterpoint to the desiccated, singular "I." It is

> die Lehre vom realen Ich, von den realen Seelen, und beginnt mit der Zertrümmerung der Vorstellung vom eingekapselten beziehungslosen wirkungslosen Ich, mit der Zertrümmerung des ganz abstrakten, schwindelhaft aufgeblasenen Individuums.
>
> (the doctrine of the real I, of real Souls and begins by dismantling the idea of an encapsulated, disconnected, ineffective I, by dismantling the wholly abstract, bogus, bloated individual).[69]

The final "I" on the page – that which appears in "dies sage *ich* nur andeutend" – is given in the context of this dismantling, in terms of the preliminary hints toward connections that would be forever precluded by an isolated psyche.

Two Girlfriends Commit Murder by Poisoning, 1924: The Criminal Underworld as Metabolic Landscape

Both "Blick auf die Naturwissenschaft" and *Freundinnen* deal with an inherent mismatch between question and answer. For Döblin, the real issues raised by the trial of Elli Klein and Margarete Nebbe were not in any way resolved or even understood by the experts who testified at it. The flock of esteemed sexologists and psychiatrists who analyzed the women and the more than 600 letters they had exchanged in the course of their relationship assumed that a unidimensional, causal logic was at play. They cited Elli's insufficiently developed uterus, the dangers of a post-war industrial modernity, the "masculinizing" influences of the Great War for women on the homefront, and a wholesale denigration of Wilhelmine family values. Even the most cursory examination of *Freundinnen* makes Döblin's allergy to causal logic patently clear. To Döblin's eyes, the answers supplied by the trial's experts and later taken up by the popular press were inadequate because they mistakenly imagined that the "how" of their response was the same as the "how" of the question posed. In reality – or, rather, in Döblin's formulation of it – "how did this happen?" had little to do with the niceties of biological

68 Döblin, "Blick," p.1138
69 Döblin, "Blick," p.1138.

or sociological explanations (essentially equivalent to the unsatisfying mathematical answer in "Blick"). An adequate retelling of the case would need to recover the original, more meaningful (and more general) set of questions that the trial raised. *Freundinnen*, then, is a retrial that tries to imagine the sorts of questions that could be asked within the shifting, unstable interconnections of the case.

In *Freundinnen* Döblin returns to this idea of a necessary interconnectivity branching between individuals. He mentions in the epilogue that he visited the tavern where the two women first met and the apartment where they lived. He discusses speaking with them personally. In returning to the physical locations involved in the case, he claims to discover a symbiosis both between people and the "Wohnungen, Häusern, Straßen, Plätzen" (apartments, houses, streets, squares) they inhabit.[70] Human beings cannot be understood in the absence of one another or in isolation from the physical coordinates of their lives. Döblin's movement through the streets of Berlin is – like the story that precedes it – part of the process of demolishing the disconnected "I," of rescuing it from bloated abstraction and of reinstalling it in concrete, collectively accessible, experience. It is the "investigation of an investigation" and Döblin would have the reader believe that these interconnections are not his own invention, but simply a happy discovery, properly revealed by a narrative that forecloses empiricism's trite reductions.[71]

How should one speak in this domain? What would the proper type of inquiry look like? Writing for the morning edition of the *Berliner Börsen-Courier* in one of the earliest reviews of Döblin's text, Leo Greiner remarked that:

> Die Tatsachen sind einfach, ihre psychische Basis ist sehr breit, überlagert von jeder "Weltmasse" von der Döblin spricht, ein Über- und Gegeneinander von Beziehungen, Abstossungen, Verknäuelungen. Döblin gibt sich wieder, aus einem Zentrum heraus: wie sie sich knüpfen, fliehen, ballen.
>
> (The facts are simple, their psychological basis is very extensive, [they are] overlaid by each "world-mass" of which Döblin speaks, a stacking and conflict of correlations, repulsions, entanglements. Döblin reproduces them, from the center outward: as they are knotted around one another, as they shun one another, as they agglomerate.)[72]

---

70 Döblin, *Freundinnen*, p.114.
71 Herzog, *Crime Stories*, p.47.
72 Greiner, review of *Die beiden Freundinnen*, p.156.

Greiner's 1925 remarks are important for two reasons. First, Greiner is attentive to the ways in which the personages involved became secondary to the workings of their own psychological drives. There are no characters in the text, at least not in the traditional sense. Second, Greiner picks up on Döblin's own use of language, and the extensive set of influxes, ejections, flows, streams and discharges that organize so much of the "plot" of intersecting drives in a highly fluid landscape.

The interactions between Elli and her husband, even in the early months of their marriage are characterized by assimilations and rejections. As their sexual encounters increasingly come to involve physical violence, Elli's husband Link (who lacks a first name for most of the the text), "eröffnet ein neues Seelengebiet in ihr" (opens a new soul-region) in her.[73] He ruthlessly beats her, tears her clothing, dumps buckets of water on her, and, at one point attempts to throw her naked from the apartment window. She flees Berlin to her family home in Braunschweig several times in the course of the story.

The physical intensity of the relationship is paralleled by formative changes within Elli's psychological state. Throughout the course of the book Elli alternately embraces the paternalistic family setting of her parents' house in Braunschweig and the sexual freedom and purportedly loose morals of post-war city life, largely characterized by her relationship to Margarete. The external world of Berlin and the new, dark, "Seelengebiet" opened in Elli by Link are made to resemble one another: Elli's psychological darkness is suggestively juxtaposed with the urban environment. The city is a dangerously female space, ultimately centered around the little community of Elli, Margarete, and Margarete's mother. By contrast the family home in Braunschweig is (predictably) the domain of the old order, and the old (childish, responsible, heterosexual) Elli.

Link baits Elli and drives her further into this new "finstere Haßsphäre" (dark hate-sphere).[74] Their violent intercourse expresses itself both as the wrenching open of the dim region in the bowels of Elli's soul, and through Link's need to "dig" (wühlen) in Elli's body. He claims that he wishes to literally eat her feces:

---

73 Döblin, *Freundinnen*, p.23. The idea of extreme abuse and brutalization within the domestic sphere is in line with circulating ideas, particularly advocated by Magnus Hirschfeld (who testified at the trial), that the war, combined with the time soldiers spent in the trenches, resulted in perverse, pathological forms of sex, in part through the influence of military brothels. Döblin's decision to push these concerns into the background and to foreground the development of a new psychic domain in Elli needs to be thought of as a conscious choice not to write the story which had already been told by Hirschfeld.

74 Döblin, *Freundinnen*, p.48.

> Er mußte in ihrem Leib wühlen, Sinnlichkeit aus jeder Hautfalte herausfühlen. Er hatte den Drang, sie unbildlich, fast körperlich zu verschlingen. Es war kein bloßes Wort, wenn er ihr in der wilden Verschlingung sagte: er müsse ihren Kot haben, er müsse ihn essen, verschlucken.
>
> (He wanted to dig in her body, to draw sensuality from every fold of skin. He had the impulse to physically devour her, in the most literal sense. It was no mere turn of phrase, when – in the midst of their wild entanglements – he said he wanted to eat her feces).[75]

When Elli first comes into contact with Margarete Bende a shift occurs in her entire psychological perspective: her inner timbre changes.[76] Two drives begin to work in her: an untamable hatred for Link, which she wishes to pour out, and a passionate love for Margarete. The two forces gush over her soul: "Diese beiden überstarken, verkoppelten Gefühle strömten eine Faszination über ihre Seele aus" (These two intense, interconnected feelings emitted a fascination that suffused her soul).[77] Elli's hate accords her masculine elements lending her a phallic dimension. It freely "distends, grows, seeks objects" at the very moment when Elli's desire to prove her love to Margarete lends her virility and heroism.[78] Her later fear that Link's body will burst open from the poison she is liberally adding to his food has been read in terms of bodies exploding on the field of battle during World War I.[79] It can also be seen in the context of the other swellings, distensions, and expulsions that characterize the text.

The physiological results of the poisoning mimic and concretize much of the history of Elli and Link's relationship, playing out the past stages of their troubled interactions inside Link's body.

> Die Vergiftung ergriff den Magen und Darm, machte schwere katarrhalische Entzündungen. Erbrechen, choleraartige Durchfälle traten besonders nach den großen Dosen auf. Sehr blaß und grau wurde er, der Kopfschmerz, die Neuralgien, Schwäche im ganzen Körper.
>
> (The poisoning took hold of his stomach and bowels, creating a case of gastroenteritis. Vomiting, cholera-like diarrhea typically occurred after

---

75 Döblin, *Freundinnen*, p.48.

76 Döblin, *Freundinnen*, p.54.

77 Döblin, *Freundinnen*, p.54

78 Döblin, *Freundinnen*, p.53-54: "Männlichkeit und Heroismus." See Fuechtner for discussions of the gender question here. Both Döblin's description of Elli and responses by the Weimar-era press reinforced conceptions of the post-World War I female uncomfortably crossing gender boundaries. Döblin and the press responded to Elli's past work in a munitions factory as early evidence of her masculinization.

79 Fuechtner, *Berlin Psychoanalytic*, pp.53-55.

> the largest doses. He grew pale and gray, a headache, neuralgia, weakness throughout his entire body.)[80]

If Elli's ingestions and expulsions were, at first, the object of Link's fascination, it is later she who controls, produces and desires his. The dark interior of Elli's soul (opened by the early stages of their marriage) is later reproduced by her as the *actual* opening of Link's body. An economy of sexual fluids, fecal matter, and internal drives emerges as Elli appropriates Link's interior, just as he had, earlier sexually and psychologically appropriated hers. Link's body does indeed "burst," if not in precisely the way that Elli originally feared. The vomiting and defecation create an opportunity for Elli to discharge her hate. Link's emesis is the literal mechanism for Elli to pour out the contents of her hatred.[81]

In fact, Elli's crime (das *Ver*brechen) is contained in the process of causing her husband's expulsions (das *Er*brechen): she becomes an appropriate subject for a series on contemporary crimes. One hears the echo of "Erbrechen" in later references to the women's "Verbrechen" as well as to their "Ehebruch" (adultery).[82] The series' title, *Aussenseiter der Gesellschaft: Verbrechen der Gegenwart*, should be interpreted not only as the idea of a group of individuals alienated from so-called normal society. The outsiders in Leonhard's series are not constitutively alien to Weimar culture but have been spit from its core. As such, the series mirrored the Weimar Republic's fearful fantasy that a new breed of undetectable criminals was on the rise, a new type that lacked the characteristic visual markers of traditional criminal stereotypes and could thus easily evade detection.[83] Döblin's retelling of the case plays with the cultural desire to peer into the dark and potentially threatening bowels of society and to flush out its negative elements. In *Freundinnen*, he redirects and reimagines this desire as lateral movement between conscious and unconscious states and between drives.

This idea of intake, expulsion and release suggests comparison with Freud's drive-discharge model but also draws on Nietzsche's *Will to Power* as a life-force engaged with growth, expansion, appropriation, ascendancy.[84] Döblin was familiar with Freud's drive-discharge model and he had re-

---

80 Döblin, *Freundinnen*, p.65.

81 Döblin, *Freundinnen*, p.48.

82 Döblin, *Freundinnen*, p.109.

83 Herzog, *Crime Stories*, p.47. Herzog writes, "What is ultimately at stake in this series is a redefinition of the relationship between the two main terms in the series title – outsiders and society – and the role narrative plays in establishing that relationship."

84 See #259 in Friedrich Nietzsche, *Jenseits von Gut und Böse*, Kritische Studienausgabe, ed. by Giorgio Colli and Mazzino Montinari, 15 vols., 2nd ed. (Munich: Deutscher Taschenbuch Verlag and de Gruyter, 1988), p.207-208.

viewed it in several essays immediately prior to his composition of *Freundinnen.* In December of 1922, shortly after the Seventh International Psychoanalytic Congress had convened in Berlin, and just as Elli's, Margarete's and Link's relationship was reaching its peak intensity, Döblin wrote an extensive review for *Die Neue Rundschau* on "Metapsychology and Biology." In part devoted to Freud's *Jenseits der Lustprinzips* (1920), which first introduced the theory of the drives, Döblin devoted a substantial section of the essay to an explication of drive-discharge and the workings of the stimulus-barrier.[85]

Six months later he penned two essays expanding on this idea, particularly in light of Freud's *Das Ich und das Es.* On June 10, 1923, the *Vossische Zeitung* published his "Psychoanalyse von Heute" (Psychoanalysis Today), a review of works by Ferenczi and Groddeck, alongside Freud, as well as an overview of the Seventh International Psychoanalytic Congress. A second installment on June 28 in the same paper offered a defense of psychoanalysis as an overall method. To Döblin, psychoanalysis should be thought of as a much needed "seelische Drainage, innere Kanalisation" (psychological drainage and inner sewer system) for the collective mass of men, particularly since it could release the stifled, festering energies produced by modernity.[86]

Döblin's observations were not devoid of criticism of psychoanalysis, however. He pointed out the links between Freud and the founder of psychophysics, Gustav Fechner. Freud's *Jenseits,* Döblin claimed, recalled many elements that had already been present in Fechner's work. Döblin offered the hope that the blossoming of Freud's work would bring about a similar renaissance in Fechner studies. Moreover, as he indicated in his speech on the occasion of Freud's seventieth birthday, Freud had not yet taken on the question of how psychoanalysis might relate to broader social or political questions.[87]

While there is no doubt that Döblin's was indeed "doing" psychoanalysis (as he first claimed in 1921), his essays from the 1920s indicate that this embrace was contingent on psychoanalysis's ability to reveal metapsychological elements that were already part of his own *Naturphilosophie.* The 1922 essay on "Metapsychologie und Biologie" sets up Freud alongside character typol-

85 Alfred Döblin, "Metapsychologie und Biologie," *Die Neue Rundschau,* 33.2 (Dezember, 1922), 1222-1232, reprinted in Alfred Döblin, *Kleine Schriften, Band 2,* pp.182-193 (p.186).

86 Alfred Döblin, "Praxis der Psychoanalyse," *Vossische Zeitung,* 301 (28.6.1923), 2-3; reprinted in Alfred Döblin, *Kleine Schriften, Band 2,* pp.270-274 (p.274) and p.272.

87 Alfred Döblin, "Zum siebzigsten Geburtstag Sigmund Freuds," *Vossische Zeitung* 104 (5.5.1926), reprinted in *Kleine Schriften, Band 3: 1925-1933,* ed. by Anthony W. Riley. (Olten, Freiburg im Breihgau: Walter Verlag, 1985), pp.44-55 (p.54).

ogist Ernst Kretschmer, pioneer of psychosomatic medicine Georg Groddeck, infographics artist Fritz Kahn, astrologist and popular science writer Oskar A.H. Schmitz who were working on similar issues. If the drives were important to Döblin, it was not because of their connection to psychoanalysis per se, but because they illuminated the inherent fluidity of human existence, a fluidity that Döblin had already identified as the fundamental metaphor for man's relationship to the external world.[88]

According to Döblin's *Naturphilosophie*, causality did not function as the source of events; composition did. As he wrote in *Das Ich über der Natur*, "Saltze, Säuren, Wasserstoff, Kohlenstoff, Flüssiges, Festes, elektrische Strömungen bin ich. Zu ihren Seelen neige ich mich, von ihnen komme ich, das ist mein Vater- und Mutterboden" (I am salts, acids, hydrogen, carbon, liquids, solids, electric currents. I bend toward their souls, I arise from them, this is my native soil).[89] This echoes the observation in "Das Wasser" that man (a cauldron of energy) constantly requires new waters, salts and sugars for his "metabolism and inner life." Döblin blurs the boundary between the building blocks for chemical processes and the factors contributing to psychological interiority.

It is little wonder, then, that he turned, to Fritz Kahn's infographic illustrations as a visual case study of the interconnected nature of the "I".[90] Döblin expressed admiration in "Blick" for Kahn's work, which had fist appeared as part of the illustrated series *Das Leben des Menschen* (The Life of Man), published between 1922 and 1931. Kahn and Döblin shared several common goals. Both were committed to developing a scientific vernacular, to demonstrating that new technological systems complemented the workings of the human body, and to promoting a narrative style that combined optimal terseness and precision. Both also drew on film as a key metaphor in their work. In his 1923 "Metapsychology and Biology" essay Döblin explicitly discussed Kahn's illustration, "Es wird ein Mensch gemacht!" (A Man is Made!). The title is taken from the scene in Goethe's *Faust II*, which describes the alchemical birth of Homunculus, following research by Faust's assistant Wagner (with a little help from Mephistopheles). Döblin describes the illustration as follows:

---

88 See Alfred Döblin, "Das Wasser," *Die Neue Rundschau*, 33.2 (1922), 853-858.

89 Alfred Döblin, *Das Ich über der Natur* (Berlin: Fischer, 1928), p.150.

90 Kahn's illustrations are best known through the five volumes he published as part of the popular anatomy and physiology series *Das Leben des Menschen* (The Life of Man), by the Kosmos Gesellschaft für Naturfreunde (Kosmos Society for Nature Lovers). These works appeared between 1922 and 1931 and launched him to international fame. See Cornelius Borck's introduction, "Humanist and Technical Educator," in *Fritz Kahn: Man Machine*, ed. by Uta von Debschitz and Thilo von Debschitz (New York: Springer, 2009), pp.10-19.

> Man kann nicht gut fasslicher und eindringlicher darstellen, aus welchem Teilen ein Mensch besteht, als auf dem Bilde: "Es wird ein Mensch gemacht," geschieht: zwei Hände werfen Substanzen auf einen Tiegel und giessen Flüssigkeiten hinzu; man sieht Kohle in Form von Briketts, Phosphor in Streichhölzern, Salz in Säckchen und so fort; unten in der Retorte bildet sich der Homunkulus.
>
> (One could not illustrate the parts that compose a man more intelligibly or more vividly than in the picture, "A Man is Made!" It shows the following: two hands toss substances into a crucible and pour liquids into it; one sees carbon in the form of briquettes, phosphorous in the form of matches, salt in a little bag and so on; the homunculus is formed below, in the alembic.)[91]

Unlike the mathematical equation of "Blick auf die Naturwissenschaft," the constitution of the human body in Kahn's rendering is a clear explication of complex interactions that constitute a totality (see fig. 1). Moreover, Kahn's drawing should be considered as an alternative model for the "Räumliche Darstellung der Seelenveränderung" (spatial representation of soul change) that Döblin includes as an appendix to *Freundinnen* (see fig. 2), alongside the examples of Freud's topographic diagram of the unconscious in the *Ego and Id* and Adler's model of "Personality as purpose" in *The Nervous Constitution* (1912).[92]

In Kahn's diagram, fluorine, salt, phosphorous, calcium carbonate and a long list of other substances are the primary components of man. The diagram is disproportionately split between the large number of contributing elements (occupying the upper two-thirds of the page) and the relatively small being constructed in the alembic at the bottom.[93] Kahn's diagram privileges the common material composition of all human beings over any particular individuality. Similarly, in Döblin's soul-diagrams, the "Kindlichkeit" (childishness), "Aktivität" (activity), "Perversion" (perversion), and "Grundverstimmung" (basic moodiness) form the basic component elements of the situation involving Elli, Klein, and Link.[94] Just as Kahn's hu-

---

91 Döblin, "Metapsychologie und Biologie," p.193.

92 Fuechtner has pointed out both of these as potential background sources. See both *Berlin Psychoanalytic* and "Döblin's Medical, Psychiatric and Psychoanalytical Work."

93 Whether or not Kahn intends this to be the homunculus or simply a developing human being is open to debate, though Döblin certainly reads the diagram through the Faust story.

94 I rely here on Fuechtner's translation of *Grundverstimmung* as "basic moodiness" in *Berlin Psychoanalytic*, p. 34. She has pointed out the connection to the psychoanalytical diagnosis "depressive Grundverstimmung" in her essay "Döblin's Medical, Psychiatric and Psychoanalytical Work," p.126.

man being is constructed from the synthesis of multiple basic material elements, so too is the "Spatial Representation" built around the interlocking dynamics of drives presented as being common to all. The characters in *Freundinnen* share component motivations, if in dramatically different proportions. Döblin's diagram, like Kahn's, dispenses with physiological differentiation and gender distinctions. Döblin's circularly rendered soul-states might either recall the vaguely brain-shaped outline of Freud's topographical model or the fractal diagrams populating the pages of Hausdorff's *Mengenlehre.* Although Freud and Adler are certainly sources for the Döblin diagrams, their work is less directly concerned with the psyche's connections to material composition that so clearly interested Döblin during this period. Kahn's "Es wird ein Mensch gemacht!" visually connects Döblin's claims in his short essays of the 1920s that discuss human beings' material constitution – their salts, sugars, and acids, their constellations, equalizations, levelings-off – with the materialist elements of the soul diagrams that conclude *Freundinnen.*

Murder by Other Means

Freundinnen is neither a simple projection of the drive-discharge model onto Elli and Margarete's story nor a one-to-one mapping of Nietzsche's ideas on expulsion and appropriation in the will to power. It reflects Döblin's own modifications and synthesis of both. Overall, *Freundinnen* rejects orienting itself toward any single explanatory mechanism or school of thought, in favor of exploring a *Naturphilosophie* that both framed larger works from the period like *Berge Meere und Giganten* and *Manas,* and reached back toward a Romantic philosophy of wholeness. Against this expanded backdrop the murder weapon in *Freundinnen* is not the poison after all, nor are Elli and Margarete traditional *Giftmischerinnen.* The title, which refrains from referring to the women as such, is nevertheless deliberately misleading. If one can speak of a murder weapon at all here, it is the drives themselves. Contrary to the Pitaval tradition, the text turns out to be the far larger story of how psychiatrists, sexologists, and the popular press misunderstood their object of inquiry, by imagining it to be composed of isolated, monolithic elements and subject to linear causality. Focusing exclusively on what "caused" the women to plot the murder of their spouses, the court, press and public failed to reflect the inapplicability of their own line of questioning.

Döblin's early Nietzsche essays, his reflections in "Blick" on the impossibility of an isolated "I," and his *Naturphilosophie* have all found their way into *Freundinnen,* and contribute to its metabolic landscape. The Berlin Program's demand for a "Tatsachenphantasie" is a call to think of writing as a form of permeability, a coming to the world. In "Blick auf die

Naturwissenschaft" Döblin speaks of the risk in failing to heed that call. While the tale of Elli and Margarete's struggle may well be paradigmatic of Weimar gender politics, or of changing perceptions of the criminal underworld, or of heightened concerns about the need for collective social prophylaxis, on a deeper level it narrates the fundamental connections that bind inside and outside, the part to the whole, the individual to the collective. Its multiple connotations suggest the limits of linear empirical description and the need for thinking in terms of dynamic interconnections.

Fig. 1: "A Man is Made!"

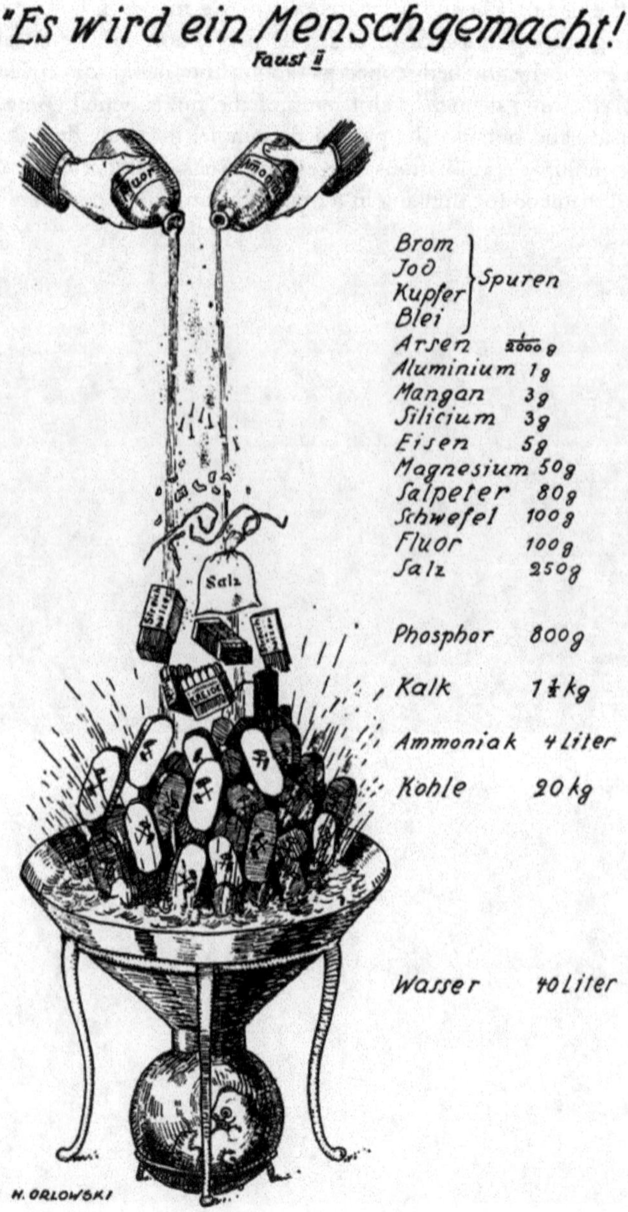

Fig. 2: "Spatial Representation of Soul Change"

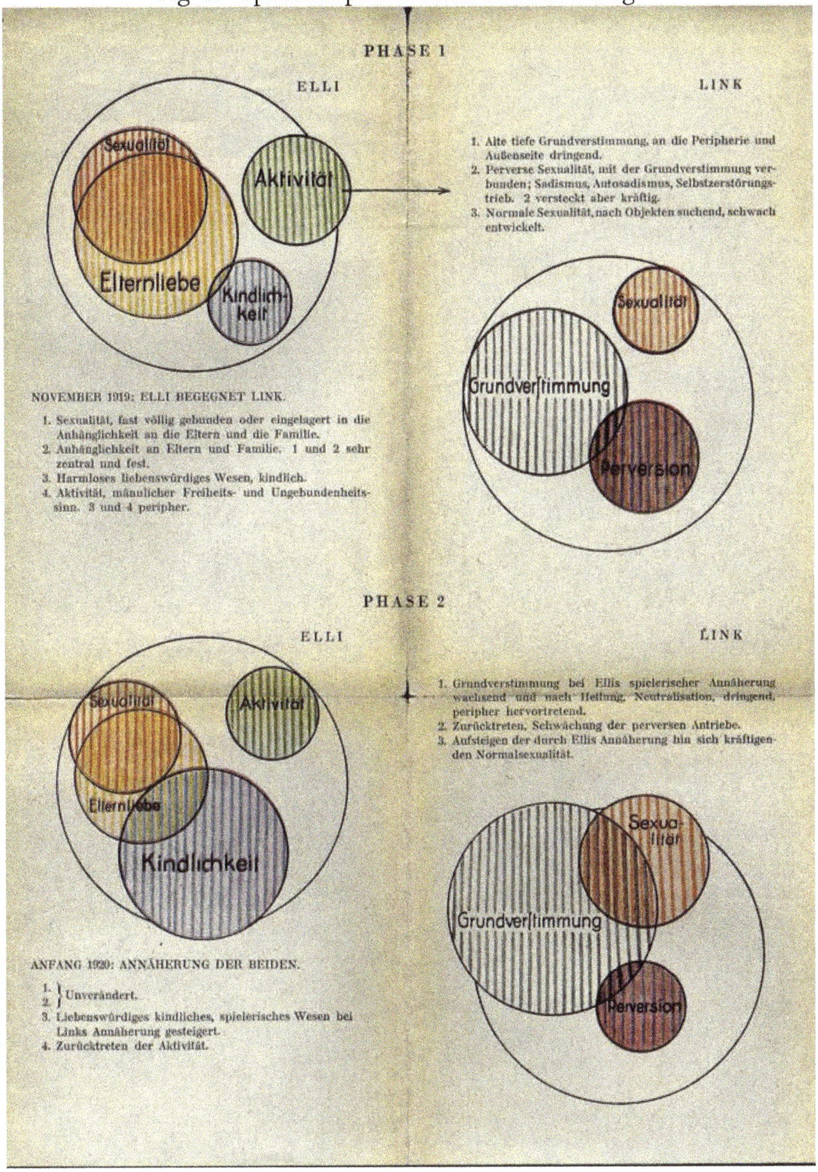

The *Lebenskraft*-Debate Recast: The Posthuman and Radical Mediation

Agency in the Anthropocene: Goethe, Radical Reality, and the New Materialisms

Heather I. Sullivan

Abstract: Our current era has been termed the Age of the "Anthropocene," or the human-inflected geological era. This essay addresses the implications of human impact on the Earth as a form of "radical reality" by addressing the broad spectrum of human and non-human agency. The analysis follows a three-step process: it begins with an introduction to the new materialisms and distributed agency in contrast to Howard Tuttle's notion of "radical reality" based on human consciousness. It then explores the agency of nature's "vibrancy" in the debate occurring early in the Anthropocene (during Goethe's lifetime) between "vitalism" and "mechanism." Finally, I use this context to explore Goethe's optics as a view that, like the new materialisms, is grounded in the interactivity of human and non-human energies. I juxtapose Tuttle's notion of radical reality with the new materialisms via Goethe in order to explore the broader implications of human and non-human agency in the age of the Anthropocene. Goethe offers convenient access into the Anthropocene with surprisingly prescient insights into what we now see as ecological enmeshments within nature's systems.

We are in the Age of the Anthropocene, or the human-inflected geological era, as the Nobel Laureate in atmospheric chemistry, Paul Crutzen, announced in 2000.[1] Since then, many scholars have adopted the term to describe the scientifically traceable impact of human activity across the entire planet since the Industrial Revolution. With such wide-spread traces and changes attributable to human beings, our actions loom ever larger and our agency to guide our future seems ever more profound. Debates rage concerning whether we should engineer the globe intentionally to counter the damage, or to try to work within the parameters in which we have evolved along with our co-species and thus limit our impact. The sway of human activity on reality seems ever more radical. Yet, at the same time, thinking "globally" means that our actual individual agency appears diminished, particularly when compared to traditionally humanist assumptions of a self-determining rational individual. Hence we face the complex dilemma of navigating between the vast collective impact of human beings as a species on the surface of the earth and climate and an acknowledgement of our limited individual agency in the scale of these circumstances.

In addressing this striking disparity of scale in the context of our ecological enmeshment, I consider our agency as a "distributed" force; that is, it

1 Cf. Rob Nixon, *Slow Violence and the Environmentalism of the Poor* (Cambridge: Harvard UP, 2011).

not so much driven by singular individuals making well-considered choices, but rather by a plurality of interactions and impulses from very large groups and a multitude of discourses, other beings of all kinds and on many scales, and, also, by our physical and cultural environments. Indeed, "matter" itself has an "agentic capacity" that influences our daily choices. We exist within the movements and meshes of living and non-living energies, from viruses and bacteria to weather and economics, even while our talents for technology and cultural constructions have shifted the planet's flows and climate. Agency in the age of the Anthropocene is complex and kaleidoscopic, distributed and global. This understanding of agency is one of the framing theses of the "new materialisms" for which reality emerges from the combined energies of vibrant matter (from quantum level to the cosmic), bodies, things, and cultural discourses.

To carry out this study, I examine the works of an author writing at the Anthropocene's dawn, one who expresses the shifting views on the body-mind-environment interface at the time when the radical planetary changes that we are now experiencing broadly were beginning to gain velocity: Johann Wolfgang von Goethe (1749-1832). Goethe, in fact, spoke of modernity's increasing speed as "veloziferisch," or devilishly fast. He is also acclaimed as the exemplary author of the modern, self-determining subject who creates his own destiny. From *Werther* to *Faust*, his literary figures stride through history and their eponymous texts creating worlds and their futures. Or at least that is the standard account of Goethe as major German author and shaper of cosmopolitan European modernity. In contrast to that view, I suggest that Goethe's vision – as we note when viewing both his science and literature together – documents figures who may *believe* that they choose their own fates yet actually *engage in reciprocally determining exchanges* with their companions and their physical surroundings in a manner best described as distributed agency. In other words, Goethe presents human reality as an enmeshment within historical culture, local communities, global interactions, and, not to be forgotten, multiple scales of natural forces. Discussing how these entanglements among culture, physical nature, and intellectual world-building interact as part of human "reality" presents a challenge for studies of the environment and the human being alike.

In this essay, I address the question of the "physically" radical reality of the age of the Anthropocene as documented in the new materialisms, and compare it to the "mentally" radical reality of human consciousness proposed by Howard Tuttle and explored in this volume.[2] Tuttle writes that the composition of reality is inevitably altered by the presence of our conscious-

---

2 Howard Tuttle, *Human Life is Radical Reality: An Idea Developed from the Conceptions of Dilthey, Heidegger, and Ortega y Gasset* (New York: Lang, 2005).

ness perceiving and shaping itself and its surroundings. I juxtapose these types of radical reality through the lens of agency: who and what has it, how or if it can be directed, and what agency in the age of the Anthropocene means for humanity's "radical reality." In his *Human Life is Radical Reality*, Tuttle explores how reality and the very matrix of our existence is impacted by human consciousness. As human subjects, we are always (only) within the range of our consciousness, though we extend our "reality" outwards, and so shape ever more of our environment accordingly. Tuttle declares radical reality to be the idea that in order for something to be real for human beings,

> it must somehow appear with and be kept in view of at least one human life. Ortega's claim that it is the destiny of human beings to 'humanize the world' is neither an anti-environmental notion nor an anthropocentric one. Human beings cannot avoid understanding circumstances in terms of the nature and understanding they actually possess. Even the idea of a non-anthropomorphic environmentalism is in fact a humanization of circumstances by human life itself. We cannot avoid imparting meanings and values that are not human meanings and values. (Tuttle, p.178)

Tuttle labels our perspective as inevitably human, in that we can see the world only through our own eyes and cultural systems. It is not just our frame that is altered in the contact, but reality itself is shaped in both directions by our "humanizing."

Tuttle's concrete assessment of human reality as being human shares with mechanistic views a belief that we human beings can escape our frame – bodily, material, ecological – and "transcend." Tuttle's radical reality is the human ability to have a "voluntary, transcending construction" that is different from our "biological nature" (Tuttle, p.141). Tuttle thus maintains the difference between human reality and the rest of physical reality:

> The issue here is the distinction between our life as a voluntary, transcending construction, as radical reality, and our life as our biological nature. Our historical and fabricated being and our biological being are not identical. Our self-transcending and historical being is a denaturalized existence which cannot be discerned by exclusive reference to our anatomy, sensations, or physiology [...]. While our biological existence, of course, is a necessary condition of our life as embodied, it is not a sufficient condition to explain our life as radical reality, and it is not anything radically real. Our human being is both natural and extra-natural; we are thus 'ontological centaurs.' Our body, of a given and fixed nature, does its work automatically, through biological laws of growth and decay, but our extra-natural life as radical reality is not given ready-made or realized according to laws; it is achieved in historical time, an existence which accumulates and fabricates itself toward its own future. (Tuttle, p.141)

Tuttle maintains a line between our "biological existence" and our "self-transcending being," something I question through both Goethe and the new materialists. Certainly, our reality is "achieved in historical time, an existence which accumulates and fabricates itself toward its own future," yet unlike Tuttle, I follow the new materialist's assertion that this type of existence is pervasive in matter's widespread creative forces in many forms (Tuttle, p.141). In fact, there is increasing evidence that "reality" is more a cacophonous and creative symphony of co-emergence including the tones from both mind and matter. As John McCarthy writes in his 2006 discussion of Goethe and complexity theory in *Remapping Reality: Chaos and Creativity in Science and Literature (Goethe-Nietzsche-Grass)*: "Perhaps the creative act alone is real. Perhaps all principles of reality are ultimately derivative of the one principle of creative convergence and divergence of inner and outer spaces, of matter and mind."[3]

It is significant, therefore, that Tuttle prevaricates most promisingly, despite generally maintaining the problematic dichotomy of "human biology" versus "transcendent ability," when he emphasizes the requirement of "relations" that allow the existence of the two things related: significantly, there is therefore no "I," or human subject, *without its world*. We may see the world through human eyes and exist in a "human reality" which is like no other reality, yet this does not happen in a vacuum; the two are fully interdependent. Tuttle claims:

> Human life remains inseparable from its confrontation with a world which exists as something serviceable for the mediation of what hinders or advances it. Neither the 'I' nor its circumstances can be authentically understood as independent 'things in themselves' which are each independent of the others. [...] *(T)he 'I' and its circumstances are always bonded together in a dialectical bipolor relation.* This polarity is neither the pure 'outside world' of realism nor the pure 'inside world' of idealism. (Tuttle, pp.47-48, emphasis mine)

For Tuttle, experience is always a polarity of things and viewers. His polarized view is, however, one based on a contrast between "human life" and the world "as something serviceable" for it. This tension remains. He nevertheless emphasizes a bipolar relationship between the individual and its "circumstances" that each depends on the other. With the emphasis on the bipolar tension, Tuttle provides fertile ground for our study of the new materialisms and questions of agency and the environment in the Anthropocene.

---

3 John A. McCarthy, *Remapping Reality: Chaos and Creativity in Science and Literature (Goethe-Nietzsche-Grass)* (Amsterdam: Rodopi, 2006), p.70.

In addressing agency and our physical and mental enmeshment the scopes of "reality," this paper follows a three-step process: it begins with an introduction to the new materialisms and distributed agency. I then briefly explore the agency of nature's "vibrancy" in the debate occurring early in the Anthropocene (the age of Goethe is the dawn of this era) between "vitalism" and "mechanism" in terms of Tuttle's notion of radical reality. Finally, I use this context to explore Goethe's optics as a view that, like the new materialisms, is grounded in the interactivity of human and non-human energies. In sum, I juxtapose the notion of radical reality with the new materialisms via Goethe in order to explore the broader implications of human and non-human agency in the age of the Anthropocene.

The "new materialisms" emerge from a wide range of scientific work, the social sciences, and the humanities. Unlike traditional Marxist materialism that emphasizes economic production as part of a (seemingly inevitable) teleological development into rather utopian social systems, the new materialisms concentrate on the interrelationships of human beings, bodies, and cultures in terms of patterns of matter and energy, autopoiesis, and complex systems including weather, the dissipative structures described by Ilya Prigogine, and all living beings (and economics, too). This view emphasizes above all the agentic capacities of matter in many forms and not just the human will. Agency in this context thus indicates the ability to impact and alter the surroundings, broadly speaking. Obviously, there is a wide spectrum of agentic capacity. The flows of matter and energy with such agentic capacity include the powerful weather patterns such as tornadoes and hurricanes, the bodily processes of all living things, and the long-term alterations to inorganic forms as well, including geological forces, the impact action of solar energy, and such effects as the fluctuations of electricity in the power grid. These all take on a "life of their own," so to speak, that is outside our control. Major works in the new materialisms include Stacy Alaimo's 2010 *Bodily Natures* and her 2008 jointly edited volume with Susan Hekman, *Material Feminisms*; Karen Barad's 2010 *Meeting the Universe Halfway*; Jane Bennett's 2010 *Vibrant Matter: A Political Ecology of Things*; Diana Coole's and Samantha Frost's 2010 volume, *New Materialisms: Ontology, Agency, and Politics*; Andrew Pickering's 1995 *The Mangle of Practice: Time, Agency, & Science*; and Serenella Iovino's and Serpil Oppermann's literary studies in "material ecocriticism.[4]

4 Stacy Alaimo, *Bodily Natures: Science, Environment, and the Material Self.* (Bloomington: Indiana UP, 2010), *Material Feminisms*, ed. by S. Alaimo and Susan Hekman (Bloomington: Indiana UP, 2008); Karen Barad, *Meeting the Universe Halfway: Quantum Physics and the Entanglement of Matter and Meaning* (Durham: Duke UP, 2007); Jane Bennett, *Vibrant Matter: A Political Ecology of Things* (Durham: Duke UP, 2010); *New Materialisms: Ontology, Agency, and Politics,* ed. by Diana Coole and Samantha Frost (Durham:

These works bridge the power of discourses and physical matter, thereby contextualizing human activity within broader cultural and natural forces.

The performative aspects of emerging reality are not limited to human linguistic capacity and our consciousness, since our bodily immersion and co-entanglements with light and the machinations of our intestinal bacteria, for example, are constant. It is not just discourses shaping our bodies and behaviors: the type of music playing in the background at the mall apparently impacts our shopping choices, the amount of daily sunlight can influence our moods, and, more ominously, the hormones introduced into our bodies from factory farming practices may be altering the early onset of puberty. The list of significant factors altering – literally – our bodies and minds grows ever larger in the Anthropocene's industrial coating of anthropogenic matter across the globe. The impact of such medicinal and toxic elements is explored at great length in Alaimo's *Bodily Natures*; in this essay I look primarily at the question of light as a part of our bodily environment. One cannot argue, after all, that we exist independently from light. We are products of solar energy, and it is an ecologically relevant exercise to consider the implications for our agency.

I cannot do full justice to the range of ideas from the new materialisms here; in brief, I concentrate on the questions of agency with an eye towards our participation in broader flows of energy and matter (the downgrading of human agency into contextualized interactions) juxtaposed with the immense impact of humanity during the age of the Anthropocene (upgrading our agency as a species). Barad gives us solid ground for thinking about matter's agency: her research in quantum mechanics demonstrates concretely how matter is "agentive" and how the "linguistic turn" in scholarship has tended to neglect matter itself.[5] Building on the physics of Niels Bohr, Barad presents a "posthumanist performative" approach to "understanding technoscientific and other naturalcultural practices that specifically acknowledges and takes account of matter's dynamism" (Barad, p.135). She works with optics – like Goethe, but in her case also with electrons and quantum particles. This leads her to see matter not as "reflective" (of a subject) but rather as "refractive" (dispersed with specific patterns) and agentic in its

---

Duke UP, 2010); Andrew Pickering, *The Mangle of Practice: Time, Agency, & Science* (Chicago: U of Chicago P, 1995). See especially the literary application of the new materialisms as "material ecocriticism" by Serenella Iovino and Serpil Oppermann in their essay "Material Ecocriticism: Materiality, Agency, and Models of Narrativity," in *Ecozon@*, 3.1 (2012), 75-91, and their volume on *Material Ecocriticism* (Bloomington: Indiana UP, 2014).

5 Cf. especially her chapter on "Agential Realism: How Material-Discursive Practices Matter," pp.132-185.

movements: "Matter is neither fixed and given nor the mere end result of different processes. Matter is produced and productive, generated and generative. Matter is agentive, not a fixed essence or property of things" (Barad, p.37). In fact, she writes that:

> Matter's dynamism is generative not merely in the sense of bringing new things into the world but in the sense of bringing forth new worlds, of engaging in an ongoing reconfiguring of the world. Bodies do not simply take their places in the world. They are not simply situated in, or located in, particular environments. Rather, "environments" and "bodies" are intra-actively co-constituted. (Barad, p.170)

In short, Barad presents what I consider an "ecological" view from quantum mechanics, one that puts human beings into matter's mix of bodies and environments constantly reconfiguring each other at varying levels. According to Barad: "[T]he phenomena produced are not the consequences of human will or intentionality or the effects of the operations of Culture, Language, or Power. Humans do not merely assemble different apparatuses for satisfying particular knowledge projects; they themselves are part of the ongoing reconfiguring of the world" (Barad, p.171). This is one aspect of what the new materialisms, even the seemingly abstract studies in quantum mechanics, offer current ecocritical explorations of the human-environment interfaces: reconfigurations and diffraction are the patterns in which we exist and in which we shape our surroundings. In this context, the toxic wastes and disasters ensuing in the wake of our "reconfigurations" take on a life of their own. Radioactivity, for example, lives on past any master narratives we may tell (as Rob Nixon notes in *Slow Violence and the Environmentalism of the Poor*, depleted uranium used in many recent American munitions has a half-life of 4.51 billion years, thus at a scale that one might only label cosmic).[6]

Barad contextualizes our agency within scales such as the quantum that we are not typically used to seeing in environmental contexts. Bennett, on the other hand, works with more familiar scales of bodies and humanly visible objects, yet she similarly declares that agency is distributed. "Agency is," she writes, "distributed across a mosaic" (Bennett, p.38). She also notes, however, that human agency is like a bicyclist "riding a bicycle on a gravel road. One can throw one's weight this way or that, inflect the bike in one direction or toward one trajectory of motion. But the rider is but one actant operative in the moving whole" (Bennett, p.38).

Energies and vitalities are a "swarm," or an assemblage, and not part of a binary of active subject working with passive matter. Bodies are therefore

---

6 Nixon, *Slow Violence* (2011), p.201.

imbricated in other systems of bodies and matter where they act alongside and with others: "bodies enhance their power *in* or *as a heterogeneous assemblage*" (Bennett, 23, original emphasis). One may perceive each bodily motion as an autonomous decision but that is more a willful overlooking of context than an accurate perception. Hence, even our most conscious actions are actually, in Bennett's terms that are typical of the new materialisms, "distributed." "What this suggests for the concept of *agency* is that the efficacy or effectivity to which that term has traditionally referred becomes distributed across an ontologically heterogeneous field, rather than being a capacity localized in a human body or in a collective produced (only) by human efforts" (Bennett, p.23). For Bennett, distributed agency is at a bodily level, such as the ingestion and digestion of food that our bodies actively consume thereby engaging the eaten and the eater in an "eating encounter" in which bodies are shown to be but temporary congealments of a materiality that is a process of becoming" (Bennett, p.49).

All living things from human beings to birds, animals, plants, and microbes are enmeshed within manifold processes and at multiples scales. Additionally relevant are the food substances inherent to life, solar energy, and the ever more omnipresent particles of industrial debris covering the planet since the explosion of the Anthropocene. These scales do not eradicate human agency but rather contextualize it within our own species and discourses as well as our relation to other species with whom we co-exist. In this sense, we are not quite ourselves but rather scales of collectivity coagulated into fleshy bodies. This materiality sounds alienating, but it elucidates a vision of "shared reality." As Bennett writes:

> Vital materiality better captures an "alien" quality of our own flesh, and in so doing reminds humans of the very *radical* character of the (fractious) kinship between the human and the nonhuman. My "own" body is material, and yet this vital materiality is not fully or exclusively human. My flesh is populated and constituted by different swarms of foreigners. The crook of my elbow, for example, is a "special ecosystem, a bountiful home to no fewer than six tribes of bacteria […]." (Bennett, p.112, original emphasis)

"Radical reality" in Bennett's terms is based on the "kinship between the human and nonhuman"; and indeed, it is that ecological context to which I compare Tuttle's "radical reality" focusing instead on human consciousness. In short, thinking of the environment in terms of the new materialisms means finding new grounds on many different scales of our human-matter interface. These do not begin and end with the bodily boundaries, but rather our bodies are part of the co-emerging flows.

Such ideas about flow and distributed agency are not entirely new, of course, and one finds these concepts throughout Western culture and in many cultures across the globe. For this particular discussion of human and non-human, relevant background can be found the eighteenth- and nineteenth-century *Lebenskraft*-related debate between "vitalism," as expounded upon by Paracelsus in the sixteenth century in which the four elements are themselves active agents, and the "mechanistic" explanations describing all non-human life as a machine as per Descartes and Newton. In terms of agency, this dichotomy was fierce: the vitalistic explanations suggested that a life force akin to a kind of wide-spread agency animated the world, whereas the mechanistic views described nature as "mere matter," a machine moving unceasingly according to the laws of physics utterly devoid of agentic capacity.[7] In the mechanistic view, only human beings and the divine have agency, since our "soul" separates us from matter and links us to the divine. Interestingly, the scientific view drew an ultimate line between passive matter and the immaterial (human) soul and so maintained an ultimate distinction between human beings and the rest of the world, whereas vitalism more scientifically, at least in terms of contemporary quantum physics and ecology, included humanity in the mix of the vibrant material world.

In vitalism, our minds and bodies are more than simple vessels for our souls, though the explanations for matter's energy were often fanciful or spiritually based. During the Age of Goethe, romantic "Naturphilosophie" sought various unifying principles for the life force, mind, soul, and even body, thereby bridging the two sides and revealing the faulty structure of the dichotomy.[8] Many of these discussions circled around the problem of agency; that is, the issue of how life shapes itself and acts on its own terms (or as mechanistic enactments of divine directive), and how this relates to human activities. As Andrew Cunningham and Nicholas Jardine write in the introduction to *Romanticism and the Sciences*, this "age of reflection" sought a self-understanding based on the "unity of spiritual process in nature and ourselves."[9] The more scientific and mechanistic perspectives rejected vitalism's holistic and spirit-based harmonies while also fueling a dualistic and seemingly paradoxical tendency to assert that our bodies are determined by mate-

7 Cf. Hartmut Böhme, *Natur und Subjekt* (Frankfurt am Main: Suhrkamp, 1988); Val Plumwood, *Environmental Culture: The Ecological Crisis of Reason*, (London: Routledge, 2006); and Carolyn Merchant, *The Death of Nature: Women, Ecology and the Scientific Revolution*, (New York: Harper Collins, 1989), esp. pp.117-126 and 235.

8 Cf. Kate Rigby, *Topographies of the Sacred: The Poetics of Place in European Romanticism*, (Charlottesville: U of Virginia P, 2004).

9 "Introduction: The Age of Reflexion," in *Romanticism and the Sciences*, ed. by Andrew Cunningham and Nicholas Jardine (Cambridge: Cambridge UP, 1990), pp.1-9 (p.3).

rial laws yet our souls are free. Thus, mechanistic thinkers claimed that there is an absolute difference between the human being as a soul-carrier or a rational mind, and all other materiality and living beings. In other words, the hard-edged scientific knowledge cemented an ultimate (and non-scientific) distinction between humanity and the rest of the organic world.

This mechanistic view is part of the traditional materialism that understood human beings in isolation from both our environment and our "companion species," as Donna Haraway describes the species with which human beings co-evolved including our fellow animals, but also bacteria and plants, etc.[10] As we now realize with greater ecological knowledge, there are very troubling implications associated with the belief that human beings are radically separate from rest of the biosphere; indeed, the mechanistic model neglecting the ecological context of the human species is radically insufficient. Both Nixon in *Slow Violence and the Environmentalism of the Poor* and Thomas Heberlein in *Navigating Environmental Attitudes* have written poignantly about the disastrous social and environmental impact of practices that pretend humanity and ecology are two entirely separate categories.[11] Although the vitalist-mechanistic debates continue in various forms today, the blind spots of both views render the dichotomy ineffective for explaining human participation in Earth's realities, particularly in terms of what we now know from such fields as ecological science, quantum mechanics, non-equilibrium thermodynamics, complexity theory, animal studies, and all of these in relationship to artistic creativity. One might more productively think non-dualistically about creativity, mind, matter and reality. As McCarthy writes:

> Once we begin to distance ourselves from the binary mode and think complementarily – even holistically – we begin to notice linkages previously unsuspected. We also come to appreciate the fact that nonlinearity is actually the rule in the real world, whether in cloud formations, gypsy moth populations, particle physics, our heart rhythms and brain waves – or in enduring works of art. (McCarthy, p.271)

In other words, creativity appears to be a shared capacity of bodies, clouds, and our minds. In the rest of the essay, I work through the implications of interwoven agencies for the radical reality of our consciousness in conjunction with the autopoieitic creativity and vibrant "becoming" of the physical, ecological, reality of planet Earth through Goethe's optical lens.

---

10 Cf. Donna Haraway, *The Companion Species Manifesto: Dogs, People, and Significant Otherness* (Chicago: Prickly Paradigm, 2003).

11 Thomas Heberlein, *Navigating Environmental Attitudes* (New York: Oxford UP, 2012).

This philosophical strand emphasizing emergence from relations *preceding the things related* resonates with Barad's views on quantum mechanics. From the bridging work of the romantics and Goethe continuing through contemporary ideas in complexity, chaos theory, quantum mechanics, non-equilibrium systems, networks, environmental thinking, postmodern literature and the arts, etc., we find models of "relationality" that overcome such dichotomies.[12] McCarthy echoes and updates Goethe in terms of chaos theory with the assertion that: "At the center of rumination is the creative act in both nature and the arts. The chief hypothesis is the belief that science and the arts are *interrelated via deep structures*" (McCarthy, p.14, emphasis mine). These deep structures in both the natural world and in the arts are shared energies and, according to the new materialisms, they express agentic capacities beyond the human. This is not revived vitalism based on spirituality or a "vital force," but rather a contextualization of human abilities within a living, growing environment constantly shaped by complex flows of energy and matter. In this sense, reality is an ongoing emergence of relations amongst creative patterns. The radical reality of humanity is not dissolved into these motions but rather a supplemental extension infused with, and related to, other earthly processes.

Goethe's work reflects similar ideas regarding human activities. He was never entirely bound to a dichotomy of realism versus idealism, and, indeed, his science and literature bridge such artificial divides. Particularly in his optical works but also throughout his science, as Ingrid Dzialas writes, Goethe sees human beings engaging with "elementary natural phenomena," like colors, in terms of *interactive* processes, rather than as an "active" mind processing "passive" nature.[13] Taming the mind to avoid hasty assumptions and

---

12 I explore these in terms of non-equilibrium thermodynamics elsewhere; see Heather I. Sullivan, "Affinity Studies and Open Systems: A Nonequilibrium, Ecocritical Reading of Goethe's *Faust*." In: *Ecocritical Theory: New European Approaches*, eds. Axel Goodbody and Kate Rigby (Charlottesville: U of Virginia Press, 2011), pp.243-255.

13 In Ingrid Dzialas's *Auffassung und Darstellung der Elemente bei Goethe*, (Berlin: Ebering, 1939), she describes Goethe's sense of relationship to the elements as an "Einheitsgefühl." Of air, she writes that Goethe "weiß genau, daß er von der Atmosphäre abhängig ist – und er will es auch nicht anders. Wie ein mystisches Erlebnis durchschauert ihn diese Erkenntnis. Er hat das Gefühl, als ob die ganze ihn umgebende Welt in ihn eindringe und mit ihrem Sein sein eigenes Sein erfülle. Bei keinem Element hat Goethe das tatsächliche Ineinandergehen, die Einheit zwischen Mensch und Außenwelt so stark empfunden wie bei der Luft. Eine engere Verbindung kann es nicht geben. In der Luft und von der Luft leben wir und reagieren im höchsten Grade auf sie" (p.43). (Goethe 'knows precisely that he is dependent on the atmosphere – and he does not wish it to be otherwise. Like a mystical experience, this knowledge thrills him. He feels as if the entire surrounding world infiltrates him and fills his own essence with its essence. With no other element did Goethe perceive so

to allow the natural phenomena to emerge on their own terms, to renounce the false belief in mastery, is one of Goethe's primary goals in all of his science and one that permeates his literary work as well. Dennis Sepper's book-length comparison of Goethe's work to Newtonian systems clearly outlines the Goethean methodology in which one "ought to let things speak for themselves."[14] Those who are aware and sensitive enough to allow themselves to see and flow with nature's patterns will most readily be able to explain them, since nature's patterns and our own being and vision are deeply related, as McCarthy and Astrida Tantillo also explain.[15] Goethe's work from early in the Anthropocene offers a view exploring our shared material participation in physical reality like all living things, even as he struggles simultaneously to maintain the sense that we human beings are also absolutely unique, as is typical to humanism. In other words, Goethe provides an excellent example of work that connects the traditional, subject-centered humanism, and ideas similar to those of the contemporary work in the "new materialisms," that emphasize non-human forms of agency or agentic capacity as the context for our own bodily and intellectual environments and actions.

Goethe's figures in *Werther* engage with, and are shaped by, cultural and natural forces outside their control, including class expectations, thunder storms, and flooding rivers; in *Faust* with literary traditions, cosmic pacts between the Lord and Mephistopheles, witches' brew, sexual desires, water nymphs, and the elements of fire, water, air, and the earth; and in *die Novelle* and *Das Märchen* with lions and tigers, gold, unruly giants, uncontrolled shadows, mysterious vegetable debts, and the antics of will o' wisps, to mention a few. We human beings are in the mix of elements, in other words. Gernot Böhme describes how Goethe inscribes the human being from *within nature*, so that we, too, are "Naturwesen":

> Der Erkennende, der Mensch, ist Naturwesen, und er kennt deshalb die Natur von innen heraus, wir würden heute sagen: aus der Teilnehmerperspektive. Die-

---

strongly the concrete infusion, the unity of human and outer world, as he did with air. There cannot be a closer connection. In the air and from the air we live and react to the greatest extent.')

14 Dennis Sepper, *Goethe contra Newton: Polemics and the Project for a New Science of Color*, (Cambridge: Cambridge UP, 1988), p.3.

15 Cf. McCarthy, esp. his chapter 5 on *Faust*; and Astrida O. Tantillo, *The Will to Create: Goethe's Philosophy of Nature* (Pittsburgh: U of Pittsburgh P, 2002).

> ser Gedanke ist allerdings bis heute nicht ausgedacht und in seinen Erkenntnismöglichkeiten nicht im entferntesten ausgeschöpft.[16]
>
> (The perceiver/knower, the human being, is a natural being, and s/he thus knows nature from the inside out. Today, we would call this seeing from the participant's perspective. This thought has not, through today, been fully thought through and has not been by any means exhausted in its full potential for perceptual/knowledge possibilities" [translation mine].)

Furthermore, in Goethe's many scientific writings, he details a very clear emphasis on the interactive processes of the human "subject" and the observed world or object. The observer must temper her impulses and let "nature" or, in the case of Goethe's optical treatise, *Zur Farbenlehre*, the colors and light, emerge on their own terms. The elements of color and light, as well as water, fire, earth, and air and other non-human factors have significant influence in Goethe's works, so much so that they appear to have "agentic" capacities. One might say, for example, that water determines much of Faust's fate as he battles against the sea with his dike, uses the water nymphs to accomplish victory against the "other Kaiser," and when he faces the putrid swamp in his final moments. Even his final "ascent" into the heavens looks a lot like the flow of water rising with warm air, as I discuss elsewhere.[17] With this kind of distributed agency among humans and non-humans, Goethe's works are similar to the new materialisms and their emphasis on the active impact, or "vibrancy" and creativity of matter, as Jane Bennett describes it, in which we partake rather than lead from "outside."

For Goethe, our sensory connection to the material world, "nature," most strongly arises from our visual perception, from sight. In his optical works, *Zur Farbenelehre* and *Beiträge zur Optik*, Goethe's description of vision is representative of the interface of mind and matter. Goethe criticizes Newton's famous optics for numerous reasons including Newton's (correct) theory that white light contains all the spectrum of colors, which Goethe believes was based on "faulty logic." That is, Goethe (with, in this case, significant insight) insists that colors emerge only when light interacts with objects and eyes; whereas Newton sees the colors in terms of light itself (it is a matter of light's pure essence). Goethe's colors are specifically *interactive*:

---

16 Gernot Böhme, "Natur hat weder Kern noch Schale. Goethes Methode der Naturbetrachtung," in *Goethe: Ungewöhnte Ansichten*, ed. by Karl Richter and Gerhard Sauder (St. Ingbert, Germany: Röhrig UP, 2001), pp.9-21 (p.15).

17 Heather I. Sullivan, "Ecocriticism, the Elements, and the Ascent/Descent into Weather in Goethe's *Faust*," *Goethe Yearbook*, 17 (2010), 55-72.

> Newton begeht hierbei den Fehler, den wir schon früher gerügt haben, und den er durch sein ganzes Werk begeht, daß er nämlich das prismatische Bild als ein fertiges unveränderliches ansieht, das es doch eigentlich immer nur ein werdendes und immer abänderliches bleibt.[18]
>
> (Newton makes hereby the mistake that we earlier rebuked, and that he makes throughout his entire work; namely that he sees the prismatic image as final and unchanging. It, however, only ever occurs as something always becoming and always changing [translation mine].)

Colors and all of nature – including human beings – are always "becoming" and changing according to Goethe. Science must work within this framework of emergence rather than in an ultimate divide with an objective, outside observer and solid, fixed object. Things in Goethean terms have a more agentic capacity than Newton's view of light as an object of physics with specific and straightforward characteristics. Goethe rejects specifically the mechanistic model here, as Walter Heitler asserts in his discussion of the *Farbenlehre*.[19]

Additionally, Goethe tends to describe nature, eyes, and colors as interactive, intertwined relations rather than isolated objects. Tuttle's "relations" between the "I" and the world are similar, but his stress of the humanly inflected "radical reality" leans more heavily towards our consciousness than to the broader scope of physical interactivity. Goethe describes such interactivity at many levels and in many forms, including the polarity of dark and light, and the co-shaping of eyes and colors, all of which are always experienced in conjunction with objects. He sees nature and eyes as fully interrelated, and the form of human perception as emerging from the fundamental polarity of dark and light. Newton thinks of light itself, as something in isolation, rather than in terms of its relations to the rest of the world as Goethe preferred: "Newton scheint vom Einfacheren auszugehen, indem er sich bloß an's Licht halten will; allein er setzt ihm auch Bedingungen entgegen so gut wie wir, nur daß er denselben ihren integrierenden Anteil an dem Hervorgebrachten ableugnet" (FL, p.304). (Newton appears to proceed from the simpler [starting point] in that he wants to address only light. However, he also subjects it to conditions just as we do, only that he denies their assimilative impact on the result" [translation mine]).

---

18 Johann Wolfgang von Goethe, *Zur Farbenlehre*, ed. Manfred Wenzel (Frankfurt am Main: Klassiker, 1991), p.337. Future references to this work are made parenthetically, labelled "FL."

19 Walter Heitler, "Goethean Science," in *Goethe's Way of Science: A Phenomenology of Nature*, eds. David Seamon and Arthur Zajonc (New York: State U of New York P, 1998), pp.55-69.

Note that Goethe insists we include our own impact on the process and results in the scientific understanding. In this, he speaks with contemporary quantum physics and the new materialists who see an emergent reality shaped by material and human interactions. Frank Schweitzer speaks in this regard of Goethe's "aktives Mitwerken des Erkennenden"[20] ["active participation of the perceiver/knower"]:

> Der Erkenntnisprozeß ist also durchaus auf das aktive Mitwirken des Erkennenden angewiesen. Erkenntnis hier ist kein Akt bloßer Konstruktion, sondern schließt auch einen nicht-konstruierbaren, *nur erlebbaren* Teil mit ein, der sich auf Genuß und Empfindung bezieht." (Schweitzer, p.390)
>
> (The perceptual/knowing process is related fully to the active participation of the perceiver/knower. Perception/knowledge is here no act of simple construction, but rather includes also a piece that is not able to be constructed, only able *to be experienced*, and that relates to enjoyment and sensation" [translation mine]).

To ignore the act of perception and observation in science is, in Goethe's phrase, an example of Newton's "Gewalt des Selbstbetruges" and "Unredlichkeit" (violence of self-deception and unspeakableness) (FL, p.315).

Goethe condemns Newton's refusal to acknowledge the role of the observer and the broader context. In McCarthy's terms, his "main objection to Newtonian optics, for example, was Newton's failure to include the observing subject in his calculations and the conditions of the experiment itself. Above all, Goethe criticizes a perceived tendency toward microreduction which leads to the loss of the broader context of the isolated part studied."[21] Michael Böhler similarly stresses Goethe's rejection of the Newtonian division between "Mensch und Natur," describing how Goethe believed that Newtonian methods are destructive and cannot lead "zur Wahrheit" (to truth).[22] Goethe critiques the isolated and static quality of a Newtonian uni-

---

20 Frank Schweitzer, "Naturwissenschaft und Selbsterkenntnis," in *Goethe und die Verzeitlichung der Natur*, ed. Peter Matussek (Munich: Beck, 1998), pp.383-398 (p.390).

21 John A. McCarthy, "'The Pregnant Point': Goethe on Complexity, Interdisciplinarity, and Emergence," in Herbert Rowland, ed., *Goethe, Chaos, and Complexity* (Amsterdam: Rodopi, 2001), pp.17-31 (p.26).

22 Michael Böhler writes: "Es ist nun diese der wissenschaftlichen Methodik eigene Trennung von Mensch und Natur sowie der Isolation des Subjekts und des Intellekts von ihrem ursprünglichen Zusammenhang, es ist diese abstrahierende ‚Scheidekunst', welche Goethe für höchst verderblich hält. Denn auch wenn sie nur instrumentellen Charakter haben sollte, auch wenn sie nur Mittel zum Zweck der Wissensvermehrung wäre--: für Goethe bedeutet das Verfahren einen verhängnisvollen künstlichen Bruch

verse, devoid of becoming and without human impact on the process of observation.

If Newton sees the process of perception as final and complete, it is, in contrast, only so for a brief moment in Goethe's vision. In fact, Goethe sees eyes as themselves creative. The ability to become, develop, and create is distributed among colors, eyes, colored things, lights, and human beings. This is not to assign agency randomly among objects and organs, nor to assert that we determine the full scope of creativity. Instead Goethe asserts a co-creativity of humans and nature. The eyes and the thing perceived participate in exchanges based on fundamental relatedness as interactive processes. Goethe writes of the eyes' creativity: "Dieses Organ [das Auge] ist immer in der Disposition, selbst Farben hervorzubringen, und genießt einer angenehmen Empfindung, wenn etwas der eignen Natur Gemäßes ihm von außen gebracht wird" (This organ (the eye) is always capable of bringing forth colors itself, and it enjoys a pleasant sensation when something compatible with its own nature is presented to itself.) (FL, pp.247-248; translation mine). The eyes and the colors, in other words, are fully interrelated; we do not come to nature as an outsider but rather as part of the natural systems, or as Böhme states, with a "Teilnehmerperspektive" (participatory perspective). Sepper comments on the interactive process of perception typical to Goethe's science, and notes that it is fully based on this sense of relationships and coordination with the phenomena emerging as part of the visual process itself. Hence the Goethean process is predominantly experiential (that is, empirical), not theoretical. Goethe, he notes, understands truth as residing "less in propositions than in experience, less in statements about nature than in the adequacy of one's approach to it" (Sepper, p.184). Goethe strives for an "adherence to a way (method) that corresponds to and amplifies the encounter of human beings with nature" (Sepper, p.184).

In that we are now working within an acknowledgement of the Anthropocene, and the highly "amplified" encounter of human beings and the rest of the world, this attention to how human beings correspond to nature – that is, how we resonate with and against our vibrant surroundings –

---

im Zusammenhang zwischen Mensch und Natur." (It is this separation typical to the scientific methodology that divides human from nature and that isolates the subject and the intellect from its original context, it is this abstracting 'art of division,' which Goethe considers highly destructive. In that even when it supposedly has only an instrumental characteristic, or only serves as means to an end for the goal of increasing knowledge, this process means for Goethe a fateful, artificial break of the connection between humans and nature.) "Naturwissenschaft und Dichtung bei Goethe," in *Goethe im Kontext: Kunst und Humanität, Naturwissenschaft und Politik von der Aufklärung bis zur Restauration*, ed. Wolfgang Wittkowski, (Tübingen: Niemeyer, 1984), pp.313-339 (p.333).

is all the more relevant. Indeed, Sepper's phrase of "amplifying the encounter" takes on a new, more disturbing meaning in this context. The more we achieve "correspondence" and so understand our enmeshment in the world, the more, it seems, that we also attempt to "amplify" energy flows and gradients in ways that benefit us in the short term. Our very success is an amplification that cannot continue without limits, ecologically, and without exponential impact on the biosphere. Goethe's insights are that we "amplify" from within instead of "mastering from without." This may not change our behavior but it certainly helps us understand it.

The human agent can attain a high level of accuracy in perception and assessment of the colors and natural world by acknowledging the creativity and interactivity of the things around her. This "accuracy" pertains to perception for Goethe, but not only there. It is, in fact, his formula for great art: to see the world as it develops and "becomes" and to seek to replicate and converge one's own genius with respect to "nature's" creative acts. This strategy holds both for the individual education and for the artist. Self-restraint is not so much a monastic exercise of deprivation as it is an ability to merge creative energies with the surrounding forces and patterns, and to benefit from their wisdom rather than only to impose ourselves onto the world: whether our surroundings be composed of fellow thinkers, nature's creativity, or matter's basic forms in plant growth, the shimmering colors, or the flows of water and the movements of the earth. As Frederick Amrine notes, this process of creative perception/interaction is as much about the metamorphosis of things as it is the "metamorphosis of the scientist."[23] Above all, one must attempt to line up one's vision with the "natural order," which agentically emerges if one only will allow it to do so instead of imposing oneself blindly onto the world (as Newton does, according to Goethe's vituperative critiques). Goethe stresses repeatedly that one must carry out scientific and poetic studies in a natural order. Newton fails, because he carries out "seine Versuche nicht in einer natürlichen Ordnung, sondern auf eine künstlich verschränkte Weise" (his experiements not in a natural order, but rather an aritificially limited manner) (FL, p.422; translation mine). This natural order is a key to his work: the sequence of descriptions and development of ideas is central.

Because of Goethe's typical emphasis on "natural order," the structural sequence of the *Farbenlehre* is itself of great significance. Goethe divides his treatise into three parts: the explanation of colors themselves; the critique of Newton's optics; and the historical development of optics and color studies.

---

23 Frederick Amrine, "The Metamorphosis of the Scientist," in *Goethe's Way of Science: A Phenomenology of Nature*, ed. by David Seamon and Arthur Zajonc (New York: State U of New York P, 1998), pp.33-54.

Each section contextualizes the other. In the first section on the colors, he also has three parts: the physiological, or colors belonging to the "eye": the physical colors relating to the interaction of light on the surface of an object such as shimmering mother of pearl; and the chemical colors deriving from the color-absorbing matter of the objects seen. Goethe begins therefore with the eye (or human brain, as we now realize) engaging with light, then moves to interaction of light with surfaces and objects, and closes with unchanging colors of objects as they appear when light shines on them. Each step is part of our engagement among eyes, light, and/or objects rather than of things themselves in isolation.

Above all, Goethean optics are based on relations and interactions: "In der ganzen sinnlichen Welt kommt alles überhaupt auf das Verhältnis der Gegenstände untereinander an, vorzüglich aber auf das Verhältnis des bedeutendsten irdischen Gegenstandes, des Menschen, zu den übrigen" (In the entire sensory world, everything depends entirely on the relationship of the objects among each other, particularly however on the relationship of the most meaningful of earthly objects, the human being, to the others.) (FL, p.83; translation mine.). Repeatedly, he stresses the interactions and "relations," or the "Beziehungen": "Immer bleibt es aber auch hier die Hauptsache, daß die *Beziehungen* wahrhaft *eingesehen* werden" (The main issue here is always that the *relations* are *truthfully* seen) (FL, p.182). Here is where he draws the major distinction between his own optics and those of Newton who sees only colors and not light acting in space with objects, not to mention the actual process of visual perception.

I am not arguing that Newton was wrong, but rather that his view, as part of the mechanistic divide between human beings and the environment, is itself in need of context, at least in terms of the new materialisms and ecological thinking. In this sense, Goethe's somewhat (in)accurate yet revolutionarily contextualizing optical studies are worth recalling. For Goethe, Newton's inward focus eliminates the equally significant interactions and relations that place humanity – the observers – in the polarity of light and dark. Similarly, light is always contextualized in relation to the surroundings and not a stand-alone entity. Goethean science, in sum, posits exchanges and polarities as the basis of the physical world, and we as perceiving beings in the realm of light are fully within these processes. There is some continuation of traditional humanism in the sense that Goethe understands human beings as the "most meaningful earthly object" ("der bedeutendste irdische Gegenstand"), and yet Goethe maintains an emphasis on our perception and awareness of reality in terms of light, colors, and the other aspects of the physical world. Our agency is not isolated from the workings of solar energy, in other words. Goethe's entire optical opus rests upon the assertion that only in a relational understanding of interactions does reality emerge. His polarity is not passive objects manipulated and seen by active subjects but rather active matter across the spectrum. In terms of radical reality, this shifts the emphasis from the human consciousness to the interactive exchanges of many forces cohering into forms – diffraction patterns, waves, and spirals – of distributed agency.

Goethe thus provides us with an access point for understanding both the changing ideas of modernity at the beginning of the Anthropocene in terms of the human-material interfaces, and a bridge to the contemporary discussions in the new materialisms. Not just vitalistic or animistic musings, Goethe's science (re)animates the world with both human and non-human agencies. His interdisciplinary views based on life-long scientific interests and towering literary accomplishments provide a bridge to the new materialisms and their similarly cross-disciplinary notion of vibrant nature. I utilize the study of "radical reality" to help us rethink what exactly we mean with "environment" and reality broadly. In particular, the study of the era 1770-1830 is a study of the early Anthropocene: the very era when science disassociates human beings from the material world while simultaneously enabling the much more rapid impact on the Earth's ecological systems with the industrial revolution and the emerging age of fossil fuels. For this reason, my focus on ecocriticism and Goethe does not favor scenic landscapes and charismatic mega-fauna (or their devastation, or even environmental justice) but rather emphasizes the effort to disturb what we take to be our "human" and separate foundation from the rest of the planet. The task is to study how our bodies, minds, and cultures are imbricated into material systems and vice versa. Perhaps paradoxically, the more knowledge we gain about our utter dependency on ecological functioning, the more we assert that we can technologically transcend in our capacity as "master" who functions outside the system. Tuttle's assertion that we can "transcend" our biological body is troubling in this context. It is therefore an important exercise to imagine and explore both narratives and scientific studies that document just how fully we are enmeshed *within* the solar-powered biosphere and part of the light-life-energy-matter cycles therein. That optics and color studies – especially the role of light – can thus be "ecological" is particularly relevant in the age of fossil fuels now known as the Anthropocene.

In conclusion, it must be said that Goethe is not – avante la letter – a strictly "green" or environmental writer per se. Many of his texts, like *The Sorrows of Young Werther*, do include what we now term nature writing that express clear enthusiasm for nature's wonders. But nature writing itself does not necessarily prompt a re-thinking of humanity's relation to ecological systems. Indeed, Goethe's position at the brink of the Anthropocene meant that he could only partly see the extent of coming changes. Nevertheless, his *Farbenlehre* clearly takes a stance immersing us materially and physically in our world of light in a way that resonates with the new materialists' efforts to reshape our understanding of agency as a joint event of both the human and non-human. This Goethe does from "within" the world, as Böhme's term "Teilnehmerperspektive" suggests. Analogous to the new materialisms, Goethe's optical studies prompt us to look beyond the sublime scenery, interesting animal species, and water-air-soil nexus apparent in contemporary ecology studies. Thus he offers convenient access to understanding the early Anthropocene. With the goal of making visible a broad spectrum of the diffracted and distributed agencies actually in play, this essays offers an alternative to a concept of radical reality based solely on human consciousness. The radical reality of the Anthropocene consists of the many proliferating human and non-human agentic capacities – caused, in part, by the release of fossil fuels into the atmosphere – that radiate around the planet. We are inextricably bound up with that radically creative reality.

Lebenskraft, Radical Reality, and Occidental Medicine: How Science is Leading us back to a Holistic View

Monica Ledoux

Abstract: Analyzing a human being through her/his constituent parts, physiology has led to significant breakthroughs in research. However, this approach is reaching its limits as one becomes ever more aware that essential holistic factors such as multicausal illnesses, the importance of the microbiome, and psychological factors are not addressed. A holistic approach to medicine reemerged in the nineteenth century under different guises, such as homeopathy, and with varying degrees of scientific grounding. More recent initiatives, such as translational microbiome research, multidisciplinary consultations and a new focus on patient/doctor communication, are now gaining momentum, boosted by new scientific discoveries. It is likely that a more holistic approach to studies of human well being in twenty-first century medicine will offer opportunities for better understanding how the body responds to its environment.[1]

The Focus on Anatomy

In Renaissance Europe Andreas Vesalius (1514-1564), a professor at the University of Padua, became convinced that autopsy was a necessary exercise for the genuine understanding of human anatomy. He considered empirical, direct observation to be the only reliable method of investigation, which represented a dramatic break with previous medical practice. His dissecting of cadavers introduced a period of rapid advances in knowledge of the anatomy and development of surgical procedures. The introduction of the microscope in laboratory techniques in 1674 by Anton van Leeuwenhoek (1632-1723) allowed for enhanced knowledge of ever-smaller structures such as blood corpuscles, lymph cells, nerve cells, and bacteria. These advances eventually led to such subspecialties as hematology, immunology, neurology, and microbiology. Moreover, scientific progress led to classifying diseases in medical textbooks. Diseases were listed as oncologic, inflammatory, infectious, and genetic and were assigned specific causes. Since then, medical science has progressed in all domains by assessing a medical problem according to the affected areas. This approach allowed for significant knowledge gains about human anatomy and physiology, until the amount of knowledge grew so large that medicine itself had to be compartmentalized into more manageable specialties: cardiology, hepatology, dermatology,

1 Please note that this chapter retains the medical science practice of listing source citations as references to the numbered items in the works cited included at the end of this chapter.

urology, ophthalmology, etc. with doctors taking care of a single organ: heart, liver, skin, kidneys, or eyes.

Despite all the stunning advantages of scientific advances for patient treatment, one downside became visible: while doctors concentrated on facts concerning the particular organ they were treating, they ran the risk of loosing sight of the patient as a person with his/her individual way of functioning and feeling. In the past, this led to such absurdities as caretakers in hospitals referring to a hospitalized patient as "the liver in room 5" because they could not remember his name or professors discussing a patient's chart in a hospital room with interns without realizing that the concerned person was not in his bed.

Renewed Holism

In the last third of the twentieth century, signs of a movement toward a holistic understanding of the multifarious causes of an illness began to reappear in papers presented at scientific meetings. One new approach, for example, was to view an illness as being simultaneously oncologic and genetic, inflammatory and psychosomatic. Researchers theorized, moreover, that in certain diseases more than two causal factors or triggers might be involved, as in the case of eczema which is inflammatory and often has a genetic predisposition (e.g., filaggrin gene) that flares up with stress and thus has psychosomatic symptoms too. Moreover many patients can react to environmental factors as well. Clearly, human organisms are complex and not reducible to their constituent parts.

In medical textbooks, disease continued to be classified as infectious, oncologic, and autoimmune, along with a notation of their reductionist cause. In more and more chapters a second or third causal factor was mentioned. What if we were to move away from a single organ/single factor disease toward a view of human beings existing in a somewhat fragile equilibrium determined by a multitude of internal and external factors? Increasingly, medical science is beginning to move away from a focus on a single organ as well as the simplified search for a single factor that leads to a specific disease and is returning to a more holistic view of the human body as an assemblage of interconnected systems. The reductionist search for a specific cause of a disease potentially blinds the researcher to the nature of the interactive factors involved in the appearance of a "malfunction."

As noted, seventeenth-century Dutchman Anton van Leeuwenhoek perfected the microscope that allowed the discovery of microorganisms. Since then, science has gained insight into a multitude of microorganisms along a broad scale, the best-known ones being bacteria, fungi, and viruses. They are largely considered infectious and are studied with the goal of finding treatments. Historically, we have cultivated them, which allowed us only a glimpse of microbial reality. However, microbes do not all multiply in lab conditions and even when they do, many are outcompeted by faster growing organisms. With standard laboratory bacterial cultures, we can scarcely culture 1% of bacterial species present on our body.[7, 29] However, a recent technological advance has revolutionized what we know about microorganisms and their relationship to the total human body: gene sequencing. Until recently we needed to culture bacteria, and the success of finding even one species would largely depend on the choice of the culture medium; we now can obtain a mapping of our microbial landscape. Sequencing not only detects a large variety of bacteria, but also quantifies how many of which species are present at a certain location in the body.

The insights garnered from metagenomics are impressive. Our body, which might have seemed an agglomerate of human cells occasionally attacked by a microorganism, is in fact largely outnumbered by microorganisms on the ratio of 1:10. Most of these microorganisms live in perfect harmony together. The ensemble of all these microorganisms populating our body is called the microbiome. Every organ system seems to have its particular composition of different microorganisms, its own microbiome. The microbiome's original definition describes it as an "ecological community of commensal, symbiotic, and pathogenic microorganisms that literally share our body space."[19]

How important is this symbiosis to us? Do these microorganisms share our body space without us actually needing them? Initial studies point in the direction of a deep interconnectedness. These microorganisms are not sprinkled on us like some bothersome dust, but literally share the space with our human cells, for example in different layers of our skin. The results of our effort to eradicate them were revealed in tests on mice: when raised in a germ free lab, the test mice developed a disturbed immune system (e.g, defective gut-associated lymphoid tissue and mesenteric lymph nodes, reduced epithelial expression of immune molecules, and improper T-cell differentiation.)[5, 9, 20, 21] Yet despite such advances in research, the generally accepted conceptual framework in medical science has not experienced significant change.

Earlier Holistic Approaches

In long nineteenth-century Germany, one alternative voice was that of Samuel Hahnemann (1755-1843), the founder of homeopathy. For him, *Lebenskraft* (vital force) was an essential notion. Thus he writes in paragraph 9 of his book *Organon der Heilkunst* (Organon of the Art of Healing):

> Im gesunden Zustande des Menschen waltet die geistartige, als Dynamis den materiellen Körper (Organism) belebende Lebenskraft (Autocratie) unumschränkt und hält alle seine Teile in bewundernswürdig harmonischem Lebensgange in Gefühlen und Tätigkeiten, so daß unser inwohnende, vernünftige Geist sich dieses lebendigen, gesunden Werkzeugs frei zu dem höhern Zwecke unsers Daseins bedienen kann.[11]
>
> (In the healthy condition of man, the spirit-like vital force (autocracy), the dynamis that animates the material body (organism), rules with unbounded sway and sustains all the parts of the organism in admirable, harmonious, vital operation, as regards both sensations and functions, so that our indwelling, rational mind can freely employ this living, healthy instrument for the higher purpose of our existence.)[12]

Hahnemann, who graduated with a medical degree with honors, was disturbed by the deleterious side effects of eighteenth-century classical medicine with its frequent bloodletting and high incidence of post-surgical infections. He believed many patients stood a better chance of survival if they did not have too close contact with doctors. As a result he developed alternative treatment principles. His central postulate, *similia similibus curentur*, stated that "to cure mildly, rapidly, certainly, and permanently, choose, in every case of disease, a medicine which can itself produce an affection similar to that thought to be cured!"[12] In order to find these medications he adopted newer scientific methods of his time, turned to chemical analysis, experimented with different procedures and documented their effects on healthy volunteers, including himself. When he observed symptoms similar to a disease after intake of a composite, he concluded that this would cure the disease, even more so in a highly diluted form. For the description of symptoms he insisted on a holistic approach, writing: "In these investigations, the ascertainable physical constitution of the patient (especially when the disease is chronic), his moral and intellectual character, his occupation, mode of living and habits, his social and domestic relations, his age, sexual function, etc., are to be taken into consideration." These symptoms were the expression of *Lebenskraft*: "it is only the vital force, deranged to such an abnormal state, that can furnish the organism with its disagreeable sensations, and incline it to the irregular processes which we call disease."[12] While his followers seem to have largely abandoned the search for scientific proof in the two hundred years that followed, they clearly maintained the importance of the concept of *Lebenskraft*, reflecting it in the typical homeopathic patient-caregiver interview, which the many different homeopathic schools and approaches existing today have in common.

There is still no evidence-based explanation to affirm the effectiveness of Hahnemann's medical theory. No firm evidence exists for claiming that his globuli (homeopathic remedies) had superior results over a placebo. On the other hand, his holistic approach in patient interviews is largely appreciated by patients today. Studies of doctor/patient communication indicate a clear influence of inclusive communication on the patient's health outcome. Thus, it supports the value of a holistic approach and suggests an intimate mind-body connection. As for the lack of proof regarding the advantage of globuli over placebos, the lack of verifiable evidence does not mean that they are without effect, as placebos have been proven to lead to surprising, positive results in health outcomes. Because those results are often attributed to non-specific factors, one can again argue in favor of an intimate mind/body connection.

Thus, while homeopathy could not demonstrate superior, specific results for homeopathic remedies over placebos,[18, 27] the holistic approach of homeopathic caregivers is generally appreciated by patients. Or as David Reilly, the head of a homeopathic hospital in Glasgow, puts it in a 2005 interview, the therapeutic relationship is not simply an interesting aside in the history of healing; it is seminal: "I again would say what complementary medicine has been screaming out over these last two decades is the fact that these holistic factors of engagement are critically important, are of central importance."[28] Studies of patient/caregiver communication and so-called context effects confirm the positive influence of doctors' and patients' quality of interaction on health outcomes.[8, 22, 31] Hence, they strongly suggest the influence of the mind on the body and, consequently, the advantage of the homeopathic interviewer's approach.

While homeopaths continue to try to prove their globulis' superior effectiveness over placebo, the latter are not without effect. Placebo studies also point out the importance of non-specific factors. For example, patients who underwent thorax surgery in connection with lung cancer needed significantly less pain medication (34% less) after their surgery, when they received an intravenous saline solution and were told it was a potent painkiller compared to patients with the same infusion who were told nothing.[25] Thus, the patient's expectations influence analgesia. The list of studies showing how the mind influences the body is long, going as far as describing a better therapeutic outcome if the caregiver believes in a placebo's success.

Aware of what patient-caregiver-communication can contribute in a time when many diseases like obesity, diabetes, and heart attacks are linked to lifestyle, Dr. Reilly teaches others how to transform the therapeutic relationship from one that treats problems with beta-blockers, antidepressants, antibiotics, antiarrhythmics, antifungals etc. to one that activates the patient's own internal healing capacities that are linked to something akin to the notion of *Lebenskraft*. The absence of scientific proof for such a linkage has not prevented homeopathic clinics and university departments for alternative medicine from blossoming in many countries such as Germany and Switzerland despite strong opposition in other countries such as Denmark. In contrast to Europe, where heated discussions about the relative value of empirically tested medical science versus patient-centered, non-invasive therapies take place, in Chinese medicine and Indian Ayurveda both approaches coexist harmoniously. More and more homeopathic hospitals accept a broad range of treatment, by combining classical and homeopathic medicine.

Another example of a holistic voice was Hahnemann's friend, Christoph Wilhelm Hufeland (1762-1836), founder of macrobiotics. In his main work on the art of prolonging human life, eating habits play an important part. He describes in detail the quantity and quality of recommended food intake, such as avoiding meat, big quantities, and quick and irregular meals.[13] Many of his recommendations are still found in modern macrobiotics, which, similar to homeopathy, could not be verified by evidence-based scientific methods. Nonetheless, his holistic approach did produce positive results. Because he considered *Lebenskraft* to be the central force in an interconnected network of moral, psychological, nutritional, environmental and socioeconomic factors, health was a measure of *Lebenskraft*, and the best doctor could not help if external factors like work conditions or poverty were not rectified. Thus, he cared for the poor pro bono, wrote books about how to treat them without making things worse, and helped establish Berlin's first charity hospital. He accomplished all this while also serving as personal physician to the Prussian king and teaching at the University of Berlin. He used his considerable influence to push for national hygiene laws, advocated health policies for schools, and proposed a social health insurance network that formed the basis of Bismarck's public health system seventy years later, in 1871. However intuitive Hufeland's interconnected patient approach was, its benefits seem clear.

Radical Reality and the Microbiome

About 100 years later in 1964, Bell's theorem proved mathematically that certain quantum correlations, unlike all other correlations in the universe, cannot arise from any local cause. He thus introduced a radically new reality of quantum entanglement that posited interconnectedness on the nano scale.[32] Moreover, in the late twentieth century, gene-sequencing technology also highlighted interconnectivity as a model for medical science and microbiology. The human body, already recognized as a complex set of organisms and structures, became even more complexly interrelated with the advent of insight into the microbiome. The human body is deeply entangled with billions of bacteria that prove instrumental in maintaining human health. Recent research provides growing evidence on how the symbiosis of the human and the microbiome is vital to us. Indeed, as noted above, our bodies prove to be just 10% human and 90% bacterial in terms of the number of cells. Researchers point to the role that a disturbed, unbalanced gut microbiome plays in the development of diseases of immunological deregulation[3, 4, 16, 17, 24] and demonstrate that certain bacteria are active antitumor agents in combatting bladder and colon cancers.[1, 10, 15]

Are all these microorganisms really important, or do only some of them, the most frequently occurring ones, play a role in maintaining our health? For example, considering the skin microbiome, one finds three abundant species: Propionibacterium, Staphylococcus, and Corynebacterium. Together with these abundant species, one discerns many families of bacteria that each make up less than 1% of the total flora of that particular location. Let us call them minority species. They are not well studied and were not previously known to colonize the skin. Nevertheless, they could be linchpins of the skin ecosystem, just like metagenomic studies of soil ecosystems have shown that several low-abundance fungal species are actually highly active in essential decomposition processes.[2, 7] In the case of eczema and wound healing, the interaction between bacterial species and our body again underscores the complex interactive and fragile network of smaller complex units.

What do these new scientific insights mean for the traditional classification of infectious diseases? As we discover that infectiousness is linked to genetics, the traditional scheme for classifying infectious diseases will have to be revised. For example, tuberculosis is now viewed to be genetically determined, because some people seem more protected than others.[14] Similar insights have been made into the interaction of certain viruses in oncogenesis, the induction of tumors, such as the human papilloma virus (HPV) known to trigger cervical cancer. HHV-8, a herpes virus can cause Kaposi's sarcoma, a systemic disease with cutaneous lesions, affecting middle aged men of Mediterranean descent, Africans, immunosuppressed and AIDS patients. Kaposi's violaceous skin lesions became famous through Tom Hanks in the film *Philadelphia*. Similarly, HHV-8 also causes the rare primary effusion lymphoma and Castleman's disease, equally characterized by an overgrowth of cells in the lymphatic system.[6, 23, 26, 30] Being a recent development, gene technology is just beginning to help us understand to what degree our microbiome influences the entire interactive organ, nerve, and skeletal systems to maintain a healthy balance in the human body. In the twentieth century, classical medicine more or less timidly introduced genetic or psychological cofactors as causing certain pathological symptoms. After scarcely twenty years of microbiome research we can already appreciate the human body as a fragile, symbiotically interconnected network, in which a disequilibrium of one small, complex unit impacts the others in ways we are just beginning to fathom.

Continued advances in understanding the microbiome will undoubtedly induce us to view the patient more emphatically as situated in an equilibrium of multiple ecosystems. To reiterate, these systems are not only internal to the human body but also external and include social psychology as well as the physical environment. Social and physical parameters impact the human body more than previously thought. All internal and external factors taken together should prompt us to rethink our way of looking at the individual patient and her/his symptoms. Thus, caregiver and patient will need to communicate better in order to gain a full picture of physical and mental health. This more comprehensive approach must occur despite the ever- increasing time constraints imposed in many places by the healthcare system on healthcare providers. Increased communication between doctors treating the same patient and multidisciplinary care structures will bring significant improvements.

As the center of diagnosis and treatment, the patient must be the main source of information and be a partner in therapeutic decisions. Evidence is mounting that the recentering of the patient within a complex system of interaction has already begun to optimize healthcare outcomes. Such is the case, for example, with probiotic treatments for allergies in early childhood and in pregnant woman. Ultimately, caregivers will see disease classifications change to accommodate new treatment options that focus attention on reestablishing equilibrium within the body and between the body and the wider environment. While none of this is tied directly to the historical *Lebenskraft* debate, the emphasis on enhanced vitality in the ailing subject is part and parcel of the vital force necessary to the quality of life.

Works Cited

1. H. Akaza, A. Iwasaki, M. Ohtani, N. Ikeda, K. Niijima, I. Toida, and K. Koiso, "Expression of Antitumor Response. Role of Attachment and Viability of Bacillus Calmette-Guerin to Bladder Cancer Cells," *Cancer,* 72 (1993), 558-63.
2. P.Baldrian, M. Kolarik, M. Stursova, J. Kopecky, V. Valaskova, T. Vetrovsky, L. Zifcakova, J. Snajdr, J. Ridl, C. Vlcek, and J. Voriskova, "Active and Total Microbial Communities in Forest Soil Are Largely Different and Highly Stratified During Decomposition," *Isme j,* 6 (2012), 248-58.
3. B. Bjorksten, P.Naaber, E. Sepp, and M. Mikelsaar, "The Intestinal Microflora in Allergic Estonian and Swedish 2-Year-Old Children," *Clin Exp Allergy,* 29 (1999), 342-6.

4. B. Bjorksten, E. Sepp, K. Julge, T. Voor, and M. Mikelsaar, "Allergy Development and the Intestinal Microflora During the First Year of Life," *J Allergy Clin Immunol,* 108 (2001), 516-20.
5. D. Bouskra, C. Brezillon, M. Berard, C. Werts, R. Varona, I. G. Boneca, and G. Eberl, "Lymphoid Tissue Genesis Induced by Commensals through Nod1 Regulates Intestinal Homeostasis," *Nature,* 456 (2008), 507-10.
6. E. Cesarman, Y. Chang, P.S. Moore, J. W. Said, and D. M. Knowles, "Kaposi's Sarcoma-Associated Herpesvirus-Like DNA Sequences in Aids-Related Body-Cavity-Based Lymphomas," *N Engl J Med,* 332 (1995), 1186-91.
7. Y. E. Chen, and H. Tsao, "The Skin Microbiome: Current Perspectives and Future Challenges," *J Am Acad Dermatol,* 69 (2013), 143-55.
8. Z. Di Blasi, E. Harkness, E. Ernst, A. Georgiou, and J. Kleijnen, "Influence of Context Effects on Health Outcomes: A Systematic Review," *Lancet,* 357 (2001), 757-62.
9. P.G. Falk, L. V. Hooper, T. Midtvedt, and J. I. Gordon, "Creating and Maintaining the Gastrointestinal Ecosystem: What We Know and Need to Know from Gnotobiology," *Microbiol Mol Biol Rev,* 62 (1998), 1157-70.
10. J. M. Grange, O. Bottasso, C. A. Stanford, and J. L. Stanford, "The Use of Mycobacterial Adjuvant-Based Agents for Immunotherapy of Cancer," *Vaccine,* 26 (2008), 4984-90.
11. Samuel Hahnemann, *Organon Der Heilkunst* (Heidelberg: Karl F. Haug Verlag, 1992), p.105.
12. Samuel Hahnemann, *Organon of Medicine.* Fifth Edition (Philadelphia: Boericke and Tafel, 1901), pp.38, 49, 53, 89
13. Christoph Wilhelm, Hufeland, *Makrobiotik Oder Die Kunst Das Menschliche Leben Zu Verlaengern* (Stuttgart: U. F. Macklot, 1826), pp.7, 8, 396-425, 59-71.
14. A. Jepson, A. Fowler, W. Banya, M. Singh, S. Bennett, H. Whittle, and A. V. Hill, "Genetic Regulation of Acquired Immune Responses to Antigens of Mycobacterium Tuberculosis: A Study of Twins in West Africa," *Infect Immun,* 69 (2001), 3989-94.
15. D. A. Juckett, C. F. Aylsworth, and J. M. Quensen, "Intestinal Protozoa Are Hypothesized to Stimulate Immunosurveillance against Colon Cancer," *Med Hypotheses,* 71 (2008), 104-10.
16. M. Kalliomaki, S. Salminen, H. Arvilommi, P.Kero, P.Koskinen, and E. Isolauri, "Probiotics in Primary Prevention of Atopic Disease: A Randomised Placebo-Controlled Trial," *Lancet,* 357 (2001), 1076-9.
17. M. Kalliomaki, S. Salminen, T. Poussa, H. Arvilommi, and E. Isolauri, "Probiotics and Prevention of Atopic Disease: 4-Year Follow-up of a Randomised Placebo-Controlled Trial," *Lancet,* 361 (2003), 1869-71.

18. J. Kleijnen, A. J. de Craen, J. van Everdingen, and L. Krol, "Placebo Effect in Double-Blind Clinical Trials: A Review of Interactions with Medications," *Lancet,* 344 (1994), 1347-9.
19. Joshua Lederberg, McCray, Alexa T., "Ome Sweet 'Omics-- a Genealogical Treasury of Words," *The Scientist* 2001, 15(7):8.
20. A. Lundin, C. M. Bok, L. Aronsson, B. Bjorkholm, J. A. Gustafsson, S. Pott, V. Arulampalam, M. Hibberd, J. Rafter, and S. Pettersson, "Gut Flora, Toll-Like Receptors and Nuclear Receptors: A Tripartite Communication That Tunes Innate Immunity in Large Intestine," *Cell Microbiol,* 10 (2008), 1093-103.
21. A. J. Macpherson, and N. L. Harris, "Interactions between Commensal Intestinal Bacteria and the Immune System," *Nat Rev Immunol,* 4 (2004), 478-85.
22. S. Meryn, "Improving Doctor-Patient Communication. Not an Option, but a Necessity," *Bmj,* 316 (1998), 1922.
23. N. Munoz, F. X. Bosch, X. Castellsague, M. Diaz, S. de Sanjose, D. Hammouda, K. V. Shah, and C. J. Meijer, "Against Which Human Papillomavirus Types Shall We Vaccinate and Screen? The International Perspective," *Int J Cancer,* 111 (2004), 278-85.
24. L. Nylund, R. Satokari, J. Nikkila, M. Rajilic-Stojanovic, M. Kalliomaki, E. Isolauri, S. Salminen, and W. M. de Vos, "Microarray Analysis Reveals Marked Intestinal Microbiota Aberrancy in Infants Having Eczema Compared to Healthy Children in at-Risk for Atopic Disease," *BMC Microbiol,* 13 (2013), 12.
25. A. Pollo, M. Amanzio, A. Arslanian, C. Casadio, G. Maggi, and F. Benedetti, "Response Expectancies in Placebo Analgesia and Their Clinical Relevance," *Pain,* 93 (2001), 77-84.
26. M. Schiffman, P.E. Castle, J. Jeronimo, A. C. Rodriguez, and S. Wacholder, "Human Papillomavirus and Cervical Cancer," *Lancet,* 370 (2007), 890-907.
27. A. Shang, K. Huwiler-Muntener, L. Nartey, P.Juni, S. Dorig, J. A. Sterne, D. Pewsner, and M. Egger, "Are the Clinical Effects of Homoeopathy Placebo Effects? Comparative Study of Placebo-Controlled Trials of Homoeopathy and Allopathy," *Lancet,* 366 (2005), 726-32.
28. D. Spence, E. Thompson, D. Reilly, J. Mawer, and C. Ellis, "Homeopathy," in *Case notes Programme 8* ed. by M. Porter (radio science unite, radio 4: BBC, 2005).
29. J. T. Staley, and A. Konopka, "Measurement of in Situ Activities of Nonphotosynthetic Microorganisms in Aquatic and Terrestrial Habitats," *Annu Rev Microbiol,* 39 (1985), 321-46.
30. M. R. Staudt, Y. Kanan, J. H. Jeong, J. F. Papin, R. Hines-Boykin, and D. P.Dittmer, "The Tumor Microenvironment Controls Primary Effusion Lymphoma Growth in Vivo," *Cancer Res,* 64 (2004), 4790-9.
31. M. A. Stewart, "Effective Physician-Patient Communication and Health Outcomes: A Review," *Cmaj,* 152 (1995), 1423-33.
32. H. Wiseman, "Physics: Bell's Theorem Still Reverberates," *Nature,* 510 (2014), 467-9.

Epilogue: *Lebenskraft* Legacies

John A. McCarthy

> It is a pleasurable undertaking to examine nature and self simultaneously without doing violence either to nature or to the human spirit, but rather by bringing the two into balance via a gentle exchange of their reciprocal influence (Goethe, *Maxims and Reflections*).[1]

A theory is merely a way of organizing ideas that seem to make sense of the world. In fact, Goethe had defined theory as being nigh synonymous with careful scrutiny (*anschauen*).[2] "Kein Phänomen," he declared apodeictically in an unpublished maxim, "erklärt sich an und aus sich selbst; nur viele, zusammen überschaut, methodisch geordnet, geben zuletzt etwas, das für Theorie gelten könnte."[3] (No phenomenon is explainable in and through itself alone; only when many phenomena are considered conjointly and methodically ordered can they ultimately lead to something that might be considered theory.) *Lebenskraft*, or as early embryologist Caspar Friedrich Wolf (1735-94) termed it in his 1759 dissertation, the *vis essentialis*, was just such a theory. But the term of course also evolved in its use over 150 years.[4] It evolved as a result of empirical observations of the self in conjunction with its environment. Generally, essential force would connote the formative impulse inherent in any organism. Today we might think of it as an epigenetic force analogous to the so-called elective affinities in chemistry or the com-

---

1 Johann Wolfgang von Goethe, *Maximen und Reflexion*, in Goethe, *Werke*, Hamburger Ausgabe in 14 Bänden (München: deutscher Taschenbuch Verlag, 1982), vol. XII, p.434 (#500): "Es ist ein angenehmes Geschäft, die Natur zugleich und sich selbst zu erforschen, weder ihr noch seinem Geiste Gewalt anzutun, sondern beide durch gelinden Wechseleinfluss miteinander ins Gleichgewicht zu setzen." All translations are my own unless otherwise noted.

2 Johann Wolfgang von Goethe, "Vorwort," *Zur Farbenlehre. Didaktischer Teil* (1808), in Goethe, *Werke* (1982), XIII, p.317. Cf. John A. McCarthy, "The 'Pregnant Point': Goethe on Complexity, Interdisciplinarity, and Emergence," in: *Goethe, Chaos, and Complexity*, ed. by Herbert Rowland (Amsterdam and New York: Rodopi, 2001), pp.17-33 (pp.28-29).

3 Goethe, *Maximen und Reflexionen*, XII, p.434 (#500).

4 Caspar Friedrich Wolff, *Theoria generationis. Editio nova augta et emendata* (Halle a.d. Saale: Christian Hendl, 1774), Pars I: De generatione plantarum, §4, p.2. http://www.archive.org/details/theoriagenerati00wolfgoog (accessed 7 Dec. 2014).

plementary attraction and repulsion forces of gravity. To deploy an even more current term, we can construe the *vis essentialis* as an attempt to describe complex emergent processes. "Inherent in a relatively amorphous seminal structure that can mould itself into an organism," the essential force is a deep micro-structure that defies exact empirical determination.[5] As the contributions to this volume strongly suggest, the *vis essentialis* is ultimately rooted in the ancient concept of πνεῦμα (*pneúma*), the vital, unifying force in the universe which is related to Aristotle's view of the soul and body as irreducibly interconnected, whereby the soul becomes manifest only in the activity of the body.

It was not so much that the human body and the world more broadly had changed; rather, the way one thought about the nature of *Lebenskraft* changed with ever deeper insights into internal, electro-chemical workings of the body and its environment. If natural philosophers initially approached scientific inquiry deductively, that is, predisposed to view every thing, creature, and plant as God willing it to be so (also known in some quarters as Intelligent Design), they could only describe what they found in that preordained theological context. If, alternatively, a scientist proceeded strictly according to the empirical-analytic method, dissecting for example the material parts of an animal the better to understand how they combine to form the vital whole just deconstituted, that researcher's inductive method would contribute, perhaps unwittingly, to an understanding of life as an emergent phenomenon. Inductive arguments can lead to new ideas and thus expand our knowledge of the world in a way impossible for deductive arguments that set out to prove rather than test the original hypothesis.

A notable difference in the two approaches, therefore, is that the former deductive example was marked historically by a starting point anchored in spiritual transcendence, whereas the latter modern approach is rooted in material processes themselves and is characterized by a "deep naturalism."[6] Modern science recognizes that the latter is inherently more complex. Emergence – conceived as a process whereby larger systems, patterns, and regularities arise through interactions among smaller or simpler "systems" that themselves do not exhibit such properties (here one might think of interacting micro-organisms, interacting molecules, interconnected neurons) – is intimately related to the so-called deep naturalism that Bell discerned in

---

5 François Duchesneau, "'Essential Force' and 'Formative Force': Models for Epigenesis in the 18th Century," in: *Self-Organization and Emergence in Life Sciences,* ed. Bernard Feltz, Marc Crommelinck, and Philippe Goujon (Dordrcht NL: Springer, 2006), pp.171-86 (pp.172-73).

6 Cf. Matthew Bell, *Goethe's Naturalistic Anthropology: Man and Other Plants* (Oxford: Clarendon Press, 1994), pp.189-205, esp. pp.190, 194.

Goethe's scientific method. The causal relations in emergent phenomena are not simple; they are highly intricate because they can occur across different scales and via multiple, interconnected feedback loops. In regard to this dominant view in contemporary natural science, physicist Richard Haglund notes: "Emergence in physical systems refers precisely to those properties of the solutions [i.e., the mathematical valuation of phenomena] which are not obvious from the equations. [...] [T]his emergent perspective would be substantially more consistent with Goethe's view of both Nature and the human role of the observer."[7] Hence, we can conclude that emergent structures – revealed via the inductive method – are more than the sum of their parts.[8]

From this perspective, we might identify three milestones in reconceptualizations of deep naturalism. First, Darwin's *The Origin of Species* (1859). This treatise, as David Sloan Wilson argues in his *Evolution for Everyone* (2007), occasioned an epochal shift. With its principles of variation, adaptation, and selection, it introduced nothing less than a third way of thinking about the relationship between the individual and the whole, humans and their environment.[9] It remains a mainstay of twentieth-century biological science.

Second is Barbara McClintock's (1902-92) pioneering work on gene translocation in maize which revealed that the "hereditary" genetic code was not fixed and inalterable as then presumed. In asking what the phenomenon of the so-called "jumping genes" meant for the corn itself, she ultimately realized that she was asking the wrong question.[10] Perhaps the jumping genes were genetic "parasites" analogous to viruses, albeit at the genetic level? While her work had progressed significantly from the 1930s to the 1950s, her insights were largely ignored. They were so revolutionary that the necessary conceptual framework had not yet been established in the profession.[11]

7 Haglund, Richard F., jr, "Visualization and Emergence in Contemporary Physics," in *Goethe, Chaos, and Complexity*, ed. by Herbert Rowland (Amsterdam and New York: Rodopi, 2001), pp. 57-68 (p.60; see also p.68).

8 Cf. John A. McCarthy, "Emergence, Horizons, Continuance," in: J.A. McCarthy, *Remapping Reality: Chaos and Creativity in Science and Literature* (Amsterdam & New York: Rodopi, 2006), pp.323-35.

9 David Sloan Wilson, *Evolution for Everyone: How Darwin's Theory Can Change the Way we Think About Our Lives* (New York: Delacorte Press, 2007), pp. 17-18.

10 A jumping gene is a fragment of nucleic acid such as a transposon that can become incorporated into the DNA of a cell; it is a transposable element that can change its position within the genome, sometimes causing major changes. Barbara McClintock received the Nobel Prize in 1983 for discovering it.

11 Rudolf Hausmann, *Die Entdeckung des Lebens. Wege und Irrwege großer Forscher* (Darmstadt: Wissenschaftliche Buchgesellschaft, 2009), pp.138-41.

This professional blindness occurred although Darwin's theory of evolution had long been mainstream. Even though Gregor Mendel (1822-84) had applied the Darwinian paradigm to his study of pea plant characteristics and was able to establish many of the rules of heredity (the so-called Mendelian inheritance), the initial reception of his ideas was similarly muted until his rediscovery in the twentieth century. His American rediscoverer, Rollins A. Emerson, was head of the Department of Plant Breeding at Cornell University, where McClintock was a graduate student and subsequently held an appointment as instructor. With Emerson's support, McClintock helped assemble a group of plant breeders and cytologists to pursue projects in the new field of cytogenetics. All this provided the framework for McClintock's significant advances in cytogenetic techniques.[12]

Third, with her symbiotic theory on the origin of organelles critical to life such as mitochondries and chloroplasts, Lynn Margulis (1938-) reshaped the way researchers conceptualized cell biology. Classical genetic theory had taught that genetic information was anchored in the cell nucleus and limited neo-Darwinian evolutionary theory to the mechanism of mutations which impacted the nucleus and to the mutations that the mechanism of selection produced. Margulis thought of these operations differently. She thought not in terms of linear cause and effect, but rather of a symbiotic relationship. The cells of more complex organisms such as eukaryotes, she theorized, did not result from mutation and selection but from integration. Like McClintock before her, Margulis at first encountered polite silence. Her first publication was repeatedly rejected by top-tier professional journals until a lower-tier one finally accepted it in 1967. The logical consequence of Margulis's findings regarding bacterial integration (that also explained, among other things, photosynthesis) is to view all life forms as micro-elements in the macro-system of Mother Earth, the classical Gaia hypothesis.[13]

Indeed, the human genome project itself is just another logical step in the evolution of the search for that which binds things together. No longer viewed as a *vis vitalis* or *vis essentialis*, *Lebenskraft* could in light of the foregoing be recast as an emergent symbiotic relationship of collaborative elements that support one another. On the one hand, it is related to the genetic information contained in living cells. This coded information – and the way it

---

12 Pray, L. & Zhaurova, K. "Barbara McClintock and the discovery of jumping genes (transposons)," *Nature Education*, vol. 1, no. 1 (2008), 169. See also http://www.nobelprize.org/nobel_prizes/medicine/laureates/1983/mcclintock-bio.html (accessed 31 Dec. 2014).

13 Hausmann, pp.142-43.

can be decoded and recoded – is crucial. Hausmann remarks at the conclusion of his survey of momentous advances in the biological sciences:

> Doch folgender Aspekt drängt sich auf: Die fundamentale Einsicht, dass genetisch festgelegte, informationstragende Makromoleküle mit ihren Wechselwirkungen die Triebfeder aller Lebensprozesse sind, ist eine Erkenntnis des letzten Jahrhunderts (Hausmann, pp.151-52).
>
> (Yet the following aspect demands attention: namely, the fundamental insight that genetically determined, information-bearing macromolecules with their inherent interactions are the driving force of all life processes; this is a [major] recognition of the last century.)

Indeed, the microbiome resident in the human body could be seen as additional layers of coding that implant themselves in the host body and interact with the 10% of us that is human. In an expanded sense, then, microbial interactions on the nano level might be seen paradoxically as part and parcel of the Anthropocene.

Moreover, this rather surprising thought extension is a logical application of the historical concept of *Lebenskraft* to contemporary biological science. While one might greet such a thought with skepticism and even incomprehension, challenging accepted paradigms has traditionally met with initial resistance. Some might react like a German friend, a biological chemist by training, who wondered why it would even occur to one to ask: what does it all mean? Why seek a ghost in the machine?[14] Why not simply be satisfied in knowing how the body and its cell structure work and how that knowledge might be applied in altered form? For example, by growing human organs from stem cells? While such an application might in fact be the next great development in biochemistry, investigating the nature of *Lebenskraft* in the twenty-first century goes beyond the practical. If stem-cell research allows us to create "from scratch," then we humans will have replaced the Divine Creator and His finished universe. We will have become in an epigenetic sense our own creator or *alter deus*. This much we know: the human organism evolved from the environment and reacts to it. And now human consciousness is also seemingly caught up in feedback loops with the material basis of life.[15]

---

14 Arthur Koestler, *The Ghost in the Machine* (London & New York: Hutchinson & Macmillan, 1967).

15 Thus the new science of memetics analogous to genetics. See, e.g., Richard Brodie, *Virus of the Mind: The New Science of the Meme* (Carlsbad CA etc.: Hay House, 2009; first published Seattle WA: Integral Press, 1996) and Francis Heylighen & Klaas Chielens, "Cultural Evolution and Memetics," in: *Encyclopedia of Complexity and System Science*, editor-in-chief Robert A. Meyers, 11 vols. (New York & Heidelberg: Springer,

In sum, then, this collection of essays takes its cue from the science of structure that examines relationships among the constituent parts of a whole. Radical reality denotes the complexity of interaction involved in the evolution of life forms and their maintenance in and through the total environment of the earth. While rooted in the ancient concept of Gaia, empirical and rational inquiry only gradually evolved into process science and process philosophy. Process art is the logical extension of the former in reaction to a paradigm of emergence. It is, as we can now aver, the re-enactment of deep naturalism. In commemorating the fiftieth anniversary of the publication of Thomas Kuhn's influential *The Structure of Scientific Revolutions* (1962) – that studied how new paradigms only gradually gained acceptance in scientific circles – science historian Andrew Pickering observed that the history of science argued for the inexhaustibility of the world we inhabit: it is "a Taoist world of endless, decentered, unpredictably emergent flows, transformations and becomings [...] not at all the sort of place that requires us to inhabit one specific world rather than another."[16]

This inexhaustible flux and the potential for multiple organisms to inhabit one and the same space informs in one way or another the literary and aesthetic explorations in these pages of the symbiotic relations between the different "worlds" of inorganic matter, organic body, mind, and soul. *Lebenskraft* has been the rhizomatic red thread running through each domain, connecting them in subtle and nuanced ways. Hence, this project can also be viewed as participating in the discourse of the posthuman. In his reflections on the lineaments of an emergent culture – an eternally emergent culture, it must be stressed, where everything is always immediate and mediated – Ihab Hassan was surely correct to assert: "We need first to understand that the human form – including human desire and all its external representations – may be changing radically, and thus must be revisioned."[17] That radical revisioning entails not only an extension of human consciousness into the cosmological end of the scale but also into the minutest, internal workings of the body and world we inhabit on the nano scale of our investigative gaze.

2009). Accessed online 4 Jan. 2015. http://pespmc1.vub.ac.be/Papers/Memetics-Springer.pdf.

16 Andrew Pickering, "The World Since Kuhn," in: *Social Studies of Science*, vol. 42, no.3 (2012), 467-73 (pp. 468-69).

17 Ihab Hassan, "Prometheus as Performer: Toward a Posthumanist Culture?" in: *The Georgia Review*, vol. 31, No. 4 (Winter 1977), 830-850 (p.843). Cf. Also N. Katherine Hayles, *How we Became Posthuman: Virtual Bodies in Cybernetics, Literature, and Informatics* (Chicago & London: U of Chicago P, 1999), esp. pp. 247-92.

To return to the source of my concluding reflection: in the early nineteenth century, Goethe could only intuit this kind of intimately complex interaction when he penned the maxim cited in the masthead about the gratifying collaboration between subject and object, or when he spoke of a "delicate empiricism," suggesting that deeper understanding of the relationship between subject and object is reserved for a future, more advanced age:

> Es gibt eine zarte Empirie, die sich mit dem Gegenstand innigst identisch macht und dadurch zur eigentlichen Theorie wird. Diese Steigerung des geistigen Vermögens aber gehört einer hochgebildeten Zeit an.[18]
>
> (There is a delicate empiricism which makes itself utterly identical with the object, thereby becoming true theory. But this enhancement of our mental powers belongs to a highly advanced age.)[19]

There are many signs that we are drawing close to, if not already in, that highly advanced age, one marked by enhanced mental powers where one is inclined to view empiricism as actual theory, not just as method. The era increasingly speaks of a "system of importances" in human interactions with the world (Tuttle) and foregrounds performance over cognition (Pickering's "dance of agency"). In any event, the present era invites us to be more aware of the interactions of bodies, environments, minds, and souls in terms of performed radical reality.

---

18 Goethe, *Maximen und Reflexionen*, Hamburger Ausgabe, vol. XII, p.435 (#509).

19 Johnann Wolfgang von Goethe, *Works*, Surhkamp Edition in 12 Volumes, ed. by Victor Lange, Eric A. Blackall, and Cyrus Hamlin, Vol. 12: *Scientific Studies*, ed. and translated by Douglas Miller (New York: Suhrkamp Publishers, 1983), vol. XII, p.307.

Select Bibliography

Select Bibliography of Works Cited

Ahern, Daniel, *Nietzsche as Cultural Physician* (University Park, PA: Pennsylvania State UP, 1995).

Alaimo, Stacy, *Bodily Natures: Science, Environment, and the Material Self* (Bloomington: Indiana UP, 2010)

—, and Susan Hekman, eds., *Material Feminisms* (Bloomington: Indiana UP, 2008)

Aldersey-Williams, Hugh, *Anatomies: A Cultural History of the Human Body* (New York and London: W.W. Norton, 2013).

Amrine, Frederick. "Goethean Intuition." *Goethe Yearbook*, 8 (2011), 35-50.

—, "The Metamorphosis of the Scientist," in *Goethe's Way of Science: A Phenomenology of Nature*, ed. by David Seamon and Arthur Zajonc (New York: State U of New York Press, 1998), 33-54.

Anz, Thomas, "'Gesund' und 'krank.' Kriterien der Kritik im Kampf gegen die literarische Moderne um 1900," in *Ethische contra ästhetische Legitimation von Literatur*, ed. by Walter Haug and Wilfried Barner (Tübingen: Niemeyer, 1986), 240-250.

—, *Gesund oder krank? Medizin, Moral und Ästhetik in der deutschen Gegenwartsliteratur* (Stuttgart: Metzler, 1989)

Arz, Maike. *Literatur und Lebenskraft: Vitalistische Naturforschung und bürgerliche Literatur um 1800* (Stuttgart: Verlag für Wissenschaft und Forschung, 1996).

Babich, Babette, "Nietzsche's Critique of Scientific Reason and Scientific Culture," in in *Nietzsche and Science*, ed. by Gregory Moore and Thomas Brobjer (Hampsire UK: Ashgate, 2004), 133-54.

Barad, Karen, *Meeting the Universe Halfway: Quantum Physics and the Entanglement of Matter and Meaning* (Durham: Duke UP, 2007)

Barkhoff, Jürgen, "Goethes Ehrfurchtsgebärden in den *Wanderjahren*," in *Anthropologie und Literatur um 1800*, ed. by J.B. and Eda Sagarra (Munich: iudicium, 1992), 161-186

—, *Magnetische Fiktionen: Literarisierung des Mesmerismus in der Romantik* (Stuttgart and Weimar: Metzler, 1995)

Baumgarten, Alexander Gottlieb, *Metaphysik*, trans. by Georg Friedrich Meier (Halle: Carl Hermann Hemmerde, 1766)

Bell, Matthew, *Goethe's Naturalistic Anthropology: Man and Other Plants* (Oxford: Clarendon Press, 1994)

Benjamin, Walter, "Kriminalromane, auf Reisen," in *Gesammelte Schriften*, ed. by Rudolf Tiedemann and Hermann Schweppenhäuser, 7 vols (Frankfurt am Main: Suhrkamp, 1972-1989), IV.1, 381-382.

Benjamin, Walter, *Ursprung des deutschen Trauerspiels* (1928), (Frankfurt am Main: Suhrkamp Verlag, 1963)

Bennett, Jane, *Vibrant Matter: A Political Ecology of Things* (Durham: Duke UP, 2010)

Boerhaave, Herman and Peter Shaw, *A New Method of Chemistry: Including the History, Theory, and Practice of the Art*, 2 vols (London: T. Longman, 1741)

Boschung, Urs, "Lebenslauf [Albrecht von Haller]," in *Albrecht von Haller: Leben - Werk – Epoche*, ed. by Hubert Steinke, Urs Boschung, and Wolfgang Pross (Göttingen: Wallstein, 2009), 15-82.

Bowler, Peter J. , *The Eclipse of Darwinism: Anti-Darwinian Evolution Theories in the Decades around 1900* (Baltimore: Johns Hopkins UP, 1983)

—, *The Non-Darwinian Revolution: Reinterpreting a Historical Myth* (Baltimore: Johns Hopkins UP, 1988)

Blumenbach, Johann Friedrich, *Anfangsgründe der Physiologie, aus dem Lateinischen übersetzt und mit Zusätzen vermerhrt von Joseph Eyerel* (Vienna: Wappler, 1789)

Böhme, Hartmut, "Natur hat weder Kern noch Schale. Goethes Methode der Naturbetrachtung," in *Goethe: Ungewöhnte Ansichten*, ed. by Karl Richter and Gerhard Sauder (St. Ingbert, Germany: Röhrig UP, 2001), 9-21

—, *Natur und Subjekt* (Frankfurt am Main: Suhrkamp, 1988)

Body, N.O., *Aus eines Mannes Mädchenjahren* (Berlin: Gustav Rieckes Buchhandlung, 1907)

—, *Memoirs of a Man's Maiden Years*. Trans. Deborah Simon (Philadelphia: U of Pennsylvania P, 2006)

Bölsche, Wilhelm, *Die naturwissenschaftlichen Grundlagen der Poesie. Prolegomena zu einer realistischen Ästhetik. Mit zeitgenössischen Rezensionen und einer Bibliographie der Schriften Wilhelm Bölsches*, ed. by Johannes J. Braakenburg (Munich: dtv, 1976), 1-65.

Bonds, Mark Evan, "Idealism and the Aesthetic of Instrumental Music at the Turn of the Nineteenth Century," *Journal of the American Musicological Society*, 50 (1997), 387-420.

Botsch, Walter. *Die Bedeutung des Begriffs Lebenskraft für die Chemie zwischen 1750 und 1850*. Diss. Stuttgart, 1997.

Bottigheimer, Ruth B. *Grimms' Bad Girls and Bold Boys: The Moral and Social Vision of the Tales* (New Haven: Yale U P, 1987)

Brandis, Joachim Dietrich, *Versuch über die Lebenskraft* (Hannover: Hahn'sche Buchhandlung, 1795)

Brecht, Bertolt, "Über die Popularität des Kriminalromans" in *Gesammelte Werke*, ed. by Elisabeth Hauptmann in collaboration with Rosmarie Hill, 20 vols (Frankfurt: Suhrkamp, 1967), XIX, 450-458.

Brobjer, Thomas H., "Nietzsche Reading and Knowledge of Science: An Overview," in *Nietzsche and Science*, ed.by Gregory Moore and Thomas Brobjer (Hampsire UK: Ashgate, 2004), 21-50.

Brodie, Richard, *Virus of the Mind: The New Science of the Meme* (Carlsbad CA etc.: Hay House, 2009; first published Seattle WA: Integral Press, 1996)

Broman, Thomas Hoyt, *The Transformation of German Academic Medicine, 1750-1820* (Cambridge, U.K.: Cambridge UP, 1996)

Brown, Richard S. G., "Nietzche that Profound Physiologist," in *Nietzsche and Science*, ed.by Gregory Moore and Thomas Brobjer (Hampsire UK: Ashgate, 2004), 51-70

Büchner, Ludwig, *Force and Matter: Empirico-Philosophical Studies Intelligibly Rendered*, trans. J.F. Collingwood (London, 1870)

—, *Kraft und Stoff. Empirisch-Naturphilosophische Studien in allgemein verständlicher Sprache*, 13th ed (Leipzig: Theodor Thomas, 1874)

Butler, Judith, *Bodies that Matter: On the Discursive Limits of "Sex"* (New York and London: Routledge, 1993)

Cardigos, Isabel, *In and out of Enchantment: Blood Symbolism and Gender in Portuguese Fairytales* (Helsinki: Suomalainen Tiedeakatemia, 1996)

Carson, Ronald, "The Hyphenated Space: Liminality in the Doctor-Patient Relationship," in *Stories Matter: The Role of Narrative in Medical Ethics*, ed. by Rita Charon and Martha Montello (New York and London: Routledge, 2002), 171-182.

Chemero, Anthony, *Radical Embodied Cognitive Science* (Cambridge, MIT, 2009)

Claßen, Isabella, *Darstellung von Kriminalität in der deutschen Literatur, Presse und Wissenschaft 1900-1930* (Frankfurt: Land, 1988)

Coole, Diana and Samantha Frost, eds., *New Materialisms: Ontology, Agency, and Politics* (Durham: Duke UP, 2010)

Cowan, Michael, *Cult of the Will: Nervousness and German Modernity* (University Park, PN: The Pennsylvania State UP, 2008)

Coulter, Harris, *Divided Legacy: A History of the Schism in Medical Thought*. Vol. 2. *Progress and Regress: J.B. Van Helmont to Claude Bernard* (Washington, D.C.: Wehawken, 1977)

Crawford, Claudia, *The Beginnings of Nietzsche's Theory of Language* (Berlin: de Gruyter, 1988)

Cunningham, Andrew and Nicholas Jardine, eds., *Romanticism and the Sciences* (Cambridge: Cambridge UP, 1990)

Darwin, Charles, *The Expression of the Emotions in Man and Animals*, ed. by Francis Darwin (London: Murray, 1904) [1st edn. 1872], 25-82.

—, *The Origin of Species by Means of Natural Selection* (London: John Murray,1859)

Darwin Erasmus, *Zoonomia or The Laws of Organic Life*, vol I (London: J. Johnson, 1794)

De Almeida, Hermione, *Romantic Medicine and John Keats* (New York and Oxford: Oxford U P, 1991)

Debschitz, Uta von and Thilo von Debschitz, eds., *Fritz Kahn. Man Machine /Maschine Mensch* (Vienna: Springer, 2009)

Decker, Hannah S., *Freud in Germany: Revolution and Reaction in Science, 1893-1907* (New York: International Universities Press, 1977)

de Pater, C., "Experimental Physics," in *Leiden University in the Seventeenth Century: An Exchange of Learning* ed. Thomas H. Lunsingh Scheurleer, and G.H.M. Posthumus Meyjes (Leiden: Brill, 1975), 309-27.

Diamond, Milton, "Pediatric Management of Ambiguous and Traumatized Genitalia," *The Journal of Urology*, 162 (1999), 1020-1028.

—, and H.K. Sigmundson, "Sex Reassignment at Birth: Long Term Review and Clinical Implications," *Archives of Pediatrics and Adolescent Medicine*, 151 (March 1997), 298-304.

—, and Hazel Glenn Beh, "Surgical Treatment of Infants with Ambiguous Genitalia:□ Deficiencies in the Standard of Care and Informed Consent," *Michigan Journal of Gender & Law*, 7.1 (2000), 1-63.

Döblin, Alfred, *Die beiden Freundinnen und ihr Giftmord,* (Berlin: Die Schmiede, 1924)

—, "Blick auf die Naturwissenschaft," *Die Neue Rundschau* 34.2 (1923), 1132-1138.

—, *Das Ich über der Natur* (Berlin: Fischer, 1928)

—, "Gedächtnisstörungen bei der Korsakoff'schen Psychose,"Inaugural Dissertation zur Erlangung der medizinische Doktorwürde, Medizinische Fakultät der Albert-Ludwig-Universität zu Freiburg i.B., 1905.

—, "An die Geistlichkeit," *Neue Rundschau* 30 (1919), 1270-1277.

—, *Kleine Schriften, Band I: 1902-1921*, ed. by Anthony W. Riley (Olten, Freiburg i. Br.: Walter Verlag, 1985)

—, *Kleine Schriften, Band 2: 1922-1924,* ed. by Anthony W. Riley (Olten, Freiburg i. Br.: Walter Verlag, 1985)

—, *Kleine Schriften, Band 3: 1925-1933*, ed. by Anthony W. Riley (Olten, Freiburg i. Br.: Walter Verlag, 1985)

—, "Metapyschologie and Biologie," *Die Neue Rundschau* 33.2 (December, 1922), 1222-1232 (repr. in Döblin, *Kleine Schriften, Band 2: 1922-1924*, 182-193)

—, "Zu Nietzsches Morallehren," (1903), *Nietzsche und die Deutsche Literatur*, ed. by Bruno Hillebrand, Band 1: Texte zur Nietzsche-Rezeption 1873-1963 (Munich: Deutscher Taschenbuch-Verlag; Tübingen: Niemeyer, 1978), 331-358 (repr. in Döblin, *Kleine Schriften, Band I: 1902-1921*, 29-55)

—, "Praxis der Psychoanalyse," *Vossische Zeitung* 301 (28.6.1923), 2-3 (repr. in Döblin, *Kleine Schriften, Band 2: 1922-1924*, 270-274)

—, "An Romanautoren und ihre Kritiker. Berliner Programm.," *Der Sturm,* 158-159 (May 1913), 17-18.

—, "Zum siebzigsten Geburtstag Sigmund Freuds," *Vossische Zeitung* 104 (5.5.1926) (repr. in Döblin, *Kleine Schriften, Band 3: 1925-1933*, 44-55)

—, "Das Wasser," *Die Neue Rundshau*, 33.2 (1922), 853-858.

—, "Der Wille zur Macht als Erkenntnis bei Freidrich Nietzsche," (1902), *Nietzsche und die Deutsche Literatur*, ed. by Bruno Hillebrand, Band 1: Texte zur Nietzsche-Rezeption 1873-1963 (Munich: Deutscher Taschenbuch-Verlag; Tübingen: Niemeyer, 1978), 315-330 (repr. in Alfred Döblin, *Kleine Schriften, Band I: 1902-1921*, ed. by Anthony W. Riley (Olten, Freiburg im Breihgau: Walter Verlag, 1985), 13-29)

Dollenmayer, David B. *The Berlin Novels of Alfred Döblin* (Berkeley: U of California Press, 1988)

Dollinger, Roland, "Alfred Döblins Naturphilosophie in den Zwanziger Jahren," in *Philosophia naturalis: Beiträge zu einer zeitgemässen Naturphilosophie,* ed. by Thomas Arzt, Roland Dollinger, Maria Hippius (Würzberg: Königshausen und Neumann, 1996), 135-150.

—, "Technology and Nature: From Döblin's *Berge Meere und Giganten* to a Philosophy of Nature," in *A Companion to the Works of Alfred Döblin*, ed. by Roland Dollinger, Wulf Koepke and Heidi Tomann Tewarson (Rochester, NY: Camden House, 2004), 93-110.

Dreger, Alice, "A History of Intersexuality: From the Age of Gonads to the Age of Consent," *The Journal of Clinical Ethics*, 9.4 (Winter 1998), 345-355.

—, *Hermaphrodites and the Medical Invention of Sex* (Cambridge: Harvard UP, 1998)

Du Bois-Reymond, Emil, "Glaubensbekenntnis über das Wesen der sogenannten Lebensvorgänge und die dabei tätigen Kräfte" (1848), in *Untersuchungen über tierische Elektrizität* (Berlin, 1848), vol. I, p.xxxvi.

—, "Über die Lebenskraft. Aus der Vorrede zu den 'Untersuchungen über tierische Elektrizität' vom März 1848," in *Reden von Emil du Bois-Reymond in zwei Bänden*, ed. by Estelle du Bois-Reymond (Leipzig: Veit & Comp., 1912), II, 1-26.

Duchesneau, François, "'Essential Force' and 'Formative Force': Models for Epigenesis in the 18th Century," in *Self-Organization and Emergence in Life Sciences,* ed. Bernard Feltz, Marc Crommelinck, and Philippe Goujon (Dordrcht NL: Springer, 2006), 171-86.

Dundes, Alan, "Nationalistic Inferiority Complexes and the Fabrication of Fakelore: A Reconsideration of Ossian, the *Kinder- und Hausmärchen*, the *Kalevala*, and Paul Bunyan," in *Papers I-IV [and] Plenary Papers: The 8th Con-*

gress for the International Society for Folk Narrative Research, Bergen, June 12th-17th 1984, vol. 1, ed. Reimund Kvideland and Torunn Selberg (Bergen: International Society for Folk Narrative Research, 1984), 154-71.

Dzialas, Ingrid, *Auffassung und Darstellung der Elemente bei Goethe* (Berlin: Ebering, 1939)

Egger, Irmgard, *Diätetik und Askese: Zur Dialektik der Aufklärung in Goethes Romanen* (Munich: Fink, 2001)

Eibl, Karl, "Darwin, Haeckel, Nietzsche: Der idealistisch gefilterte Darwin in der deutschen Dichtung und Poetologie des 19. Jahrhunderts. Mit einer Hypothese zum biologischen Ursprung der Kunst," in *Fritz Mauthner: Sprache, Literatur, Kritik*, ed. by Helmut Henne and Christine Kaiser (Tübingen: Niemeyer, 2000), 87-108.

Elder, Sace, *Murder Scenes. Normality, Deviance and Criminal Violence in Weimar Berlin* (Ann Arbor: U of Michigan Press, 2013)

Ellis, John M., *One Fairy Story Too Many: The Brothers Grimm and Their Tales* (Chicago: U of Chicago Press, 1983)

Engstrom, Eric, *Clinical Psychiatry in Imperial Germany: A History of Psychiatric Practice* (Ithaca: Cornell UP, 2003)

Fausto-Sterling, Anne, *Sexing the Body: Gender Politics and the Construction of Sexuality* (New York: Basic Books, 2000)

Feuerbach, Johann Paul Anselm von, ed., *Aktenmäßige Darstellung merkwürdiger Verbrechen*, (Gießen, 1828)

Fink, Gottfried Wilhelm *Kurze Geschichte der Homöopathie: Entdeckung und Entfaltung der sanften Heilkunde und Medizin* (Leipzig: Bohmeier, 2009)

Finkelstein, Gabriel, *Emil du Bois Reymond: Neuroscience, Self, and Society in Nineteenth-Century Germany* (Cambridge MA: MIT U P, 2013)

Fischer, Ernst Peter, "Going Bravely into the Unexplored," *Max Planck Research: Science Magazine of the Max Planck Society*, 1 (2006),15-18.

Fischer, Jens-Malte, "Dekadenz und Entartung – Max Nordau als Kulturkritiker des Fin de Siècle," in *Fin de Siècle. Zu Literatur und Kunst der Jahhrundertwende*, ed. by Roger Bauer and Eckhard Heftrich (Frankfurt a.M.: Klostermann, 1997), 93-111.

Fore, Devin, "Döblin's Epic: Sense, Document and the Verbal World Picture," *New German Critique* 99 (Fall, 2006), 171-207.

Forkel, Johann Nikolas, *Allgemeine Geschichte der Musik*, 2 vols. (Leipzig: 1788)

Frazer, James George. *The Golden Bough: A Study in Magic and Religion* abridged edition (New York: Macmillan, 1956)

Northrup Frey, "Polemical Introduction," in N.F., *Anatomy Of Criticism Four Essays* (Princeton U P, 1957), 3-29.

Fuchs, Hanns, Richard Wagner und die Homosexualität (Berlin: Behr, 1902)

Fuechtner, Veronika, Berlin Psychoanalytic: Psychoanalysis and Culture in Weimar *Republic Germany and Beyond* (Berkeley: U of California Press, 2011)

—, "Döblin's Medical, Psychiatric and Psychoanalytical Work" in *A Companion to the Works of Alfred Döblin*, ed. by Roland Dollinger, Wulf Koepke and Heidi Tomann Tewarson (Rochester, NY: Camden House, 2004), 111-139.

Gaderer, Rupert, *Poetik der Technik: Elektrizität und Optik bei ETA Hoffmann* (Freiburg: Rombach Verlag KG, 2009)

Galvani, Luigi, *Commentary on the Effects of Electricity on Muscular Motion*, ed. by Margaret Glover Foley (Norwalk: Burndy Library, 1954)

Gantenbein, Urs Leo, "Similia Similibus: Samuel Hahnemann und sein Schatten Paracelsus," *Nova Acta Paracelsica* 13 (1990): 293-328.

Gay, Volney P., ed., *Neuroscience and Religion: Brain, Mind, Self, and Soul* (Lanham & Boulder: Rowman & Littlefield, 2009)

Gay, Volney P., "Neuroscience and Religion: Brain, Mind, Self, Soul," in *Neuroscience and Religion: Brain, Mind, Self, and Soul*, ed.by Volney P. Gay (Lanham & Boulder: Rowman & Littlefield, 2009), 1-18.

Carlson, Stephan, "The Neuroscience of Religious Experience: An Introductory Survey," in *Neuroscience and Religion: Brain, Mind, Self, and Soul*, ed.by Volney P. Gay (Lanham & Boulder: Rowman & Littlefield, 2009), 153-73.

Gigante, Denise, *Life: Organic Form and Romanticism* (New Haven: Yale UP, 2009)

Gilman, Sander, "Preface: Whose Body Is It, Anyway? Hermaphrodites, Gays, and Jews in N.O. Body's Germany," in N.O. Body, *Memoirs of a Man's Maiden Years*, trans. by Deborah Simon (Philadelphia: U of Pennsylvania Press, 2006), vii-xxiv.

Goethe, Johann Wolfgang von, *Wilhelm Meisters Wanderjahre*, ed. by Gerhard Neumann and Hans-Georg Dewitz, in J.W. von Goethe, *Sämtliche Werke, Briefe, Tagebücher und Gespräche*, ed. by Hendrik Birus, Dieter Borchmeyer et al, 40 vols (Frankfiurt a.M.: Deutscher Klassiker Verlag, 1987-2013), X (1993)

Goethe, Johann Wolfgang von, *Zur Farbenlehre*, ed. Manfred Wenzel in J.W. von Goethe, *Sämtliche Werke, Briefe, Tagebücher und Gespräche*, ed. by Hendrik Birus, Dieter Borchmeyer et al, 40 vols (Frankfiurt a.M.: Deutscher Klassiker Verlag,), XXII.1 (1991)

—, *Werke*, Hamburger Ausgabe, ed. by Erich Trunz, 14 vols (Munich: Beck, 1982-2008)

—, *Works*, Surhkamp Edition in 12 Volumes, ed. by Victor Lange, Eric A. Blackall, and Cyrus Hamlin, Vol. 12: *Scientific Studies*, ed. and translated by Douglas Miller (New York: Suhrkamp Publishers, 1983)

Goldmann, Stefan, *Christoph Wilhelm Hufeland im Goethekreis: Eine psychoanalytische Studie zur Autobiographie und ihre Topik* (Stuttgart: M&P, 1993)

—, "Von der Lebenskraft zum Unbewussten: Konzeptwandel in der Anthropologie um 1800," in *Homöopathie und Philosophie & Philosophie und Homöopathie,* ed. by Rainer G. Appell (Eisenach: Bluethenstaub, 1998), 149-74.

Gouk, Penelope, "Raising Spirits and Restoring Souls: Early Modern Medical Explanations for Music's Effects," in *Hearing Cultures: Essays on Sound, Listening, and Modernity*, ed. by Veit Erlmann (Oxford: Berg, 2004), 87-105.

Grosz, Elizabeth, *Volatile Bodies: Toward a Corporeal Feminism* (Bloomington: Indiana UP, 1994)

Gray, Paul, *Plato's Ghost: The Modernist Transformation of Mathematics* (Princeton: Princeton UP, 2008)

Greiner, Leo, "Untitled Review of *Die Beiden Freundinnen*," *Berlin Böursen-Courier* (29.3.1925), reprinted in *Alfred Döblin in Spiegel der zeitgnössischen Kritik*, ed. by Ingrid Schuster and Ingrid Bode (Bern and Munich: Franke Verlag, 1973), 155-156.

Grimm, Jacob, *Deutsche Mythologie*, photomechanical reproduction of the 4th edn, ed. by Elard Hugo Meyer, 3 vols (Graz: Akademische Druck- und Verlagsanstalt, 1953)

—, *Deutsche Rechtsaltertümer*, 4th edn, ed. by Andreas Heusler and Rudolf Hübner, 2 vols (Leipzig: Dieterich'sche Verlagsbuchhandlung Theodor Weicher, 1899)

Grimm, Jacob and Wilhelm, *Altdeutsche Wälder*, vol. 1 (Kassel: Thurneissen, 1813).

—, *Briefe der Brüder Grimm an Savigny*, ed. by Wilhelm Schoof (Berlin: Schmidt, 1953)

—, *Briefwechsel zwischen Jacob und Wilhelm Grimm aus der Jugendzeit*, 2nd ed., ed. by Herman Grimm and Gustav Hinrichs (Weimar: Hermann Böhlaus Nachfolger, 1963)

—, *Deutsches Wörterbuch*, Vol. 4, 2nd edition (Leipzig: S. Hirzel, 1877)

—, eds., *Hartmann von Aue: Der arme Heinrich; Aus der Straßburgischen und vatikanischen Handschrift herausgegeben und erklärt durch die Brüder Grimm* (Berlin: Realschulbuch, 1815)

—, *Kinder- und Hausmärchen: Ausgabe letzter Hand mit den Originalanmerkungen der Brüder Grimm*, 7th edn, ed. by Heinz Rölleke, 3 vols (Stuttgart: Reclam, 2001)

Grimm, Wilhelm, *Kleinere Schriften*, ed. by Gustav Hinrichs, vol. 4 (Berlin: Ferdinand Dümmler, 1882)

Grusin, Richard, "The Radical Mediation of Life in the Anthropocene" (22 October 2013), C21 Blog: http://www.c21uwm.com/2013/10/23/the-radical-mediation-of-life-in-the-anthropocene/

Haglund, Richard F., jr, "Visualization and Emergence in Contemporary Physics," in *Goethe, Chaos, and Complexity*, ed. by Herbert Rowland (Amsterdam and New York: Rodopi, 2001), 57-68.

Hahnemann, Samuel, *Gesammelte kleine Schriften*, ed. by Josef M. Schmidt and Daniel Kaiser (Heidelberg: Haug, 2001). Eng. abridged version: *The Lesser Writings*, trans. by R.E. Dudgeon (NY: Radde, 1852)

—, *Organon der Heilkunst. Neufassung mit Systematik und Glossar von Josef M. Schmidt*, 2nd ed (Munich: Urban & Fischer, 2006). Eng. trans. *Organon of Homeopathic Medicine*, intro. by Constantine Hering (New York: Radde, 1849)

—, *Organon-Synopse: Die 6 Auflagen von 1810-1842 im Überblick*, ed. Bernhard Luft and Matthias Wischner (Heidelberg: Haug, 2001)

Haller, Albrecht von, "Dissertation on the Sensible and Irritable Parts of Animals," trans. by Oswei Temkin, *Bulletin of the History of Medicine*, 4 (1936), 656-699.

—, *Erster Umriß der Geschäfte des körperlichen Lebens für die Vorlesungen eingerichtet. Aus dem Lateinischen unter der Aufsicht des Verfassers übersetzt*, trans. by Franz Ludwig Tribolet (Berlin: Haude & Spener, 1770)

—, *The Natural Philosophy of Albrecht von Haller*, ed. by Shirley A. Roe (New York: Arno Press, 1981)

—, "Rezension: Erste Gründe einer Physiologie der eigentlichen thierischer Natur thierischer Körper," *Göttingsche Anzeigen von gelehrten Sachen* (1772), 548-553.

—, "Rezension von M. de la Mettrie *Institutions de médecine de M. de Boerhaave,*" *Göttingsche Zeitungen von gelehrten Sachen* (1745), 377-378.

—, "Über den Ursprung des Übels," in *Die Alpen und andere Gedichte*, ed. by Adalbert Elschenbroich (Stuttgart: Reclam, 1965), 53-74.

—, "Vorrede," George Louis Leclerc, Comte du Buffon, *Allgemeine Historie der Natur nach allen ihren besondern Theilen abgehandelt, nebst einer Beschreibung der Naturalienkammer Seigneur Majestät des Königes von Frankreich.* Vol I (Hamburg: Holle, 1750)

Halm, August, *Zwei Kulturen der Musik* (Stuttgart: Klett 1947)

Hamilton, John T., *Music, Madness, and the Unworking of Language* (New York: Columbia UP, 2008)

Hanslick, Eduard, *Vom Musikalisch-Schönen* (Leipzig: Breitkopf und Härtel, 1922)

Haraway, Donna, *The Companion Species Manifesto: Dogs, People, and Significant Otherness* (Chicago: Prickly Paradigm, 2003)

Harper, Catherine, *Intersex* (Oxford and New York: Berg, 2007)

Hausdorff, Felix (Paul Mongré), *Das Chaos in kosmischer Auslese: Ein erkenntniskritischer Versuch* (Leipzig: C.G. Naumann, 1898)

—, *Sant'Ilario: Gedanken aus der Landschaft Zarathustras* (Leipzig: C.G. Naumann, 1897)

Hausmann, Rudolf, *Die Entdeckung des Lebens. Wege und Irrwege großer Forscher* (Darmstadt: Wissenschaftliche Buchgesellschaft, 2009)

Hassan, Ihab, "Prometheus as Performer: Toward a Posthumanist Culture?" *The Georgia Review*, vol. 31, No. 4 (Winter 1977), 830-850.

Hayles, N. Katherine, *How we Became Posthuman: Virtual Bodies in Cybernetics, Literature, and Informatics* (Chicago & London: U of Chicago Press, 1999)

Heberlein, Thomas, *Navigating Environmental Attitudes* (New York: Oxford UP, 2012)

Heiligenthal, Peter and Reinhard Volk, eds., *Der Fall Schreber 1: um Neunzehnhundert, mit Aufsätzen von Franz Baumeyer* (Wiesbaden: Focus-Verlag, 1973)

Heinze, Sigrid, ed., *Homöopathie 1796-1996: Eine Heilkunde und ihre Geschichte* (Berlin: Edition Lit.europe, 1996)

Heitler, Walter, "Goethean Science," in *Goethe's Way of Science: A Phenomenology of Nature*, eds. David Seamon and Arthur Zajonc (New York: State U of New York Press, 1998), 55-69.

Hensler, P.G., "Rezension von Johann August Unzer, *Erste Gründe einer Physiologie der eigentlichen thierischen Natur thierischer Körper*," in *Allgemeine deutsche Bibliothek*, Bd. 16, St. 2 (1772), 502-513.

Herder, Johann Gottfried, *Werke*, ed. by Wolfgang Pross (Darmstadt: Wissenschaftliche Buchgesellschaft, 1987)

Hermann, Frank and Heinke Schmitz, *Der Verlag die Schmiede 1921-1929. Eine kommentierte Bibliographie* (Morsum/Sylt: Cicero Press, 1996)

Herzer, Manfred, *Magnus Hirschfeld: Leben und Werk eines jüdischen, schwulen und sozialistischen Sexologen* (Frankfurt/M. and New York: Campus, 1992)

Herzog, Todd, *Crime Stores: Criminalistic Fantasy and the Culture of Crisis in Weimar Germany* (New York: Berghahn Books, 2009)

Heylighen, Francis & Klaas Chielens, "Cultural Evolution and Memetics," in *Encyclopedia of Complexity and System Science*, editor-in-chief Robert A. Meyers, 11 vols (New York & Heidelberg: Springer, 2009)

Hillebrand, Bruno, ed., *Nietzsche und die deutsche Literatur*, Vol 1. *Texte zur Nietzsche Rezeption 1873-1963* (Munich: Deutscher Taschenbuch Verlag, 1978).

Hirsch, August, A.H., *Geschichte der medizinischen Wissenschaften in Deutschland*, vol. 22 of *Geschichte der Wissenschaften* (München & Leipzig: R. Oldenbourg, 1893)

Hirschfeld, Magnus, *Sexuelle Zwischenstufen. Das männliche Weib und der weibliche Mann* (Bonn: A. Marcus und E. Webers Verlag, 1918)

Hoffmann, E.T.A., *Fantasie- und Nachtstücke*, *Sämtliche Werke*, ed. Hartmut Steinecke and Wulf Segebrecht, 6 vols. (München: Winkler Verlag, 1960)

—, *E.T.A. Hoffmann's Musical Writings*, ed. by David Charlton, trans. by Martyn Clarke (Cambridge: Cambridge UP, 2004)

—, *E.T.A. Hoffmann's Musical Writings: Kreisleriana, The Poet and the Composer, Music Criticism*, ed. David Charlton (Cambridge: Cambridge UP, 1989)

Holland, Jocelyn, *Key Texts of Johann Wilhelm Ritter (1776-1810) on the Science and Art of Nature* (Leiden: Brill, 2010)

Holmes, Morgan, *Intersex: A Perilous Difference* (Selinsgrove: Susquehanna UP, 2008)

Hufeland, Christoph Wilhelm, *Die Kunst, das menschliche Leben zu verlängern* (Stuttgart: Hippokrates, 1975). English trans. *The Art of Prolonging Life* (New York: Arno Press, 1979)

—, "Mein Begriff von der Lebenskraft," *Journal der practischen Heilkunde* 6 (1798), 785-96.

Hung, Jochen, and others, eds, *Beyond Glitter and Doom: The Contingency of the Weimar Republic* (Munich, Iudicium: 2012)

Iovino, Serenella and Serpil Oppermann, "Material Ecocriticism: Materiality, Agency, and Models of Narrativity," *Ecozon@*, 3.1 (2012), 75-91.

—, eds., *Material Ecocriticism* (Bloomington: Indiana UP, 2014)

Israel, Jonathan, *Radical Enlightenment: Philosophy and the Making of Modernity 1650-1750* (Oxford UP, 2001)

James, William, *The Principles of Psychology*, 2 vols (New York: Henry Holt & Co., 1890)

Jütte, Robert, *Samuel Hahnemann: Begründer der Homöopathie* (Munich: dtv, 2005). English translation: http://www.igmbosch.de/content/language1/downloads/SamuelHahnemannTheFounderofHomeopathy.pdf

Kant, Immanuel, *Allgemeine Naturgeschichte und Theorie des Himmels* (1755), *Werke in zehn Bänden*, ed. by Wilhelm Weischedel (Darmstadt: Wissenschaftliche Buchgesellschaft, 1983), I, 227-400.

—,*Critique of Judgement* (New York: Cosimo, 2007)

—, *Universal Natural History and Theory of the Heavens Or An Essay on the Constitution and the Mechanical Origin of the Entire Structure of the Universe Based on Newtonian Principles* (1755), trans. by Ian Johnston: http://records.viu.ca/~johnstoi/kant/kant2e.htm

Karkazis, Katrina, *Fixing Sex: Intersex, Medical Authority, and Lived Experience* (Durham and London: Duke UP, 2008)

Karlavaris-Bremer, Ute, "Außenseiterinnen der Gesellschaft. Alfred Döblins Erzählung *Die beiden Freundinnen und ihr Giftmord* aus interdisziplinärer Perspektive," in *Tatsachenphantasie - Alfred Döblins Poetik des Wissens im Kontext der Moderne / Internationales Alfred-Döblin-Kolloquium Emmendingen 2007,* ed. by Sabina Becker and Robert Krause (Bern: Peter Lang AG, 2008), 265-278.

Kausch, Johann Joseph, *Psychologische Abhandlung über den Einfluß der Töne und ins besondere der Musik auf die Seele* (Breslau: Johann Friedrich Korn, 1782)

Kehlmann, Daniel, *Die Vermessung der Welt* (Reinbek bei Hamburg: Rowohlt, 2005)

Kennaway, James, *Bad Vibrations: The History of the Idea of Music as a Cause of Disease* (Farnham: Ashgate, 2012)

Kennaway, James, "Musical Hypnosis: Sound and Self from Mesmer to Brainwashing," *Social History of Medicine* 25.2 (2012), 271-289.

Kessler, Suzanne, *Lessons from the Intersexed* (New Brunswick: Rutgers UP, 1998)

Killen, Andreas, *Berlin Electropolis: Shock, Nerves and German Modernity* (Berkeley: U of California Press, 2006)

Kleist, Heinrich von, "Die heilige Cäcilie oder die Gewalt der Musik. Eine Legende," in Heinrich von Kleist, *Sämtliche Werke und Briefe in vier Bänden*, ed. by Ile-Marie Barth et al, 4 vols (Frankfurt: Suhrkamp, 1990), III, 276-313.

Knoeff, Rina, *Herman Boerhaave (1668-1738): Calvinist Chemist and Physician* (Amsterdam: Koninklijke Nederlandse Akademie van Wetenschappen, 2002)

Koehler, Peter J., "Neuroscience in the Work of Boerhaave and Haller," in *Brain, Mind and Medicine: Essays in Eighteenth-Century Neuroscience*, ed. by Harry A. Whitaker, C.U.M. Smith, and Stanley Finger (New York: Springer, 2007), 213-32.

Köhler, Rafael, *Natur und Geist: energetische Form in der Musiktheorie* (Stuttgart: Franz Steiner Verlag, 1996)

Koestler, Arthur, *The Ghost in the Machine* (London & New York: Hutchinson & Macmillan, 1967)

Kollak, Ingrid, *Literatur und Hypnose: Der Mesmerismus und sein Einfluß auf die Literatur des 19. Jahrhunderts* (New York: Campus Verlag, 1997)

Košenina, Alexander, "Juristische Fallgeschichte: Theodor Lessings *Haarmann. Die Geschichte eines Werwolfs*" (1925), *Non Fiktion*, 3 (2008), H. 1/2, 83-94.

Kottow, Andrea, *Der kranke Mann. Medizin und Geschlecht in der Literatur um 1900* (Frankfurt a.M., New York: Campus, 2006), 127-173.

Kracauer, Siegfried, "Mordprozeß und Gesellschaft," *Die Neue Rundschau*, 42 (March 1931), 431-432.

Kraus, Karl, "Max Nordau," in *Die Fackel*, No. 200, VII. Jahr, 3 April 1906.

Krohn, Rüdiger, "A Tale of Sacrifice and Love: Literary Way Stations of the *Arme Heinrich* from the Brothers Grimm to Tankred Dorst," in *A Companion to the Works of Hartmann von Aue*, ed. by Francis G. Gentry (Rochester, NY: Camden House, 2005), 223-53.

Kurth, Ernst, *Romantische Harmonik und ihre Krise in Wagners Tristan* (Hildesheim: Georg Olm, 1975)

Kutch, Lynn M. and Todd Herzog, eds., *Tatort Germany: The Curious Case of German- Language Crime Fiction,* (Rochester, NY: Camden House, 2014)

Laqueur, Thomas, *Making Sex: Body and Gender from the Greeks to Freud* (Cambridge: Harvard UP, 1990)

Leibniz, Gottfried Wilhelm, *La Monadologie*, édition annotée par Émile Boutroux (Paris: C. Delagrave, 1881)

—, *Monadologie*, ed. with introduction and notes by Hermann Glockner (Stuttgart: Reclam, 1963)

—, *Philosophical Writings*, ed. by G.H.R. Parkinson, trans. by Mary Morris and G.H.R. Parkinson, Everyman Library (London: J.M. Dent, 1997)

—, *Theodicy: Essays on the Goodness of God the Freedom of Man and the Origin of Evil*, ed. with an introduction by Austin Farrer, trans. by E.M. Huggard from C.J. Gerhardt's Edition of the Collected Philosophical Works, 1875-90 (Routledge & Kegan Paul Ltd, 1951)

Lenoir, Timothy, *The Strategy of Life* (Chicago: U of Chicago P, 1989)

Levine, George, *One Culture: Essays in Science and Literature* (Madison WI: U of Wisconsin Press, 1987)

Lichtenthal, Peter, *Der musikalische Arzt* (Vienna: Christian Friedrich Wappler und Beck, 1807)

Lightman, Alan, "The Accidental Universe: Science's Crisis of Faith," *Harpers Magazine*, December 2011, 34-40.

Lindeboom, Gerrit Arie, *Descartes and Medicine* (Amsterdam: Rodopi, 1979)

—, *Herman Boerhaave: The Man and His Work* (London: Methuen, 1968)

Link, Jürgen, "Literaturanalyse als Interdiskursanalyse. Am Beispiel des Ursprungs literarischer Symbolik in der Kollektivsymbolik," in *Diskurstheorien und Literaturwissenschaft*, ed. by Jürgen Fohrmann and Harro Müller (Frankfurt a. M.: Fischer, 1988), 284-307.

Lobel, Thelma, "Sensation." *The New Science of Physical Intelligence* (New York: Atria Books, 2014)

Lohff, Brigitte, "Die Entwicklung des Experiments im Bereich der Nervenphysiologie," *Sudhoffs Archiv: Zeitschrift für Wissenschaftsgeschichte,* 64.2 (1980), 105-29.

Lombroso, Cesare, *Der Verbrecher (homo delinquens) in anthropologischer, ärztlicher und juristischer Beziehung,* 3 vols (Hamburg: Verlagsanstalt/Königl. Buchhandlung, 1894)

Lovejoy, Arthur O., *The Great Chain of Being: A Study in the History of an Idea* (Cambridge, London: Harvard, 1964)

Martus, Steffen, *Die Brüder Grimm: Eine Biographie* (Berlin: Rowohlt, 2009)

Maiwald, Manfred, "Die beiden Freundinnen und ihr Giftmord. Juristische Betrachtungen zu einen literarischen Prozessbericht," in *Literatur und Recht: Literarische Rechtsfälle von der Antike bis in die Gegenwart*, ed. by Ulrich Mölk. (Göttingen, Wallstein Verlag, 1996), 370-382.

Mancini, Elena, *Magnus Hirschfeld and the Quest for Sexual Freedom: A History of the First International Sexual Freedom Movement* (New York: Palgrave Macmillan, 2010)

Marx, Adolf Bernhard, *Die Musik des neunzehnten Jahrhunderts* (Leipzig: Breitkopf und Härtel, 1855)

—, *Ludwig van Beethoven*, 2 vols (Berlin: Otto Janke, 1901)

Mattheson, Johann, *Der vollkommene Capellmeister* (Hamburg: 1739)

McCarthy, John A., "Energy and Schiller's Aesthetics from the Philosophical to the Aesthetic Letters," in: *Who is this Schiller Now? Essays on his Reception and Significance*, ed.by Jeffrey L. High, Nicholas Martin, and Norbert Oellers (Rochester NY: Camden House, 2011), 165-86.

—, "The Little Divine Machine: The Soul/Body Problem Revisited," in: *Neuroscience and Religion: Brain, Mind, Self, and Soul*, ed.by Volney P. Gay (Lanham & Boulder: Rowman & Littlefield, 2009), 241-269.

—, "'Live like a Man and not like a Monkey': Nietzsche's Philosophic Vitalism and Darwin's Theory of Evolution," in *The Evolution of Literature: Legacies of Darwin in European Cultures*, ed. by Nicholas Saul and Simon J. James (Amsterdam & New York: Rodopi, 2011), 255-81.

—, "Die Nietzsche-Rezeption in der Literatur 1890-1918," in *Hansers Sozialgeschichte der deutschen Literatur.* Band 7. Ed. by York-Gothart Mix. (Munich, Carl Hanser Verlag: 2000), 192-206.

—, "The 'Pregnant Point': Goethe on Complexity, Interdisciplinarity, and Emergence," in *Goethe, Chaos, and Complexity*, ed. by Herbert Rowland (Amsterdam and New York: Rodopi, 2001), 17-33.

—, *Remapping Reality: Chaos and Creativity in Science and Literature (Goethe – Nietzsche – Grass)* (Amsterdam: Rodopi, 2006)

McCracken, Peggy, *The Curse of Eve, the Wound of the Hero: Blood, Gender, and Medieval Literature* (Philadelphia: U of Pennsylvania Press, 2003)

Meier, C.A., ed., *Atom and Archetype : the Pauli/Jung letters, 1932-1958,* translated by David Roscoe with an introductory essay by Beverley Zabriskie (Princeton NJ: Princeton UP, 2001)

Merchant, Carolyn, *The Death of Nature: Women, Ecology and the Scientific Revolution* (New York: Harper Collins, 1989)

Michaelis, Christian Friedrich, "Ein Versuch, das innere Wesen der Tonkunst zu entwickeln," in *Allgemeine musikalische Zeitung* 8 (1806), 673-83, 691-6.

Michaelis, Christian Friedrich, "Über den Rang der Tonkunst unter den schönen Künsten," in *Allgemeine musikalische Zeitung* 2 (1799), 183-6.

Midgley, David, "Metaphysical Speculation and the Fascination of the Real: On the Connections between Döblin's Philosophical Writings and his Fiction Prior to *Berlin Alexanderplatz*," in *Alfred Döblin Paradigms of Modern*ism, ed. by Steffan Davies and Ernest Schonfield (Berlin: Walter de Gruyter GmbH & Co., 2009), 7-27.

Moore, Gregory and Thomas Brobjer, *Nietzsche and Science* (Hampshire UK: Ashgate, 2004)

Morus, Iwan Rhys, *When Physics Became King* (Chicago: U of Chicago Press, 2005)

Mücke, Dorothea von, "Blut und Wunder bei Achim von Arnim," in *Die Macht und das Imaginäre: Eine kulturelle Verwandtschaft in der Literatur zwischen Früher Neuzeit und Moderne*, ed. Rudolf Behrens and Jörn Steigerwald (Würzburg: Königshausen und Neumann, 2005), 143-56.

Müller, Adolf, "Ruggiero Giuseppe Boscovich," in *The Catholic Encyclopedia*, vol. 2 (New York: R. Appleton, 1907)

Müller-Seidel, Walter, "Alfred Döblin, Die Beiden Freundinnen und ihr Giftmord. Psychiatrie, Strafrecht und moderne Literatur," in *Literatur und Recht: Literarische Rechtsfälle von der Antike bis in die Gegenwart*, ed. by Ulrich Mölk (Göttingen, Wallstein Verlag, 1996), 356-369.

Müller-Sievers, Helmut, *Self-Generation: Biology, Literature, Philosophy around 1800* (Stanford: Stanford UP 1997)

Niederland, William G., *Der Fall Schreber: Das psychoanalytische Profil einer paranoiden Persönlichkeit* (Frankfurt a. M.: Suhrkamp, 1978)

Negus, Kenneth, *E.T.A. Hoffmann's Other World: The Romantic Author and his "New Mythology"* (Philadelphia: U of Pennsylvania Press, 1965)

Neubauer, John, *Bifocal Vision: Novalis' Philosophy of Nature and Disease* (Chapel Hill, NC: U of North Carolina Press, 1971)

Neuburger, Max, *Die Lehre von der Heilkraft der Natur im Wandel der Zeiten* (Stuttgart: Ferdinand Enke, 1926)

Neumann, Josef N., "Christoph Wilhelm Hufeland (1762-1836)," *Klassiker der Medizin*, ed. Dietrich von Engelhart and Fritz Hartmann (Munich: Beck, 1991), vol. I, 339-59.

Nicolai, Ernst Anton, *Die Verbindung der Musik mit der Arzneygelahrheit* (Halle: Carl Hermann Hemmerde, 1745)

Niemann, Walter, *Musik der Gegenwart* (Berlin: Schuster & Loeffler, 1921)

Nietzsche, Friedrich, *Jenseits von Gut und Böse*, in *Sämtliche Werke*, Kritische Studienausgabe, ed.by Giorgio Colli and Mazzino Montinari, 15 vols 2nd ed. (Munich: Deutscher Taschenbuch Verlag and de Gruyter, 1988), vol. IX, 9-243.

—, *Sämtliche Werke*, Kritische Studienausgabe, ed.by Giorgio Colli and Mazzino Montinari, 15 vols (Munich and Berlin: Deutscher Taschenbuchverlag and de Gruyter, 1993)

—, *Thus Spoke Zarathustra: A Book for All and None*, trans. with a Preface by Walter Kaufmann, The Modern Library (New York: Random House, 1995)

—, *Werke*, ed. by Giorgio Colli and Mazzini Montinari, 30 vols (Berlin: De Gruyter, 1969)

Rob Nixon, *Slow Violence and the Environmentalism of the Poor* (Cambridge: Harvard UP, 2011)

Noll, Alfred, ed., *Die 'Lebenskraft' in den Schriften der Vitalisten und ihrer Gegner*, Voigtländers Quellenbücher 69, 4th ed (Leipzig: R. Voigtländer, 1914)

Nordau, Max, *Entartung*, 2 vols (Berlin: Carl Duncker, 1892-1893)

—, "Evolutionistische Ästhetik," in M.Nordau, *Paradoxe* (Leipzig: Elsaesser, 1885), 280-310.

—, *Seifenblasen. Federzeichnungen und Geschichten* (Leipzig: Reclam, 1879)

—, *Mahâ-Rôg und andere Novellen* (Berlin: Alfred Scholl, 1906)

—, "Mene, Tekel, Upharsin," in M.Nordau, *Die conventionellen Lügen der Kulturmenschheit* (Leipzig: Schlicke, 1883), 1-35.

Packham, Catherine, *Eighteenth-century Vitalism: Bodies, Culture, Politics* (New York: Palgrave Macmillan, 2012)

Person, Jutta, *Der pathographische Blick. Physiognomik, Atavismustheorien und Kulturkritik 1870-1930* (Würzburg: Königshausen & Neumann, 2005)

Perthes, Nicholas and Sandra Richter, eds., *Medizinische Schreibweisen. Ausdifferenzierung und Transfer zwischen Medizin und Literatur (1600-1900)*, STSL 117 (Tübingen: Niemeyer, 2008)

Pfeiffer, Klaus, *Medizin der Goethezeit: Christoph Wilhelm Hufeland und die Heilkunst des 18. Jahrhunderts* (Cologne: Böhlau, 2000)

Pickering, Andrew, *The Mangle of Practice: Time, Agency and Science* (Chicago: U of Chicago Press, 1995)

——, "Material Culture and the Dance of Agency," in *Oxford Handbook of Material Culture Studies*, ed. by Dan Hicks and Mary C. Beaudry (Oxford UP, 2010), 191-208.

——, "The Dance of Agency," in *Ideas on the Nature of Science*, ed. by David Cayley (Goose Lane Editions, 2009), 81-91.

——, "The World Since Kuhn," *Social Studies of Science*, 42.3 (2012), 467-73.

Plumwood, Val, *Environmental Culture: The Ecological Crisis of Reason* (London: Routledge, 2006)

Porter, Roy, *The Greatest Benefit to Mankind: A Medical History of Humanity* (New York, NY: W.W. Norton, 1997)

——, "What is Disease?" *The Cambridge History of Medicine,* ed. Roy Porter (Cambridge: Cambridge UP, 2006), 71-102.

Pray, L. and K. Zhaurova, "Barbara McClintock and the discovery of jumping genes (transposons)," *Nature Education* 1.1 (2008), 169.

Preves, Sharon, *Intersex and Identity: The Contested Self* (New Brunswick and London: Rutgers UP, 2005)

Prochaska, Georg, *Commentatio de functionibus systematis nervosi* (Prague, 1784)

——, *De Usu et Functione Systematis Nervorum* (Vienna,1800)

——, *Lehrsätze aus der Physiologie des Menschen : zum Gebrauche seiner Vorlesungen* (Vienna: Wappler, 1797)

——, *The Principles of Physiology by John Augustus Unzer and a Dissertation on the Functions of the Nervous System by George Prochaska (1749-1820)*, trans. and ed. by Thomas Laycock (The Syndenham Society, London 1851)

Prosss, Caroline, "Die Kunst der Unterscheidung. Zur Darstellung der Anormalität in der Erzähliteratur des 19. Jahrhunderts," in *Bann der Gewalt. Studien zur Literatur und Wissensgeschichte*, ed. by Maximilian Bergengruen, Roland Borgards (Göttingen: Wallstein, 2009), 361-386.

Purkert, Walter, "The Double Life of Felix Hausdorff / Paul Mongré," *The Mathematical Intelligencer*, 30.4 (2008), 36-50.

Rameau, Jean-Philippe, *Traité de l'harmonie* (Paris: 1722)

Randall, Lisa, *Warped Passages: Unraveling the Mysteries of the Universe's Hidden Dimensions* (New York: HarperCollins, 2005)

Rautenberg, Ursula, *Das 'Volksbuch vom armen Heinrich': Studien zur Rezeption Hartmanns von Aue im 19. Jahrhundert und zur Wirkungsgeschichte der Übersetzung Wilhelm Grimms* (Berlin: Erich Schmidt Verlag, 1985)

Reil, Johann Christian, *Rhapsodien über die Anwendung der psychischen Kurmethode auf Geisteszerrüttungen* , 2nd edn (Halle: Curtsche Buchhandlung, 1818)

——, "Ueber das polarische Auseinanderweichen der ursprünglichen Naturkräfte in der Gebärmutter zur Zeit der Schwangerschaft und deren Umtauschung zur Zeit der Geburt, als Beytrag zur Physiologie der Schwangerschaft und Geburt," *Archiv für die Physiologie* 7 (1807): 402-501.

—, "Veränderte Mischung und Form der thierischen Materie, als Krankheit, oder nächste Ursache der Krankheitszufälle betrachtet" in *Archiv für die Physiologie* (3:1799), 424-461.

—, *Von der Lebenskraft*, ed. by Karl Sudhoff (Leipzig: Johann Ambrosius Barth, 1910). Originally published in *Archiv für Physiologie*, 1 (1796), 8–162,

Reis, Elizabeth, *Bodies in Doubt: An American History of Intersex* (Baltimore: Johns Hopkins UP, 2009)

Rémi, Cornelia, "Religion und Theologie," in *Albrecht von Haller: Leben-Werk-Epoche*, ed. by Hubert Steinke, Urs Boschung, and Wolfgang Pross (Göttingen: Wallstein, 2009), 199-225.

Ridley, Hugh, *Darwin Becomes Art: Aesthetic Visionin the Wake of Darwin, 1870-1920* (Amsterdam & New York: Rodopi, 2014)

Richards, Robert J., *The Romantic Conception of Life: Science and Philosophy in the Age of Goethe* (Chicago: U of Chicago Press, 2002)

Rigby, Kate, *Topographies of the Sacred: The Poetics of Place in European Romanticism* (Charlottesville: U of Virginia Press, 2004)

Ritte, Theodor, "Chopinitis," *Der Energetiker*, 1 (1919), 5-7.

Ritter, Johann Wilhelm, *Fragmente aus dem Nachlasse eines jungen Physikers, Key Texts of Johann Wilhelm Ritter (1776-1810) on the Science and Art of Nature*, trans. and ed. Jocelyn Holland (Leiden and Boston: Koninklijke Brill NV, 2010)

Rocca, Julius, "William Cullen (1710-1790) and Robert Whytt (1714-1766) on the Nervous System," in *Brain, Mind and Medicine: Essays in Eighteenth-Century Neuroscience*, ed. Harry A. Whitaker, C.U.M. Smith, and Stanley Finger (New York: Springer, 2007), 85-98.

Roe, Shirley A., "Anatomia animata. The Newtonian Physiology of Albrecht von Haller," in *Transformation and Tradition in the Sciences*, ed. by Everett Mendelsohn (Cambridge: Cambridge UP, 1984), 273-300.

—, "Introduction," *The Natural Philosophy of Albrecht von Haller* (New York: Arno Press, 1981)

Rölleke, Heinz, *Die Märchen der Brüder Grimm: Quellen und Studien; Gesammelte Aufsätze* (Trier: Wissenschaftlicher Verlag Trier, 2004)

—, "Weiß – Rot – Schwarz: 'Die drei Farben der Poesie': Zu Farbspielen in Grimms' 'Sneewittchen'-Märchen und anderwärts," *Fabula*, 54 (2013), 214-34.

Rosario, Vernon A, "Quantum Sex: Intersex and the Molecular Deconstruction of Sex," *GLQ: A Journal of Lesbian and Gay Studies*, 15.2 (2009), 267-284.

Rothfarb, Lee, "Hermeneutics and Energetics: Analytical Alternatives in the early 1900s," *Journal of Music Theory*, 36.1 (1992), 43-68.

Rothschuh, Karl E. *Konzepte der Medizin in Vergangenheit und Gegenwart.* Stuttgart: Hippokrates, 1978.

——, *Physiologie: Der Wandel ihrer Konzepte, Probleme und Methoden vom 16. bis 19. Jahrhundert.* Freiburg, Munich: Karl Alber, 1968.

Rousseau, George, *Nervous Acts: Essays on Literature, Culture and Sensibility* (Basingstoke: Palgrave, 2004)

Rousseau, Jean-Jacques, 'Essai sur l'origine des langues', cited in *Music and Aesthetics in the Eighteenth and Early Nineteenth Centuries*, ed. by Peter le Huray and James Day (Cambridge: Cambridge UP, 1981)

Rouvray, Dennis H., "Similarity in Chemistry: Past, Present and Future," *Topics in Current Chemistry*, 173 (1995), 1-30.

Ruston, Sharon. *Shelley and Vitality* (New York: Palgrave Macmillan, 2005)

Ryle, Gilbert, coined the phrase, "Descartes's Myth," in *The Concept of Mind* (Hutchinson, London, 1949)

Sarasin, Philipp, *Reizbare Maschinen: eine Geschichte des Körpers 1765-1914* (Frankfurt am Main: Suhrkamp, 2001)

Saul, Nicholas and Simon J. James, eds., *The Evolution of Literature: Legacies of Darwin in European Cultures* (Amsterdam & New York: Rodopi, 2011)

——, "'Once in Human Nature, a Thing Cannot be Driven Out': Evolutionary Aesthetics in Wilhelm Jensen's *The Legacy of Blood* (1869). An Early Response to Darwin," in *The Evolution of Literature. Legacies of Darwin in European Cultures*, ed. by Nicholas Saul and Simon J. James (Amsterdam and New York: Rodopi, 2011), 239-253.

——, "Raabes Geschichte(n): Realismus, Erzählen, Historie, Prähistorie," *Jahrbuch der Internationalen Raabe-Gesellschaft* (2009), 22-39.

Schafer, R. Murray, *E.T.A. Hoffmann and Music* (Toronto and Buffalo: U of Toronto Press, 1975.)

Schall, Jeffrey D., "Actions, Reasons, Neurons, and Causes," in *Neuroscience and Religion: Brain, Mind, Self, and Soul*, ed.by Volney P. Gay (Lanham & Boulder: Rowman & Littlefield, 2009), 175-98

Schatzman, Morton, *Die Angst vor dem Vater. Langzeitwirkung einer Erziehungsmethode. Eine Analyse am Fall Schreber* (Reinbek: Rowohlt. 1978)

——, *Soul Murder. Persecution in the Family* (New York: Random House, 1973)

Schelling, Friedrich Wilhelm Joseph von, *Einleitung zu seinem Entwurf eines Systems der Naturphilosophie*, ed. by Wilhelm G. Jacobs (Stuttgart: Reclam, 1988)

——, *Sämtliche Werke*, ed. K.F.A. Schelling, 14 vols (Stuttgart and Augsburg: J.G. Cotta, 1856-61)

Schiebinger, Londa, *The Mind Has No Sex?: Women in the Origins of Modern Science* (Cambridge: Harvard UP, 1989)

Schizophrenia Working Group of the Psychiatric Genomics Consortium, "Biological insights from 108 schizophrenia-associated genetic loci," *Nature*, 511 (24 July 2014), 421-27.

Schlegel, A. W., *Kritische Ausgabe der Vorlesungen*, ed. Ernst Behler, 2 vols (Paderborn: Ferdinand Schöningh, 1989-), vol. I: *Vorlesungen über Schöne Literatur und Kunst. Erster Teil: Die Kunstlehre* (1801-1802), 375-381.

Schmiesing, Ann, *Disability, Deformity, and Disease in the Grimms' Fairy Tales* (Detroit, Michigan: Wayne State UP, 2014)

Schmidt, Hans Walter, "Nerven und Musik: eine psychologisch-physiologische Betrachtung von heute," *Internationale Reform-Zeitschrift für Gesang, Klavier und Geige: Der Energetiker*, 3.8 (1922), 10-11.

Schmitz-Emans, Monika, "Magnetische Phantasien: Zur Bedeutung des Mesmerismus für Autoren der Romantik," *Der Deutschunterricht*, 57.3 (2005), 55-62.

Schönert, Jörg, ed., *Erzählte Kriminalität. Zur Typologie und Funktion von narrativen Darstellungen in Strafrechtspflege, Publizistik und Literatur zwischen 1770 und 1920* (Tübingen: Max Niemeyer, 1991)

—, *Literatur und Kriminalität. Die gesellschaftliche Erfahrung von Verbrechen und Strafverfolgung als Gegenstand des Erzählens. Deutschland, England und Frankreich 1850–1880* (Tübingen: Max Niemeyer, 1983)

Schopenhauer, Arthur, *On the Will in Nature*, trans by E.F.J. Payne (New York: Berg, 1992)

Schopenhauer, Arthur, *World as Will and Representation*, trans. by R.B. Haldane and J. Kemp (London: Routledge & Kegan Paul, 1883)

Schreber, Daniel Paul, *Denkwürdigkeiten eines Nervenkranken*, ed. by Samuel M. Weber (Frankfurt: Verlag Ullstein, Frankfurt, 1973)

Schubert, Gotthilf Heinrich, *Ansichten von der Nachtseite der Naturwissenschaft. Nachdruck der Ausgabe von 1808* (Karben: Petra Wald, 1997)

Schulte, Christoph, *Psychopathologie des Fin de Siècle. Der Kulturkritiker, Arzt und Zionist Max Nordau* (Frankfurt a.M.: Fischer 1997)

Schwanitz, Hans Joachim, *Homöopathie und Brownianismus 1795-1844: Zwei wissenschaftstheoretische Fallstudien aus der praktischen Medizin* (Stuttgart: Gustav Fischer, 1983)

Schweitzer, Frank, "Naturwissenschaft und Selbsterkenntnis," in *Goethe und die Verzeitlichung der Natur*, ed. Peter Matussek (Munich: Beck, 1998), 383-398.

Scott, Derek, "The Sexual Politics of Victorian Musical Aesthetics," *Journal of the Royal Musical Association*, 119 (1994), 91-114.

Seiler, Hanspeter, *Die Entwicklung von Samuel Hahnemanns ärztlicher Praxis* (Heidelberg: Haug, 1988)

Sepper, Dennis, *Goethe contra Newton: Polemics and the Project for a New Science of Color* (Cambridge: Cambridge UP, 1988)

Shaffer, Elinor, *The Third Culture: Literature and Science* (Berlin & New York: Walter de Gruyter, 1998)

Shapiro, Lawrence, *Embodied Cognition* (New Problems in Philosophy) (New Brunswick NJ: Routledge, 2010)

Shelley, Percy B. "Queen Mab". *Poetical Works.* Ed. Thomas Hutchison and G.M. Matthews (London: Oxford UP, 1970), 762-800.

Shildrick, Margrit and Roxanne Mykitiuk, eds., *Ethics of the Body: Postconventional Challenges* (Cambridge and London: MIT Press, 2005)

Siebenpfeiffer, Hania, *Böse Lust. Gewaltverbrechen in Diskursen der Weimarer Republik* (Köln: Böhlau Verlag, 2005)

Siemsen, Hans, "Untitled Review of *Die beiden Freundinnen*," in *Alfred Döblin in Spiegel der zeitgnössischen Kritik*, ed. by Ingrid Schuster and Ingrid Bode (Bern: Franke Verlag, 1973), 157 -158.

Simon, Hermann, "Wer War N.O. Body?," in N.O. Body, *Aus eines Mannes Mädchenjahren.* Reprint (Berlin: Edition Hentrich, 1993), 167-246

Sloterdijk, Peter, "Die Andersheilenden: Über einige Aspekte der Homöopathie im Lichte des philosophischen Therapiegedankens der Neuzeit." *200 Jahre Homöopathie: Festakt in der Paulskirche zu Frankfurt am Main. 14. September 1996* (Heidelberg: Haug, 1996), 9-38.

Smoot, George and Keay Davidson, *Wrinkles in Time* (New York: William Morrow & Co., 1993)

Söder, Hans-Peter, "Disease and Health as Contexts of Modernity: Max Nordau as Critic of Fin-de-Siècle Modernism," *GSR*, 14 (1991), 473-487.

Sokel, Walter H., "The Prose of German Expressionism," in *Expressionism as an International Literary Phenomenon*, ed. by Ulrich Weisstein (Amsterdam: John Benjamins Publishing Company, 2011), 69-88.

Sontag, Susan, *Illness as Metaphor* (New York: Farrar, Strauss and Giroux, 1978)

Splinter, Susan, "Die Anfänge der 'Elektrifikation'. Untersuchungen zur Wirkung der Elektrizität auf den menschlichen Körper 1744 an der Friedrichs-Universität zu Halle," in *"Allerhand nützliche Versuche" empirische Wissenskultur in Halle und Göttingen (1720 - 1750)*, ed. by Tanja van Hoorn and Yvonne Wübben (Hannover: Wehrhahn, 2009), 89-103.

Spreng, Michaela, *Alfred Döblins Nietzsche-Verständnis vor dem Hintergrund der philosophischen Schriften Felix Hausdorffs*, (Munich: Herbert Utz Verlag, 2014)

Stack, George J., *Lange und Nietzsche* (Berlin & New York: de Gruyter, 1983)

Stahl, Georg, *Opusculum Chymico-Physico-Medicum* (Nuremberg, 1715)

Stegmaier, Werner, "Ein Mathematiker in der Landschaft Zarathustras. Felix Hausdorff als Philosoph," *Nietzsche-Studien*, 31 (2002), 195-240.

Steig, Reinhold, and Herman Grimm, eds., *Achim von Arnim und die ihm nahe standen*, 3 vols (Stuttgart: J.G. Cotta, 1894-1904)

Steigerwald, Joan, "Treviranus' Biology: Generation, Degeneration and the Boundaries of Life," in *Gender, Race, and Reproduction: Philosophy and the Early Life Sciences in Context*, ed. Suzanne Lettow (New York State: State University of New York, 2014)

Steinke, Hubert, "Anatomie und Physiologie," in *Albrecht von Haller: Leben-Werk-Epoche*, ed. Hubert Steinke, Urs Boschung, and Wolfgang Pross (Göttingen: Wallstein, 2008), 226-54.

—, *Irritating Experiments: Haller's Concept and the European Controversy on Irritability and Sensibility, 1750-90* (Amsterdam, New York: Rodopi, 2005)

Strässle, Thomas, "'Das Hören ist ein Sehen von und durch innen': Johann Wilhelm Ritter and the Aesthetics of Music," *Music and Literature in German Romanticism*, eds. Siobhán Donovan and Robin Elliott (Rochester: Boydell & Brewer, 2004), 27-42.

—, "Johannes Kreisler im Dialog mit einem 'geistreichen Physiker': Zu E.T.A. Hoffmanns Auseinandersetzung mit Johann Wilhelm Ritter," *E.T.A. Hoffmann-Jahrbuch* 10 (2002), 96-119.

Subotnik, Rose Rosengard, *Deconstructive Variations* (Minneapolis: U of Minnesota Press, 1996)

Sullivan, Heather I., "Affinity Studies and Open Systems: A Nonequilibrium, Ecocritical Reading of Goethe's *Faust*," in *Ecocritical Theory: New European Approaches*, ed. by Axel Goodbody and Kate Rigby (Charlottesville: U of Virginia Press, 2011), 243-255.

—, and Bernhard Malkmus, eds., "The Challenge of Ecology to the Humanities: Post-Humanism or New Humanism?" in *New German Critique* (expected publication 2016)

—, "Dirty Nature and Ecocriticism: Mining, Extraction, and Solar Power in Goethe, Hoffmann, Verne, and Eschbach," *Colloquia Germanica*, 44.2 (2011), 111-131.

—, "Ecocriticism, the Elements, and the Ascent/Descent into Weather in Goethe's *Faust*," *Goethe Yearbook*, 17 (2010), 55-72.

—, "Ecocriticism, Goethe's Optics, and *Unterhaltungen deutscher Ausgewanderten*: Emergent Forms vs Newtonian 'Constructions,'" *Monatshefte*, 101.2 (2009), 151-169.

—, "The Ecology of Color: Goethe's Materialist Optics and Ecological Posthumanism," in *Material Ecocriticism*; ed. by Serenella Iovino and Serpil Oppermann (Bloomington: Indiana UP, 2014), 80-94.

——, and Dana Phillips, eds., special edition on: "Material Ecocriticism: Dirt, Waste, Bodies, Food, and Other Matter," *Interdisciplinary Studies in Literature and the Environment,* 19.3 (2012), 445-47.

Systma, Sharon, ed., *Ethics and Intersex* (Dordrecht: Springer, 2006)

Tatar, Maria, *The Hard Facts of the Grimms' Fairy Tales* (Princeton: Princeton UP, 1987)

——, *Lustmord: Sexual Murder in Weimar Germany* (Princeton, NJ: Princeton UP, 1995)

——, *Off with Their Heads! Fairy Tales and the Culture of Childhood* (Princeton: Princeton UP, 1992)

Temkin, Oswei, "Introduction [to A Dissertation on the Sensible and Irritable Parts of Animals]," *Bulletin of the History of Medicine*, 4 (1936), 651-55.

Thorson, Helga, "Masking/Unmasking Identity in Early Twentieth-Century Germany: The Importance of N.O. Body," *Women in German Yearbook*, 25 (2009), 149-173.

Tischner, Rudolf, *Geschichte der Homöopathie* (Leipzig: Schwabe, 1939)

Thomson, Ann, *Bodies of Thought: Science, Religion, and the Soul in the Early Enlightenment* (Oxford: Oxford UP, 2008)

Tripp, Edward, *Crowell's Handbook of Classical Mythology* (New York: Crowell, 1970)

Tuttle, Howard N., *Human Life is Radical Reality: An Idea Developed from the Conceptions of Dilthey, Heidegger, and Ortega y Gasset* (New York: Peter Lang, 2005)

Tuttle, Hudson, *Arcana of Nature, or the History and Laws of Creation*, 5th ed., 2 vols (Boston: W. White & Co, 1870)

——, *Geschichte und Gesetze des Schöpfungsvorganges*, trans. by H.M. Achner (Erlangen: Ferdinand Enke, 1860)

Unzer, Johann August, *Erste Gründe einer Physiologie der eigentlichen thierischen Natur thierischer Körper* (Leipzig: Weidmann, 1771)

Unzer, Johanna Charlotte, *Grundriss einer Weltweisheit für das Frauenzimmer* ed. by Heidemarie Bennent-Vahle, bearbeitete Neuaufl. der 2. Aufl. (Aachen: Ein-Fach-Verlag, 1995)

van Helmont, John Baptista, *Oriatrike or, Physick refined. The Common Errors therein Refuted, and the whole art Reformed and Rectified, Being a new Rise and Progress of Phylosophy and Medicine, for the Destruction of Diseases and Prolongation of Life*, trans. by J.C. Sometime of M.H. Oxon (London: Lodowick Loyd, 1662)

van Hoorn, Tanja, *Entwurf einer Psychophysiologie des Menschen: Johann Gottlob Krügers Grundriss eines neuen Lehrgebäudes der Artzneygelahrtheit* (1745) (Hannover-Laatzen: Wehrhahn, 2006)

—, "Nachwort," Unzer, Johann August, *Gedancken vom Schlafe und denen Träumen: nebst einem Schreiben an N.N. daß man ohne Kopf empfinden könne*, ed. by Tanja van Hoorn (St. Ingbert: Röhrig, 2004)

— and Yvonne Wübben, eds., *"Allerhand nützliche Versuche": Émpirische Wissenskultur in Halle und Göttingen (1720-1750)* (Hannover: Wehrhahn, 2009)

Verworn, Max, *Allgemeine Physiologie. Ein Grundriss der Lehre vom Leben*, fünfte, vollständig neu bearbeitete Auflage (Jena: Gustav Fischer, 1909)

Vila, Anne C., *Enlightenment and Pathology* (Baltimore: The Johns Hopkins UP, 1998)

Virchow, Rudolf, *Die Cellularpathologie in ihrer Begründung auf physiologische und pathologische Gewebelehre* (Berlin: Hirschwald, 1858)

—, "Alter und neuer Vitalismus," *Archiv für pathologische Anatomie und Physiologie und klinische Medizin*, 9 (1856), 3-55.

Wackenroder, Wilhelm Heinrich and Ludwig Tieck, *Phantasien über die Kunst für Freunde der Kunst* (Hamburg: Perthes, 1799)

Wackenroder, Wilhelm Heinrich, "Die Wunder der Tonkunst," *Sämtliche Werke und Briefe*, ed. Silvio Vietta and Richard Littlejohns, 2 vols (Heidelberg: Carl Winter Universitätsverlag, 1991)

Walch, Johann Georg and Justus Christian Hennings, *Philosophisches Lexicon*, 4th ed (Leipzig: Gleditschens Buchhandlung, 1775)

Weiler, Inge, *Giftmordwissen und Giftmörderinnen: eine diskursgeschichtliche Studie* (Tübingen: M. Niemeyer, 1998)

Weinbacher, Hannah Kristina, *Sexualmedizinisches im Werk des Arztes und Schriftstellers Alfred Döblin (1878-1957)* (München: Diss., Ludwig-Maximilians-Universität, 2011)

Weiss Adamson, Melitta, "Illness and Cure in Hartmann von Aue's *Arme Heinrich* and *Iwein*," in *A Companion to the Works of Hartmann von Aue*, ed. by Francis G. Gentry (Rochester, New York: Camden House, 2005), 125-40.

Weissmann, Adolf, *Problems of Modern Music* (London: J. Dent, 1925)

Welsh, Caroline, *Hirnhöhlen-Poetiken* (Freiburg: Rombach, 2003)

Wenner, Melinda, "Humans Carry More Bacterial Cells than Human Ones," *Scientific American* (30 Nov. 2007)

Whitehead, Alfred North, *Process and Reality: An Essay on Cosmology*, corrected edition, ed. by David Ray Griffin and Donald W. Sherburne (New York: The Free Press, 1978)

—, *Process and Reality. An Essay in Cosmology. Gifford Lectures Delivered in the University of Edinburgh During the Session 1927–1928* (New York and Cambridge UK: Macmillan and Cambridge UP, 1929)

Wiesing, Urban, *Kunst oder Wissenschaft? Konzeptionen der Medizin in der deutschen Romantik* (Stuttgart: Frommann-Holzboog, 1995)

Williams, Elizabeth A., *The Physical and the Moral: Anthropology, Physiology, and Philosophical Medicine in France, 1750-1850* (Cambridge: Cambridge UP, 1994)

Wilson, David Sloan, *Evolution for Everyone: How Darwin's Theory Can Change the Way we Think About Our Lives* (New York: Delacorte Press, 2007)

Wilson, Laura Leann, *The Portrayal of the Female Figure in Alfred Döblin's Die beiden Freundinnen und ihr Giftmord* (Lafayette, Indiana: Diss., Purdue Unviersity, 2001)

Wischner, Matthias, *Kleine Geschichte der Homöopathie* (Essen: KVC Verlag, 2004)

Wöhler, Friedrich, "Ueber künstliche Bildung des Harnstoffs," *Annalen der Physik und Chemie*, 88.2 (1828), 253–56.

Wolff, Caspar Friedrich, *Theoria generationis. Editio nova augta et emendata* (Halle a.d. Saale: Christian Hendl, 1774)

Wright, John P. and Paul Potter, *Psyche and Soma: Physicians and Metaphysicians on the Mind-Body Problem from Antiquity to Enlightenment* (New York: Oxford, 2000)

Wulffen, Erich, *Das Weib als Sexualverbrecherin. Ein Handbuch für Juristen, Verwlatungsbeamte und Ärzte. Mit kriminalistischen Originalaufnahmen* (Berlin: Langenscheidt, 1923)

Zedler, Johann Heinrich, ed., *Grosses vollständiges Universal-Lexicon aller Wissenschaften und Künste*, 64 vols (Leipzig, Halle, 1732-1750)

Zelle, Carsten, "'Ey was hat der Arzt mit der Seele zu thun?' Physiologie und Psychologie bei Albrecht von Haller und Johann Gottlob Krüger." *"Allerhand nützliche Versuche." Emprirische Wissenskultur in Halle und Göttingen (1720-1750)*, ed. by Tanja van Hoorn and Yvonne Wübben (Hannover: Wehrhahn, 2009), 21-40.

Zelle, Carsten, ed., *"Vernünftige Ärzte." Hallesche Psychomediziner und die Anfänge der Anthropologie in der deutschsprachigen Frühaufklärung* (Tübingen: Niemeyer, 2001)

Zielinski, Siegfried, "Electrification, Tele-Writing, Seeing Close Up: Johann Wilhelm Ritter, Joseph Chudy, and Jan Evangelista Purkyne," *Deep Time of the Media: Toward an Archaeology of Hearing and Seeing by Technical Means*, trans. Gloria Custance (Cambridge: The MIT Press, 2006), 159-203.

Zimmer, Carl and Benedict Carey, "A Spark for a Stagnant Search," *NYT, Science Times*, Tuesday, July 22, 2014, D1, D5.

Zimmer, Carl, "100 Trillion Connections: New Efforts Probe and Map the Brain's Detailed Architecture," *Scientific American* (January 2011)

Zipes, Jack, *The Brothers Grimm: From Enchanted Forests to the Modern World* (New York: Palgrave Macmillan, 2002)

——, ed. and trans., *The Complete Fairy Tales of the Brothers Grimm* (New York: Bantam, 2002)

Zudrell, Petra, *Der Kulturkritiker und Schriftsteller Max Nordau. Zwischen Zionismus, Deutschtum und Judentum* (Würzburg: Königshausen & Neumann, 2003)

Zumbusch, Cornelia. *Die Immunität der Klassik* (Frankfurt: Suhrkamp, 2011)

Biographical Notes on Contributors

Biographical Notes on the Contributors

Jennifer Wynne Hellwarth is Associate Professor in the Department of English and a Senior Faculty in Women's, Gender, and Sexuality Studies at Allegheny College where she has been a faculty member since 2000. She received her PhD in English Literature from the University of California, Santa Barbara, her MA in English from Stanford University, and her BA in Renaissance Studies from the University of California, Berkley. She teaches courses in Medieval and Early Modern Literature, the History of Sexuality, and Medicine and Magic in Medieval Europe and the Middle East. Her book, *The Reproductive Unconscious in Medieval and Early Modern England*, was published by Routledge Press in 2003. Her recent research focuses on the intersection of literary production and the history of medicine and sexuality in medieval and early modern Europe and the Middle East, and her current book project examines medicinal and magical remedies for managing the sexual body in Medieval Arabic, Hebraic, and European Romance.

Stephanie M. Hilger is Associate Professor of Comparative Literature and German at the University of Illinois at Urbana-Champaign. She also holds appointments in French, Gender and Women's Studies, and the European Union Center. Her research focuses on gender, class, and race in British, French, and German literature, with a particular interest in interdisciplinary approaches to literature. She is the author of *Women Write Back: Strategies of Response and the Dynamics of European Literary Culture, 1790-1805* (2009) and *Gender and Genre: German Women Write the French Revolution* (2014). She is also the editor of the *Palgrave Handbook of Literature and Medicine* (forthcoming). Her articles have appeared in such journals as *Lessing Yearbook*, *Eighteenth-Century Studies*, *Colloquia Germanica*, *Neophilologus*, *College Literature*, *French Review*, *Seminar*, *Goethe Yearbook*, and *Women in German Yearbook*. Her current book project, "Liminal Bodies: Intersexuality in Literary and Medical Discourses," for which she was awarded a faculty fellowship for study in a second discipline, investigates the literary and medical representation of individuals born with ambiguous genitalia from eighteenth-century Europe to present-day America. This interdisciplinary study explores the discursive intersection between the medical and legal *mémoire* and the autobiographical memoir and traces the related shift from the intersex individual as a silent object of medical science and legal discourse to a speaking subject in charge of his/her self-representation.

James Kennaway is a Historian of Medicine at Newcastle University. He has previously worked at Oxford, Stanford, Durham and Vienna. He has written extensively on music and medicine. His monograph *Bad Vibrations: The History of the Idea of Music as a Cause of Disease* (Ashgate) was published in 2012 and he has published many articles on related subjects in journals such as *Social History of Medicine*, *Journal of the History of Medicine* and *Allied Sciences and Current Musicology*, as well as the edited volume *Music and the Nerves, 1700-1900* (Palgrave).

Alice A. Kuzniar has been Professor of German and English at the University of Waterloo since 2008, after teaching 25 years at the University of North Carolina at Chapel Hill, with guest professorships at Princeton, Rutgers, and the University of Minnesota. Her books include *The Queer German Cinema* (Stanford UP, 2000) and Melancholia's Dog: Reflections on our Animal Kinship (University of Chicago Press, 2006). She has just completed a book entitled The Romantic Art of Homeopathy.

Monica Ledoux is Adjunct Assistant Professor of Dermatology at Vanderbilt University, Nashville TN. After completing her medical studies at the Ludwig Maximilian University in Munich, Germany, she worked at several French university hospitals, most recently the Hôpital Tarnier and Ambroise Paré in Paris. Her research centers on chronic dermatologic diseases, skin cancers and Marfan's disease. Among her publications are "Importance of Dermatological Examination of Patients with Marfan Syndrome: A Case-controlled Prospective Study" (2011), "Intertrigo prurigineux" (Pruritic intertrigo) (2010), and "Anite Streptococcique et Psoriasis en Gouttes" (Streptococcal proctitis and psoriasis guttate) (2009). Her current projects focus on microbiome and skin and the development of a doctor-patient communication-training model for dermatology residents.

John A. McCarthy is Professor of German and Comparative Literature, emeritus at Vanderbilt University. He has published widely on interconnections between philosophy and literature, literature and law, and science and literature from the eighteenth to the twentieth century. Of special note among his authored volumes are his award-winning *Crossing Boundaries: A History and Theory of Essay Writing in German 1680-1815* (1989) and the monograph on paradigm shifts in literature and science, *Remapping Reality: On Chaos and Creativity in Science & Literature* (2006). Of note among his nine edited volumes is *Disrupted Patterns: On Chaos and Order in the Enlightenment* (2000), co-edited with T.E.D. Braun. His more recent publications include "Energy and Schiller's Aesthetics from the Philosophical to the Aesthetic Letters"

(2011), "Nietzsche's Philosophic Vitalism and Darwin's Theory of Evolution" (2011), "Cognitive Mapping: Adam, Venus, and Faust" (2013), "Kirchliche Moral, privates Gefühl und bürgerliche Öffentlichkeit. Goeze, Löwen und die *Hamburgische Dramaturgie*" (2014), and "'Freie Rechtslehrer' und Rechtsreformziele in der Vernetzung von Moral und Rechtswidrigkeit" (2015). McCarthy is currently working on a monograph on the critical reception of the *Sturm und Drang* and the nature of revolutions and is also editing a volume on Shakespeare and Cultural Transfer.

Brian I. McInnis is Assistant Professor of German at the United States Military Academy, West Point. His research interests span from 1700 to the present and include the discourse of body and soul in the long eighteenth century, Lessing as a cultural critic, and the interplay of science and literature from 1700 to the present. He is currently completing a book manuscript on the body/soul discourse in novels and magazines around 1750.

Cate I. Reilly is a Ph.D. candidate in the Department of Comparative Literature at Princeton University. She has published in the *Journal of Literature and Medicine* and has been awarded fellowships from the DAAD and the Social Science Research Council (Mellon Dissertation Grant). She currently holds Princeton's Porter Ogden Jacobus prize. Her dissertation, "Naming Disorder: Psychiatry, Diagnosis and Literary Modernism in Russia and Germany, 1880 - 1929" explores modernist literature's relationship to psychiatry and psychology by focusing on transnational Russian-German exchanges.

Nicholas Saul is Professor of German at the University of Durham, England. His chief scholarly interests are Romanticism, Realisms, Classical Modernism, Romanies and literary Darwinisms. He has published essays on numerous nineteenth-century authors such as Novalis, Wilhelm Bölsche, Clemens Brentano, Goethe, Karl Gutzkow, and Ernst Haeckl and on topics ranging from aesthetic humanism to subjectivity and utopias. Among his more recent publications *German Philosophy and Literature 1700-1990* (2002), *The Role of the Romanies* (2004), *Gypsies and Orientalism in German Literature and Anthropology of the Long Nineteenth Century* (2007), *The Cambridge Companion to German Romanticism* (2009), *The Evolution of Literature* (2011), and *Realism and Romanticism in German Literature* (2013). He is currently writing a monograph on Darwinism in German Literature 1859-2008. As holder of the Christopherson-Knott Fellowship he is leading a project on "The Experience of Emergence" at the Durham Institute of Advanced Study 2014-2015.

Ann C. Schmiesing is Professor of German and Scandinavian literature at the University of Colorado at Boulder. Her research interests include disability studies and eighteenth and nineteenth-century drama, theatre history, fairy tales and folklore, and book illustration. She is the author of the books *Disability, Deformity, and Disease in the Grimms' Fairy Tales* (2015) and *Norway's Christiania Theatre, 1827-1867: From Danish Showhouse to National Stage* (2006) and has published articles on G. E. Lessing, Jacob and Wilhelm Grimm, Daniel Nikolaus Chodowiecki, Theodor Gottlieb von Hippel, Bjørnstjerne Bjørnson, Henrik Wergeland, and Johan Falkberget.

Alexis B. Smith is a doctoral candidate in the Department of German and Scandinavian at the University of Oregon. "Ritter's Musical Blood Flow Through Hoffmann's Kreisler" is drawn from her dissertation, which investigates the "location" of music in early German Romanticism and its effects on the human body. With a double Bachelor of Arts in Music and German from the University of Northern Colorado (2005), and a Master of Arts in German Literature from the University of Oregon (2008), Smith has a strong interdisciplinary interest in German literature and music. Her Master's thesis, "Behind the Structure of the Music: Adorno's Analysis of Schoenberg's *Moses und Aron*," critiques Adorno's analysis in light of Schoenberg's *Musical Idea*, questioning why the opera remains unfinished (or whether it is unfinished at all). Her research interests extend from late eighteenth- through twentieth-century German literature and philosophy in relation to music, the visual arts, and social/cultural boundaries.

Heather I. Sullivan is Professor of German, Comparative Literature, and Environmental Studies, and the faculty chair of the Sustainability Committee at Trinity University in San Antonio, Texas. She is a board member of the European Association for the study of Culture, Literature and the Environment, of the Goethe Society of North America, and chairs the translation grant committee of the Association for the Study of Literature and the Environment. Sullivan has published widely in the US and in Europe on ecocriticism, material ecocriticism, the "dark pastoral," Goethe's science and literature, the German romantics, and eco-science fiction. She guest co-edited a volume of ecocritical essays with Caroline Schaumann on "Dirty Nature" in *Colloquia Germanica* 44.2 (2012), published in 2014; guest co-edited with Dana Phillips *ISLE*'s 2012 special volume on Material Ecocriticism, and guest co-edited with Bernhard Malkmus a volume currently under review at *New German Critique* on "The Challenge of Ecology to the Humanities: Post-Humanism or New Humanism." She is author of *The Intercontextuality of Self and Nature in Ludwig Tieck's Early Works* (1997). Current projects are a co-edited volume with Caroline Schaumann on *German Ecocriticism* with a focus on texts of the Anthropocene, and a monograph on the *Dark Pastoral*, which considers global climate change literature in the Anthropocene.

Ingo Uhlig is currently Visiting Professor, Academy of Fine Arts Münster. Born 1976, he studied sociology, philosophy and ethnology in Bayreuth. He received his doctorate in media and philosophy from the Bauhaus-University Weimar in 2005 with a dissertation on Gilles Deleuze. From 2006-2014 he taught in the Department of German Studies of the Martin-Luther-University Halle-Wittenberg. He qualified for professor with a thesis on sleep in philosophy and literature, *Traum und Poiesis. Produktive Schlafzustände 1641-1810* (Wallstein Verlag, 2015). Since 2011 he is visiting professor at the Academy of Fine Arts Münster and offers courses at the Burg Giebichenstein University of Arts and Design in Halle.

www.ingramcontent.com/pod-product-compliance
Lightning Source LLC
LaVergne TN
LVHW010542300926
849162LV00002B/212

* 9 7 8 9 0 0 4 3 0 9 0 2 9 *